William Ridgeway

Origin of Metallic Currency and Weight Standards

William Ridgeway

Origin of Metallic Currency and Weight Standards

ISBN/EAN: 9783337350932

Printed in Europe, USA, Canada, Australia, Japan

Cover: Foto ©berggeist007 / pixelio.de

More available books at **www.hansebooks.com**

THE ORIGIN OF
METALLIC CURRENCY AND
WEIGHT STANDARDS

BY

WILLIAM RIDGEWAY, M.A.,
PROFESSOR OF GREEK IN QUEEN'S COLLEGE, CORK,
LATE FELLOW OF GONVILLE AND CAIUS COLLEGE, CAMBRIDGE.

ἄνθρωπος ἢ <βοῦς ἢ> γῆς ἂν εἴη μέτρον ἁπάντων.

CAMBRIDGE:
AT THE UNIVERSITY PRESS
1892

[All Rights reserved.]

PREFACE.

THE following pages are an attempt to arrive at a knowledge of the origin of Metallic Currency and Weight Standards by the Comparative Method. As both these institutions played a not inconsiderable part in the development of civilization, it seemed worth while to approach the subject from a different point of view from that from which it had been previously studied. Hitherto Numismatists when studying the Origines of Coinage had confined themselves to the materials presented to them in the earliest money of Lydia, Greece and Italy, and on the other hand the Metrologists had almost completely limited their range of observation to the systems of Babylon, Egypt, Greece and Rome. As the Comparative Method has yielded such excellent results in the study of other human institutions, I have endeavoured by its aid to get some new principles which may throw some fresh light on the first beginnings of monetary and weight systems.

The leading principle which I have here endeavoured to establish by the Inductive Method, I had already put forward in a short paper, but there are various other doctrines now published for the first time, such as the origin of the earliest Greek coin types, the origin of the earliest Greek silver coins, of the Greek Obolos, the Sicilian Litra, and Roman As, of the Mina, and its sixty-fold the Talent.

In treating of the Distribution of gold and the priority of its discovery to that of the other metals, I have been led to criticise the principles of the science of Linguistic Palaeontology, which have gained such currency in this country from Schrader's *Prehistoric Antiquities of the Aryans*, and from Dr Isaac Taylor's popular little book, *The Origin of the Aryans*. I have been led to conclude that Comparative Philology taken alone is a misleading guide in the study of Anthropology.

From the nature of this work, a certain amount of polemic was inevitable; but I trust that not a line will be found which contains anything which could be offensive to the living, or is disrespectful to great scholars now no more. I owe so much to the works of distinguished men, from whose principles I am obliged to dissent, that I feel myself almost an ingrate who assails his benefactors with the very means provided for him by their labours.

It now only remains for me to thank many friends, who have aided me and taken an interest in this work.

To Mr J. G. Frazer, Fellow of Trinity College, Cambridge, I am under obligations which I cannot adequately express in words. He has read through the proofs of the whole of this work, and there is scarcely a page which has not benefited from his most careful and acute criticism. Besides this his vast knowledge of the manners and customs of barbarous peoples has furnished me with many most valuable references, and his fine Ethnological Library has been ungrudgingly placed at my disposal. Professor W. Robertson Smith has read the proofs of those pages which deal with Semitic systems, and Prof. J. H. Middleton those treating of the Greek.

By their kind sacrifice of time and labour which have been robbed from important works of their own, the many short-comings of this book have been rendered far less numerous than they otherwise would be, but of course I alone am responsible for the manifold ones which remain.

I must also express my gratitude to Mr Head, Mr Wroth and Mr Grueber of the Coin Department of the British Museum for their kindness and courtesy in affording me every facility for studying the coins under their charge.

I have to thank the Syndics of the Cambridge University Press for having undertaken the publication of this work.

QUEEN'S COLLEGE, CORK,
Christmas Eve, 1891.

CONTENTS.

CHAPTER I.
	PAGE
The Ox and the Talent in Homer	1

CHAPTER II.
Primitive Systems of Currency	10

CHAPTER III.
The distribution of the Ox and the distribution of Gold	47

CHAPTER IV.
Primaeval Trade Routes	105

CHAPTER V.
The Art of Weighing was first employed for Gold	112

CHAPTER VI.
The Gold Unit everywhere the value of a Cow	124

CHAPTER VII.
The Weight Systems of China and Further Asia	155

CHAPTER VIII.
How were Primitive Weight Units fixed?	169

CHAPTER IX.
Statement and Criticism of the Old Doctrines	195

PART II.

CHAPTER X.

The Systems of Egypt, Babylon, and Palestine . . 234

CHAPTER XI.

The Lydian and Persian Systems . 293

CHAPTER XII.

The Greek, Sicilian, Italian and Roman Systems. Conclusion . . 304

Appendix A 389
Appendix B 391
Appendix C 394

Index . . . 407

LIST OF ILLUSTRATIONS.

FIG.		PAGE
1.	Cowrie Shell	13
2.	Wampum	14
3.	Al-li-ko-chik	15
4.	Burmese silver shell money	22
5.	Chinese hoe money	23
6.	Fish-hook money	28
7.	Siamese silver bullet money	29
8.	Silvered brass bars	30
9.	Rings found in the tombs of Mycenae	37
10.	Gold rings found in Ireland	38
11.	West African axe money	40
12.	Old Calabar copper-wire formerly used as money	41
13.	Irish bronze fibulae and West African manillas	42
14.	Ancient British Coins	93
15.	Barbarous imitation of Drachm of Massalia	111
16.	Gold Stater of Philip of Macedon	125
17.	Persian Daric	126
18.	Gold Stater of Diodotus of Bactria	126
19.	Egyptian wall painting showing the weighing of gold rings	128
20.	Regenbogenschüssel	140
21.	Chinese knife money	157
22.	Egyptian Five-Kat weight	240
23.	Lion weight	245
24.	Assyrian Duck weight	245
25.	Weights in the form of Sheep	271
26.	Coin of Salamis in Cyprus	272
27.	Bull's-head Five-shekel Weight	283
28.	Lydian Electrum Coin	295
29.	Coin of Croesus	298
30.	Coin of Eretria	306
31.	Coin of Cyrene with Silphium plant	313
32.	Coin of Cyzicus with tunny fish	316
33.	Coins of Olbia in the form of tunny fish	317

LIST OF ILLUSTRATIONS.

FIG.		PAGE
34.	Coin of Tenedos with double-headed axe	318
35.	Coin of Phanes, earliest known inscribed coin	320
36.	Archaic Coin of Samos	321
37.	Coin of Cnidus	321
38.	Coin of Thurii	322
39.	Coin of Rhoda in Spain	322
40.	Tetradrachm of Athens	325
41.	Vase from Cyrene, showing the weighing of the Silphium	326
42.	Coin of Metapontum	327
43.	Coin of Croton	328
44.	Tortoise of Aegina	328
45.	Coin of Boeotia with Shield	331
46.	Coin of Lycia	332
47.	Coin of Messana	336
48.	Aes Rude	355
49.	Bronze Decussis, with figure of Cow	356
50.	As (*Aes grave*)	361
51.	As (semi-uncial)	362
52.	As, 3rd Cent. A.D. (*Third Brass*)	362
53.	Didrachm of Corinth	362
54.	Sesterce of First Roman Silver coinage	363
55.	Didrachm of Tarentum	364
56.	Romano-Campanian coin	377
57.	Victoriatus	377
58.	Sextans (*aes grave*)	379
59.	Gold Solidus of Julian the Apostate	384
60.	Tremissis of Leo I.	385

CHAPTER I.

The Ox and the Talent in Homer.

Ἦμος δ' οὔτ' ἄρ πω Ἠώς, ἔτι δ' ἀμφιλύκη νύξ.

The object of this essay is to enquire into the origin of Metallic Currency and Weight Standards. Since August Boeckh in his metrological enquiries[1] put forth the idea that the weight standards of antiquity had been obtained scientifically, all subsequent writers with scarcely an exception have followed in the same path. This theory was undoubtedly suggested by the fact that the French Republic had established a new scientific metric system. Yet reflection might have shown scholars that even the French system was not a wholly independent outcome of science, for beyond doubt the *mètre* and *litre* and *hectare* were only varieties of older measures of length, capacity and surface, then for the first time scientifically adjusted. The discovery of certain weights of bronze and stone in the ruins of Nineveh, Khorsabad and Babylon lent force to the theory of Boeckh; the imaginations of scholars were excited by the marvellous remains of Chaldaean and Assyrian civilization which had just been brought to light by Sir A. H. Layard, and they hastened to conclude that in the mathematical science of Mesopotamia the source of all weight-standards was to be found. Egypt however put in her claim to priority, and standards based on the measurements of the Great Pyramid, or on the weight of a given quantity of Nile-water, have entered the lists against the astrologers of Chaldaea. This battle still rages hotly, Assyriologists and Egyptologists

[1] *Metrologische Untersuchungen über Gewichte, Münzfüsse und Masse des Alterthums in ihrem Zusammenhange.* Berlin, 1838.

hurling at each other statements drawn from tablets and papyri, as regards the translation of which no two of these savants are agreed. In spite of this all modern works on metrology start with the systems of Babylon and Egypt and from these they derive the systems of Greece and Italy. It would at least be more scientific to move backwards from the known to the unknown, but beguiled by the glamour of a "scientific" metrological system, scholars have turned their backs upon scientific method. Whilst our knowledge of the Assyrian and Egyptian weight systems is most imperfect, being derived from literary monuments, or from inscriptions on weights not half understood, the systems of Greece and Rome are known to us not simply from the vast literatures written in languages thoroughly intelligible, but likewise from the evidence of immense numbers of coins struck in gold and silver, by the weights of which we are enabled to check off and substantiate the literary sources.

As Greece coined money several centuries before Italy, and as its literature reaches much further back than that of Rome, it is plain that any sound enquiry into the origin of weight standards must commence with Greece. We shall therefore without further preface proceed to investigate the evidence afforded to us by the oldest Greek records.

The Homeric Talent.

In the Homeric Poems, which cannot be dated later than the eighth century B.C., there is as yet no trace of coined money. We find nevertheless in those Poems two units of value; the one is the cow (or ox), or the value of a cow, the other is the Talent (τάλαντον). The former is the one which has prevailed, and does still prevail, in barbaric communities, such as the Zulus of South Africa, where the sole or principal wealth consists in herds and flocks. For several reasons we may assign to it priority in age as compared with the Talent. In the first place it represents the most primitive form of exchange, the barter of one article of value for another, before the employment of the precious metals as a medium of currency; consequently the estimation of values by the cow is older than that by means of a Talent or "weight" of gold, or silver or copper. Again, in Homer, all values are expressed in so many oxen, as "golden arms for brazen, those worth one hundred beeves, for those worth nine beeves[1]" (*Il.* VI. 236).

The Talent on the other hand is only mentioned in Homer in relation to gold (for we never find any mention of a Talent of *silver*) and we never find the value of any other article expressed in Talents. But the names of monetary units hold their ground long after they themselves have ceased to be in actual use as we observe in such common expressions as "bet a guinea," or worth a "groat," although these coins themselves are no longer in circulation, and so the French *sou* has survived for a century in popular parlance, and the *Thaler* has lived into the new German monetary system. Accordingly we may infer that the method of expressing the value of commodities in kine, which we find side by side with the Talent, is the elder of the twain.

Was there any immediate connection between the two systems or were they as Hultsch (*Metrologie*[2], p. 165) maintains entirely independent? It is difficult to conceive any people, however primitive, employing two standards at the same time which are completely independent of each other. For instance when we find in the *Iliad*[2] that in a list of three prizes appointed for the foot-race, the second is a cow, the third is a half-talent of gold, it is impossible to believe that Achilles or rather the poet had not some clear idea concerning the relative value of an ox and a talent. Now it is noteworthy that, as already remarked, nowhere in the Poems is the value of any commodity expressed in Talents; yet who can doubt that Talents of gold passed freely as media of exchange? A simple solution of this difficulty would be that the Talent of gold represented the older ox-unit. This would account for the fact that all values are expressed in oxen, and not in Talents, the older name prevailing in a fashion resembling the usage of *pecunia* in Latin.

A complete parallel for such a practice can be still found at the present moment among some of the Samoyede tribes

---

[1] χρύσεα χαλκείων, ἑκατόμβοι' ἐννεαβοίων.
[2] *Iliad*, XXIII. 750.

of Siberia. Thus we read in the account of a recent traveller:
"He finally came to the conclusion that for the consideration
of five hundred reindeer, he would undertake the contract.
This I regarded as a very facetious sally on his part. The
reindeer however I found was the recognised unit of value,
as amongst some tribes of the Ostiaks the Siberian squirrel.
For this purpose the reindeer is generally considered to be
worth five roubles[1]." Again forty years ago Haxthausen[2] tells
us that the Ossetes, a Caucasian tribe dwelling not very far
from Tiflis, although long accustomed to stamped money,
especially on the border of Georgia, kept their accounts in cows,
five roubles being reckoned to the cow. Here then in Siberia
and in the Caucasus, in spite of a long experience not merely
of a metallic unit, but of actual coined money, we still find
values estimated in reindeer, and in cows, the older units, just
as in Homer they are stated in oxen.

We shall likewise find that when the ancient Irish borrowed
a ready made silver unit (the *uncia*) from the Romans, they
had to equate this unit to their old barter-unit the cow, just as
in modern times the wild tribes of Annam when borrowing the
bar of silver from their more civilized neighbours have had to
equate it to their native standard, the buffalo; facts in close
accord with the well known derivation of Latin *pecunia, money*
from *pecus,* English *fee* from *feoh,* which still meant cattle, as
does the German *Vieh,* and *rupee* (according to some) from
Sanskrit *rupa,* also meaning cattle.

Let us now see if we have any data to support this hypothesis. That most trustworthy writer, Julius Pollux, says
in his *Onomasticon* (IX. 60): "Now in old times the Athenians
had this (*i.e.* the didrachm) as a coin and it was called an
ox, because it had an ox stamped on it, but they think that
Homer also was acquainted with it when he spoke of (arms)
'worth an hundred kine for those worth nine[3].' Moreover in

[1] Victor A. L. Morier, *Murray's Magazine,* August, 1889, p. 181.

[2] *Trans-Caucasia,* p. 410 (Engl. trans. 1854).

[3] Pollux, IX. 73, τὸ παλαιὸν δὲ τοῦτ' ἦν Ἀθηναίοις νόμισμα καὶ ἐκαλεῖτο βοῦς, ὅτι βοῦν εἶχεν ἐντετυπωμένον. εἰδέναι δ' αὐτὸ καὶ Ὅμηρον νομίζουσιν εἰπόντα ἑκατόμβοι' ἐννεαβοίων.

the laws of Draco there is the expression, to pay back the price of twenty kine: and at the time when the Delians hold their sacred festival, they say that the herald makes proclamation whenever a gift is given by any one, that so many oxen will be given by him, and that for each ox two Attic drachms are offered: whence some are of opinion that the ox is a coin peculiar to the Delians, but not to the Athenians; and that from this likewise has been started the proverb, an ox stands on his tongue, in case any man holds his tongue for money[1]."

According to Pollux then the Attic didrachm, or at least a coin employed by the Athenians (perhaps certain coins of Euboea), was called an 'ox.' Plutarch (*Theseus*, c. 25) goes further and asserts that Theseus struck money stamped with the figure of an ox (ἔκοψε δὲ νόμισμα βοῦν ἐγχαράξας), and the Scholiast on the *Birds* of Aristophanes (1106) quotes from Philochorus, an Athenian antiquary of the third century B.C.[2], the same account of the Attic didrachms being marked with an ox.

On the other hand the highest authorities on numismatics assert that the Athenians never struck any such coins. Yet after making due allowance for the additions made by Plutarch to the more crude statement of Pollux and Philochorus, it is hard to conceive that such a belief could have arisen without some foundation, and a probable solution may be found in the fact that certain uninscribed coins, bearing the type of an ox-head, which in recent years have been assigned to Euboea, are for the most part found in Attica. We know that Eretria, and Chalcis, the great cities of Euboea, were amongst the earliest places in Greece to strike money, and it is quite possible, nay probable, that these Euboic coins formed (along with the Aeginetan didrachms) the currency in use at Athens before

[1] Cf. Aesch. *Agam*. 36; Theognis 815. Cp. τὰν ἀρετὰν καὶ τὰν σοφίαν νικᾶντι χελῶναι, a proverb (given by Pollux IX. 74) alluding to the *Tortoise* coins of Aegina; and Menander (*Al.* 1), παχὺς γὰρ ὗς ἔκειτ' ἐπὶ στόμα.

[2] ἡ γλαῦξ ἐπὶ χαράγματος ἢ τετραδράχμον, ὡς Φιλόχορος· ἐκλήθη δὲ τὸ νόμισμα τὸ τετράδραχμον τότε [ἡ] γλαῦξ· ἦν γὰρ ἡ γλαῦξ ἐπίσημον καὶ πρόσωπον 'Αθηνᾶς, τῶν προτέρων διδράχμων ὄντων, ἐπίσημον δὲ βοῦν ἐχόντων.

the time of Solon (B.C. 596). Why the name *ox* was especially recollected in after years as that of the earliest currency, we can readily understand; the name derived from the old unit of barter would at once attach itself to the coin which bore the image of the ox, and in the course of time two traditions, one that the ancient unit was the ox, the other that the first coins current at Athens bore the symbol of an ox, would merge into one, and finally patriotic feeling would ascribe the first coinage to Theseus, who was regarded as the father of so many Athenian institutions.

That, at all events, the name might be applied to a certain sum, or coin, is rendered highly probable by the fact that Draco, with true legal conservatism, retained in his code the primitive method of expressing values in oxen. Now it is evident that the term, 'price of twenty oxen' (εἰκοσάβοιον), must have been capable of being translated into the ordinary metallic currency, whether that consisted of bullion in ingots or coined money. The "cow" therefore must have had a recognized traditional and conventional value as a monetary unit, and this is completely demonstrated by the practice at Delos. Religious ritual is even more conservative than legal formula, so we need not be surprised to find the ancient unit, the ox, still retained in that great centre of Hellenic worship. The value likewise is expressed in the more modern currency. But we are not yet certain whether the two Attic drachms, which are the equivalent of the ox, are silver or gold. Now Herodotus (VI. 97) tells us that Datis, the Persian general (B.C. 490), offered at Delos three hundred *talents* of frankincense. Hultsch (*Metrol.* p. 129) has made it clear that the talent here indicated must be the gold Daric, that is the light Babylonian shekel. For if they were either Babylonian or Attic talents, the amount would be incredible. Frankincense was of enormous value in antiquity; wherefore Hultsch is probably right in assuming that in the opinion of the Persian who made the offering, the three hundred "weights" of frankincense, each of which weighed a Daric, were equal in value likewise to 300 Darics. We shall see in a moment that there was a distinct tradition that the Daric was a *Talent*, and that the Homeric one. Now the gold Daric = two Attic gold drachms; but as the cow at Delos also = two Attic drachms, and the offering of frankincense at Delos is made in *Talent*, each of which is equivalent to two *gold* Attic drachms, there is a strong presumption that this Talent is the equivalent of the ox, and that the Attic drachms mentioned by Pollux are *gold*. Besides, it is absurd to suppose that at any time two *silver* drachms could have represented the value of an ox. Even at Athens, in a time of extreme scarcity of coin, Solon, when commuting penalties in cattle for money in reference to certain ancient ordinances, put the value of the ox at *five* silver drachms[1]. Moreover it is not at all likely that the substitution of silver coin for gold of equal weight would have been permitted by the temple authorities. But we get some more positive evidence of great interest from the fragment of an anonymous Alexandrine writer on Metrology, who says[2], "the talent in Homer was equal in amount to the later Daric. Accordingly the gold talent weighs two Attic drachms." Here we can have no doubt that Attic drachms mean *gold* drachms. Are we wrong then in supposing that at Delos still survived the same dual system which we found in Homer, the Ox and the Talent? But that at Delos both were of equal value we can have little doubt. For the ox = 2 Attic drachms = 1 Daric = 1 Talent = (130 grains Troy). Who can doubt that at Delos was preserved an unbroken tradition from the earliest days of Hellenic settlements in the Aegean? Modern discovery comes likewise to our support, and we shall find that it is probable that the gold rings found by Dr Schliemann in the tombs at Mycenae were made on a standard of about 135 grs.

This identification of the ox and the Homeric Talent is of importance: for it gives a simple and natural origin for the earliest Greek metallic unit of which we read. It likewise incidentally explains the proverb, βοῦς ἐπὶ γλώσσῃ which dates

[1] Plutarch, *Solon*, c. 15.
[2] Hultsch, *Reliquiae Scriptorum Metrologicorum*, I. 301, τὸ δὲ παρ' Ὁμήρῳ τάλαντον ἴσον ἐδύνατο τῷ μετὰ ταῦτα Δαρεικῷ. ἄγει δ' οὖν τὸ χρυσοῦν τάλαντον Ἀττικὰς δραχμὰς β', γράμματα ς', τετάρτας δηλαδὴ τεσσάρας.

from a time long before money was yet coined, or even the precious metals were in any form whatever employed for currency; it possibly explains why the ox was such a favourite type on coins, without having to call to our aid recondite mythological allusions; and it clears up once for all some interesting points in Homer. In the passage of the *Iliad* (XXIII. 750 sq.) already referred to the ox is second prize, whilst an half-talent of gold is the third. The relation between them is now plain; the ox = 1 talent, and the half-talent = a half-ox.

The vexed question of the Trial Scene[1] can now be put beyond doubt. In the *Journal of Philology* (Vol. x. p. 30) the present writer argued that the two talents represented a sum too small to form the blood-price (ποινή) of a murdered man, and consequently must represent the *sacramentum* (or payment made to the Court for its time and trouble, as in the Roman *Legis actio sacramenti* described by Gaius, Bk. IV. 16), as proposed by that most distinguished scholar and jurist, the late Sir H. S. Maine[2]. We know that the two talents are equal to two oxen, but in the *Iliad*, XXIII. 705, the second prize for the wrestlers was a slave woman "whom they valued at four oxen[3]." Now if an ordinary female slave was worth four oxen (= four talents) it is impossible that two talents (= two oxen) could have formed the bloodgelt or *eric* of a freeman. Probably four oxen was not far from the price of an ordinary female slave. Of course women of superior personal charms would fetch more, for instance, Eurycleia,

"Whom once on a time Laertes had bought with his possessions,
When she was still in youthful prime, and he gave the price of twenty kine[4]."

The poet evidently refers to this as an exceptional piece of extravagance on the part of Laertes. We can likewise now

[1] *Iliad*, XVIII. 507, 8,
κεῖτο δ' ἄρ' ἐν μέσσοισι δύω χρυσοῖο τάλαντα,
τῷ δόμεν, ὃς μετὰ τοῖσι δίκην ἰθύντατα εἴπῃ.
See Appendix A for a linguistic proof that the two talents were for the Judge.
[2] *Ancient Law*, p. 375.
[3] ἀνδρὶ δὲ νικηθέντι γυναῖκ' ἐς μέσσον ἔθηκεν,
πολλὰ δ' ἐπίστατο ἔργα, τίον δέ ἑ τεσσαράβοιον.
[4] *Od.* I. 430.

get a common measure for the ten talents of gold and the seven slave women who formed part of the requital gifts proffered by Agamemnon to Achilles[1], and can form some notion of the comparative value of the prizes for the chariot race and other contests[2].

The wider question of Weight-standards in general.

But results far more important than merely the determination of the value of Homeric commodities may be obtained as regards the weight-standards of Europe and their congeners in Asia. For by taking as our primitive unit the cow or ox, we may be able to give a much more simple account of the genesis of those standards than that which hitherto has been the received one.

We have found the Homeric ox and talent identical with the didrachm or stater of the Euboic-Attic standard. All the silver coinage of Greece proper was struck either on this standard or the Aeginetic, and what is still more important for us it was on the Euboic-Attic standard alone that gold was estimated in every part of Greece. Practically the stater of this system was of the same weight as the famous Persian daric which in historical times formed the chief coin-unit of all Asia from India to the Aegean shores.

[1] *Iliad*, IX. 12 *seqq.*
[2] *Il.* XXIII. 262 *seqq.*

CHAPTER II.

PRIMITIVE SYSTEMS OF CURRENCY.

'Εξ ἀνάγκης ἡ τοῦ νομίσματος ἐπορίσθη χρῆσις.
ARISTOTLE.

LET us here propound the doctrine which seeks to obtain an explanation of the origin for weight-standards more in accordance with the facts of history and the process of development as exemplified both in ancient and modern times.

In early communities[1] all commodities alike are exchanged by bartering the one against the other. The man who possesses sheep exchanges them for oxen with the man who possesses oxen, the owner of corn exchanges his commodity for some implement or ornament of metal with the owner of the latter. The metals are only regarded as merchandise, not yet being in any degree set apart to serve as a medium of exchange in the terms of which all other commodities are valued. This is the practice which prevails in so civilized a country as China down to our own days. The only coinage which the Chinese possess is copper *cash*. According to M. le Comte Rochechouart (*Journal des Économistes*, Vol. xv. p. 103) both gold and silver are treated simply as merchandise, and there is not even a recognized stamp or government guarantee of the fineness of the metal. The traveller must carry these metals with him, as a sufficient quantity of strings

---

[1] Of course amongst the lowest races of savages such as the aborigines of Australia, even barter is almost unknown. Each man makes his own stone implements from the greenstone which is everywhere in abundance, his own clubs and boomerangs, whilst Nature supplies all his other wants.

of *cash* would require a waggon for their conveyance. Yet in exchanging silver or gold he is sure to suffer loss both from the falsity of balances and of weights and the uncertain fineness of the metal.

When in a certain community one particular kind of commodity is of general use and generally available, this comes to form the unit in terms of which all values are expressed. The nature of this barter-unit will depend upon the nature of the climate and geographical position, and likewise upon the stage of culture to which the people have attained. In the hunting stage, all the property of each individual consists in his weapons and implements of war and the chase, and the skins of wild beasts which form his clothing, and sometimes the cover of his hut or wigwam. At a later stage, when he has succeeded in taming the ox, the sheep, or the goat, or the horse, he is the owner of property in domestic animals, whose flesh and milk sustain him and his family, and whose skins and wool provide his clothing.

By this time too he has found out that it is better to make the captive whom he has taken in war into a hewer of wood and drawer of water than merely to obtain some transient pleasure from eating him after putting him to death by torture, or by wearing his skull or scalp as personal decorations.

This is now the pastoral or nomad stage.

Next comes the more settled form of life, when the cultivation of land and the production of the various kinds of cereals renders a permanent dwelling-place more or less necessary.

Property now consists not merely in slaves and domestic animals, but likewise in houses of improved construction, and large stores of grain. Man now possesses certain of the metals, gold and copper being the first to be known. How does he appraise these metals when he exchanges them with his neighbour? We shall find that he estimates them in terms of cattle, and that he at first barters them all by measures based on the parts of the human body, a method which continues to be employed for copper and iron long after the art of weighing has been invented; next he estimates his gold by certain natural units of capacity such as a goosequill, and finally fixes the amount of gold which is equivalent to a cow, by setting it in a rude balance against a certain number of natural seeds of plants. Such is the process which history tells us has taken place in the temperate regions of Asia and Europe, Africa and America. Just as it is impossible to learn the history of the growth of the earth's crust by confining our observations to one locality, and as the geologist only succeeds in gaining a true insight into the relations between the various strata by a study of the phenomena of many regions, so we shall only be able to comprehend properly the various stages in the growth of metallic currency and the origin of weight-standards by observing the facts revealed to us in various countries. Whilst in some places we shall meet with but one or two steps, in others we shall find traces of many, though often, broken strata. Like advance, however, seems impossible under the extremes of heat and cold. Hence in the latter regions the conditions of life remain almost unaltered. In the extreme north the rigour of an arctic winter forbids the keeping and rearing of domestic animals, or the cultivation of corn and vegetables. Hence the hunter form of existence remains almost unaltered. The sole or chief wealth of the people consists of the skins of the fur-bearing animals such as the seal, the beaver, the marten, or the fox, or stores of dried fish, which they exchange with traders for a few scant luxuries, or which form their own sustenance and protection against the pitiless frosts and snows.

In these regions therefore we find the skins of certain animals serving as units of account, in spite of the difference in value between those of different quality and rarity. In the Territory of the Hudson's Bay Company, even after the use of coined money had been introduced among the Indians, the skin was still in common use as the money of account. A gun nominally worth forty shillings brought twenty 'skins.' This term is the old one used by the Company. One skin (beaver) is supposed to be worth two shillings, and it represents two martens and so on. "You heard a great deal about skins at Fort Yukon, as the workmen were also charged for clothing, etc., in this way[1]." Similarly in the extreme north of Asia we find some Ostiak tribes using the skin of the Siberian squirrel as their unit of account.

The name of a small coin equal to a quarter kopeck indicates that originally the Slavs had a like form of currency. It is called *polooshka*. *Ooshka* (properly little ears) means a hare-skin, and *polooshka* means *half a hare-skin*[2].

When we turn to the torrid zone, where clothes are only an incumbrance and Nature lavishly supplies plenteous stores of fruits and vegetables, the chief objects of desire will not be food and clothing but ornaments, implements and weapons. Hence we find amongst the inhabitants of such regions in especial strength that passion for personal adornment, which is one of the most powerful and primitive instincts of the human race. Shells have from very remote times formed one of the most simple forms of adornment in all parts of the world. Shells which once perhaps formed the necklace of some beauty of the neolithic age are found with the remains of the cave men of Auvergne. Strings of cowries under their various names of *changos*, *zimbis*, *bonges* or porcelain

Fig. 1. Cowrie Shell (*Cypraea moneta*).

shells are both durable, universally esteemed, and portable, and therefore suited to form a medium of exchange, and as such they are employed in the East Indies, Siam, and on the East and West Coasts of Africa; on the tropical coasts they serve the purposes of small change, being collected on the shores of the Maldive and Laccadive islands and exported for that object. The relative value varies slightly according to their abundance or scarcity. In India the usual ratio was about 5000 to the rupee. Marco Polo found the cowry

---

[1] Whymper's *Alaska*, p. 225.
[2] Morier, *Murray's Magazine*, August, 1889, p. 181.

in use in the province of Yunnam. He says (II. p. 62, Yule's Transl.): "In Carajan gold is so abundant that they give one Saggio of gold for six of the same weight of silver. And for small change they use the porcelain shell. These are not found in the country but are brought from India." How ancient is their use in Asia is shown by the fact that Layard found cowries in the ruins of Nineveh.

Beyond all doubt the wampum belts of the North American Indians served the purpose of currency. They consisted of black and white shells rubbed down, polished and made into beads, and then strung into belts or necklaces, which were

FIG. 2. Wampum (made from the *Venus mercenaria*).

valued according to their length, colour and lustre, the black beads being the most valuable. Thus one foot of black peag was worth two feet of white peag. It was so well established as a currency among the natives that in 1649 the Court of Massachusetts ordered that it should be received as legal tender among the settlers in the payment of debts up to forty shillings[1].

[1] Jevons, *Money*, p. 24.

Nor has this employment of strings of shells as money even yet disappeared from North America. Thus Powers writes[1] of the Karoks and other tribes of California: "For money they make use of the red scalps of woodpeckers, which rate at $2.50 to $5.0 a piece, and of the dentalium shell, of which they grind off the tip, and string it on strings, the shortest pieces are worth 25 cents, and the longest about two dollars, the value rising rapidly with the length. The strings are usually about as long as a man's arm. It is called *al-li-ko-chik* (in Yarok this signifies literally Indian money) not

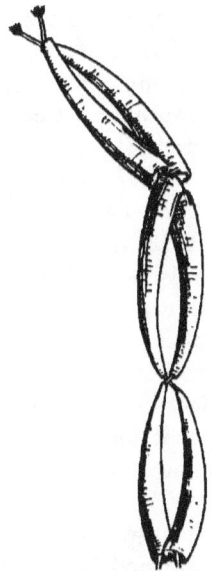

FIG. 3. Al-li-ko-chik.

only on the Klamath but from Crescent city to Eel river, though the tribes using it speak several different languages. When the Americans first arrived in the country an Indian would give 40 or 50 dollars gold for a string, but now the abundance of the supply has depreciated its value and it is principally the old Indians who esteem it." Again he writes,

[1] *Tribes of California*, p. 21.

"Some of the young bloods array their Dulcineas for the dance with lavish adornments, hanging on their dress 30, 40 or 50 dollars worth of dimes, quarter dollars and half dollars arranged in strings." This shows that the new currency of silver is treated by them in exactly the same way as the old shell strings, both of them deriving their value as media of exchange from the fact that they are the objects most universally prized as ornaments for the person.

Elsewhere the same writer observes: "Immense quantities of it (shell money) were formerly in circulation among the Californian Indians, and the manufacture of it was large and constant to replace the continual wastage caused by the sacrifice of so much on the death of wealthy men, and by the propitiatory sacrifices performed by many tribes, especially those of the coast range. From my own observations, which have not been limited, and from the statements of pioneers and of the Indians themselves, I hesitate little to express the belief that every Indian in the state in early days possessed an average of at least 100 dollars worth of shell money. This would represent the value of almost two women (though the Nishinam never actually bought their wives), or two grizzly bear skins, or 25 cinnamon bear skins or about three average ponies. The young English-speaking Indians hardly use it at all except in a few dealings with their elders or for gambling. One sometimes lays away a few strings of it for he knows he cannot squander it at the stores. It is singular how old Indians cling to this currency when they know it will purchase nothing for them at the stores; but then their wants are few, and mostly supplied from the sources of nature, and besides that the money has a certain religious value in their eyes, as being alone worthy to be offered up on the funeral pile of departed friends or famous chiefs of their tribes[1]."

Here we see how amongst the Indian tribes there was a fully developed system of inter-relations between the various objects which formed their wealth.

The horse was but a new comer into America, but he had his place soon allotted in the scale of values, being little less

[1] *Op. cit.*, p. 335.

valuable than a squaw. We cannot doubt that if the Indian had succeeded in domesticating the buffalo before the advent of the white man, it would have formed the most general unit in use, as we shall find its congeners being employed in all parts of the old world. But before the coming of the Spaniards at least one race of North America had advanced a stage beyond shell money. The Aztecs[1] of Mexico were employing a currency of gold and cacao seeds. The former in the shape of dust was placed in goose quills, which formed a natural unit of capacity, for weights were as yet unknown to the Aztecs; whilst the cacao seeds were placed in bags, each containing a specified number.

In Queen Charlotte Islands the dentalium shell was recognized as a medium of exchange by most of the coast tribes, but not so much as a medium of exchange for themselves as for barter with the Indians of the interior. With the Haidas it is still sometimes worn as an ornament though it has disappeared as a medium of exchange. The blanket of the trader has now however supplanted the *skin* as the principal unit. Not only among the Haidas but all along the coast it takes the place of the beaver-skin currency of the interior of British Columbia and of the North West Territory. The blankets used in trade are distinguished by the points or marks on the edge, woven into their texture, the best being four-point, the smallest and poorest one-point. The acknowledged unit of trade is a single two and a half-point blanket, now worth a little over $1.50 Everything is referred to this unit, even a large four-point blanket is said to be worth so many *blankets*. There is also the "Copper," "an article of purely conventional value and serving as money. This is a piece of native metal beaten out into a flat sheet and made to take a peculiar shape. These are not made by the Haidas—nor indeed is the native metal known to exist in the islands, but are imported as articles of great worth from the Chilcat country north of Sitka. Much

---

[1] Clavigero, *Hist. of Mexico*, Vol. 1. 386.

They counted the Cacao nuts by 8000 and to save the trouble of counting them they reckoned them by sacks, every sack being reckoned to contain 24,000. Cf. Prescott, *Conquest of Mexico*, Vol. 1. p. 44.

attention is paid to the size and make of the copper, which should be of uniform but not too great thickness, and should give forth a good sound when struck with the hand. At the present time spurious coppers have come into circulation, and although these are easily detected by an expert, the value of the copper is somewhat reduced and is often more nominal than real. Formerly ten slaves were paid for a good copper as a usual price, now they are valued at from forty to eighty blankets".[1] It is obvious that such costly imported articles, though now used as occasional higher units of account—much as we employ fifty-pound notes—must have had some definite use, owing to which they were so highly prized. The attention paid to their tone would lead us to conjecture that they were employed as a kind of gong, and further on we shall find certain peoples of Further Asia paying a large price in buffaloes for gongs.

Before we quit finally the northern latitudes, it is worth our while to observe the method of currency employed by the Icelanders. As metals and other products of the land were scarce in their bleak home, the stockfish (dried cod) formed naturally their chief commodity, and hence it appears on the arms of Denmark as the emblem of Iceland. There is still extant a proclamation for the regulation of English trade with Iceland issued sometime between 1413 and 1426. As, *mutatis mutandis*, it affords admirable insight into the methods by which trade was carried on between men of different nations in the emporia of the Mediterranean, and in fact everywhere else, it is worth giving it *in extenso*[2].

"I, *N. M.* do proclaim here to-day a general market between the English and the Icelandic men, who have come here with peace and fair dealing, and between the Icelandic men and the men of the islands who wish to carry on their trade here.

"First I proclaim this market on conditions of peace and lawful security between one and the other, so that each can entirely dispose of his own if he buy or if he sell. Price list in

---

[1] G. M. Dawson, 'Report on the Queen Charlotte Islands, 1878,' p. 135 B (*Geological Survey of Canada*), Montreal, 1880.

[2] F. Magnusson, *Nordiske Tidskrift for Oldkyndighed*, II. 112.

stockfish: of fish 2, 2½, or 1¾ lbs., 80 lbs. must be the equivalent of a hundred (of cloth, i.e. 129 *alens* of *vadme*, a cloth formerly used as a medium of exchange), provided the persons concerned cannot agree as to the price.

Price of (foreign) goods.	Stockfish.
48 *alen* of good and full width trade cloth	120
48 *alen* linen cloth double width	120
6 tonder (tuns) malt	120
4 do. trade flour	120
3 do. wheat	120
4 do. beer	120
1 tonde clean and clear butter	120
1 do. wine	100
1 do. pitch	80
1 do. raw tar	60
1 cask of iron, containing 400 pieces	120
⅛ tonde honey	15
⅛ do. blubber	15
½ lb. of coppers (i.e. copper cauldrons) by weight	2½
1 pair black (leather) shoes	4
1 pair of women's shoes	3
1 trade rug	30
1 "alen" timber, in planks or spars	5
⅛ tonde salt	5
½ lb. wax	5
Horse shoes of iron for 5 horses	20
Caps, knives, and other small mercer's wares, according to mutual agreement.	

"I charge all, not only the people from the country, but also the inhabitants of these islands, that ye do in no way compass any disorder or disturbance to the strangers, from the moment the guard flag is hoisted, unless they themselves allow it.

"They, who here are annoyed by word or deed, have a right to demand double idemnity therefor.

"Also I charge, and the merchants in no way the least, that they use aright the "alen" and other lawful measure for everything, as the law demands, especially as regards butter, wine and beer, flour or malt, honey or tar, so that no one deals false or with deceit with another.

"He who does so intentionally shall have sinned as greatly against the state as if he had stolen goods of like value, whereas the bargain becomes void, and damages moreover must be given to him who was deceived.

"Let us now, Ye good men, eschew all malice and trickery, riot or disturbance, quarrels and careless words: but let every man be the other's friend, without deceit.

"Prizing unity
"And old custom,
"And abiding in God's peace."

Some such proclamations were probably often made in the marts of the Aegean, such as Aegina, when Greek, Phoenician and Etruscan met for traffic under the control of some local potentate, and the protection of the god of some neighbouring shrine.

Passing to the islands of the Pacific we shall find shell money playing an important part among the primitive peoples, such as those who inhabit New Ireland, New Britain, the Pelew and the Caroline groups. It will suffice for our purpose to describe the form in which it is employed in New Britain. Mr Powell[1] tells us that the native money in New Britain consists of small cowrie shells strung on strips of cane, in Duke of York Island it is called Dewarra. It is measured in lengths, the first length being from hand to hand across the chest with arms extended, second length from the centre of the breast to the hand of one arm extended, the third from the shoulder to the tip of the fingers along the arm, fourth from the elbow to the tip of the fingers, fifth from the wrist to the tip of the fingers, sixth finger lengths. Fish are generally bought by the length in Dewarra unless they are too small. A large pig will cost from 30 to 40 lengths of the first measure (fathom) and a small one ten. The Dewarra is made up for convenience in coils of 100 fathoms or first lengths; sometimes as many as 600 fathoms are coiled together, but not often, as it would be too bulky to remove quickly in case of invasion or war, when the women carry it away to hide. These coils are very neatly

---

[1] *Wanderings in a Wild Country, or Three Years among the Cannibals of New Britain* (London, 1883), p. 55.

covered with wickerwork like the bottom of our cane chairs.... At Moko and Utuan they use another kind of money as well as this, the other being a little bivalve shell, through which they bore a hole and string it on pieces of native made twine[1]. It is also chipped all round until it is a quarter of an inch in diameter and then smoothed down into even discs with sand and pumice. Here we find strings of shells, which undoubtedly in the first instance were used for personal adornment, converted into a true currency. The simple savages whose possessions were exceedingly few and scanty, equated their fish to strings of shells which formed their only ornament, and when they got a more valuable possession in the pig, they quickly learned to appraise that animal in shell worth, just as the North American Indians learned to estimate the horse in *Wampum*. Instead of shells the natives of Fiji are said to have employed whales' teeth as currency, red teeth (which are still highly prized) standing to white ones somewhat in the ratio of sovereigns to shillings with us[2]. Passing on to the mainland of Asia we shall find that the Chinese, who in the course of ages have developed a bronze coinage of their own apart from the influences of the Mediterranean people, had in early times an elaborate system of shell money. Cowries appear in the *Ya-King*, the oldest Chinese book, 100,000 dead shell fishes being an equivalent for riches. Tortoise shell currency is also mentioned in the same book. The tortoise of various kinds and sizes was used for the greater values which would have required too many cowries. Tortoise shell is still elegantly used to express coin. Several kinds of *Cypraea* were used, including the purple shell, two or three inches long; all the shells except the small ones were employed in pairs. A writer of the second century B.C.[3] speaks of the purple shell as ranking next after the sea tortoise shells, measuring one foot six inches, which could only be procured in Cochin China and Annam, where they were used to make

[1] For shell money in the Caroline Islands cf. Kubary's *Ethnographische Beiträge zur Kenntnis des Karolinen Archipels* (Leipzig, 1889); in the Pelew Islands cf. Karl Semper, *Die Pelau Inseln* (Leipzig, 1873), p. 60; and for shell money in general cf. R. Stearn's *Ethno-conchology* (Washington, 1889).

[2] Jevons, *Money*, 25.

[3] Terrien de la Couperie, *Coins and Medals*, p. 193.

pots, basins and other valuable objects. So attached were the Chinese to these primitive coins that the usurper Wangmang restored a shell currency of five kinds, tortoise shell being the highest. From this time we hear no more of cowries in China Proper, but they left traces of themselves in the small copper coins shaped like a small Cypraea, called Dragon's eye or Ant coins[1]. It is doubtless to a similar survival that we owe those curious silver coins made in the shape of shells which come from the north of Burmah and of which there are

Fig. 4. Burmese silver shell money.

several specimens in the British Museum. They are about the size of a cowrie, and doubtless served as a higher unit in a currency, of which the lower units were formed by real shells.

In 685 B.C. in parts of China pearls and gems, gold, knives and cloth were the money, and under the Shou dynasty (1100 B.C.) we understand from ancient Commentaries that the gold circulated in little cubes of a square inch, and the copper in round, tongue-like plates by the *tchin tchu*, while the silk cloth 2 feet 2 inches wide in rolls of 40 feet formed a *piece*.

In the *Shu King*, when in 947 B.C. commutation for punishment was enacted, the culprit according to the offence was to pay 100, 200, 500 or 1000 *hwars*, or rings of copper weighing 6 *ounces*. The Chinese likewise used hoes as money, just as we shall find the wild people of Annam doing at the present hour. But in the course of time the hoe became a true currency and little hoes, such as that here figured, were employed as coins in some parts of China (*tsin*, agricultural implements). The copper knives which played so important a part in the development of Chinese coinage will be dealt

---

[1] Terrien de la Couperie, *Coins and Medals,* p. 199.

with more particularly in a later chapter. In Marco Polo's time cowries were in full use, as in the province of Yunnan[1].

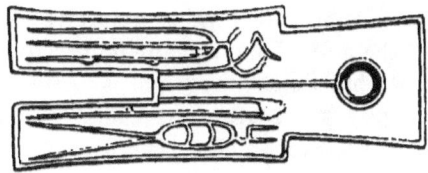

Fig. 5. Chinese hoe money.

On the borders of China and Tibet we may still find a state of things not far removed from that existing in the China of 2000 years ago[2]. The Tibetans, who in recent years employ Indian rupees, for purposes of small change cut up these coins into little pieces, which are weighed by the careful Chinese, but the Tibetans do not seem to use the scale, and roughly judge of the value of a piece of silver. Tea, moreover, and beads of turquoise are largely used as a means of payment instead of metal.

Speaking of this same region (called by him Kandu), Polo says[3]: "The money-matters of the people are conducted in this way: they have gold in rods which they weigh, and they reckon its value by its weight in *saggi*, but they have no coined money. Their small change again is made in this way: they have salt which they boil and set in a mould, and every piece from the mould weighs half-a-pound. Now eighty moulds of this salt are worth one *saggio* of fine gold." Tea seems to have taken the place of salt in modern times.

Turning next to the southern frontier of China, we shall find among the tribes of Annam a system of currency which strongly reminds us of that found in the Homeric Poems.

Among the Bahnars of Annam who border on Laos, "everything," says that excellent observer M. Aymonier, "is by barter, hence all objects of general use have a known relationship: if we know the unit, all the rest is easy. Here is the key: a

---

[1] Yule's Translation, Vol. II. p. 70.
[2] Gill, *River of Golden Sand*, II. p. 77.
[3] Yule's Translation, Vol. II. p. 45.

head, that is to say, a male slave is worth six or seven buffaloes, or the same number of pots (*marmites:* so in Homer, *Il.* XXXIII. 885, an ox is estimated at a kettle); the buffalo and the pot have the same value, which naturally varies with the size and age of the animal and the size and quality of the pot.

"A full-grown buffalo or a large pot is worth seven earthenware jars of a grey glaze, after the Chinese shape, and with a capacity of fifteen litres. One jar = 4 *muk*. (The *muk*[1] is an unit of account, but originally meant some special article.) 1 *muk* = 10 *mats*, that is to say ten of these *hoes*, which are manufactured by the Cedans, and which are employed by all the savages of this region as their agricultural implement. The hoe is the smallest amount used by the Bahnars. It is worth 10 centimes in European goods, and is made of iron[2]." Thus the buffalo is worth 280 hoes, or a little more than an English sovereign, since each hoe is worth a penny (10 centimes). The Bahnars have sheet tin $\frac{1}{2}$ millim. thick cut into pieces 11 centim. square, to be used to ornament sword-belts or to make earrings (iv. p. 390). A stick of virgin wax the size of an ordinary candle = 1 hoe, a pretty little cane hat = 2 hoes; a large bamboo hat = 2 hoes; a Bahnar knife = 2 hoes; a fine sword and sheath = 1 jar, 1 *muk*, 3 hoes; a crossbow and string = 3 hoes; ordinary arrows are sold at 30 for 1 hoe; arrows with movable heads, 20 for 1 hoe, and poisoned arrows 5 for 1 hoe; a lance-head = 3 hoes; a lance with palm handle = 4 hoes; a horse = 3 or 4 pots or buffaloes; a large elephant = 10 to 15 *heads* (slaves).

The same method of using the buffalo as the chief unit is employed by the Moïs, among whom a slave is reckoned at 10 buffaloes. Again, among tribes such as the Tjams, with whom the string of copper *cash* (or sapecs) borrowed from the Chinese, is employed as their lowest unit, a full-grown buffalo = 100 strings;[3] the Mexican *piastre* or dollar

---

[1] So the Irish *sed*, the most general name for *chattel*, originally meant simply an *ox*.

[2] *Cochin-Chine Française. Excursions et Reconnaissances*, XIII. (1877), p. 296—8.

[3] *Excursions et Reconnaissances*, XIII. No. 30 (1887), p. 296—304.

circulates freely as in China, a small pig costs 10 strings, pork by retail costs two strings per lb. (*livre*), ducks cost 1½ to 2 strings. A large caldron costs 3 buffaloes; a handsome gong = 2 buffaloes; a small gong = 1 buffalo; 6 copper platters = 1 buffalo; two swords = 1 buffalo; 2 lances = 1 buffalo; a rhinoceros' horn = 8 buffaloes; a pair of large elephants' tusks = 6 buffaloes; a small pair = 3 buffaloes. When the wild people have dealings with the more civilized peoples of the plain, who employ the Chinese cash and silver dollars, a large buffalo = 100 strings of cash, a small one = 50 strings; a fine horse = 100 strings; a she goat = a piece of cloth. The Orang Glaï have often to buy elephants' tusks, at the rate of 8 buffaloes for a pair, or 8 bars of silver (640 francs). The Szins of Kharang have often to pay a tax of a buffalo per hut, or for the whole village 10 buffaloes, the horns of which must be at least as long as their ears[1]. In Cambodia iron ingots[2] form a special kind of money. These ingots are not weighed, but they are as long as from the base of the thumb to the tip of the forefinger; they are in breadth two fingers, and one finger in thickness in the middle, tapering off to either end.

Cowries and other shells seem to have gone out of use altogether among these tribes, but we may recognize in the practice of reckoning the *cash* by the string a distinct survival of the olden time when shells were so employed. It is of great importance to note that where silver has come into use, its unit, the bar, is equated to the buffalo, the unit of barter, just as we find the Homeric gold Talent equal to the ox.

Next let us turn to India, and to the Aryans of the Rig Veda, who dwelt in the north-west of the Punjaub at the time when we first meet them. From their prayers and invocations it is easy to learn in what the wealth of this simple folk consisted. One or two examples will serve for our purpose: "The potent ones who bestow on us good fortune by means of cows, horses, goods, gold, O Indra and Vaya, may they, blessed with fortune, ever be successful by means of horses and heroes in

---

[1] M. Aymonier, *Cochin-Chine. Excursions et Reconnaissances*, Vol. I. No. 24 (1885), pp. 233 *seqq*.

[2] *Ibid.* p. 317.

battle[1]." Again, "O Indra bring us rice cake, a thousand *soma* drinks, and an hundred cows, O hero. Bring us apparel, cows, horses and jewels, along with a *mana* of gold." Yet once more: "Ten horses, ten caskets, ten garments, ten gold nuggets (*hiranya pindas*) I received from Divodāsa. Ten chariots equipped with side horses, and an hundred cows gave the Açvatha to the Atharvans and to the Pāyu." Even without further evidence than that which we have already drawn from the wild people of Annam, we might well assume that there were definitely fixed relations in value between the cows, horses, gold, rice, and cloth of the Vedic people. But absolute proof is at hand, for their close kinsmen, the ancient Persians, have left us in the Zend Avesta ample means of observing their monetary system. Thus we read in the ordinances which fix the payment of the physician that "he shall heal the priest for the holy blessing; he shall heal the master of an house for the value of an ox of low value; he shall heal the lord of a borough for the value of an ox of average value; he shall heal the lord of a town for the value of an ox of high value; he shall heal the lord of a province for the value of a chariot and four; he shall heal the wife of the master of a house for the value of a she ass; he shall heal the wife of the master of a borough for the value of a cow; he shall heal the wife of the lord of a town for the value of a mare; he shall heal the wife of the lord of a province for the value of a she camel; he shall heal the son of the lord of a borough for the value of an ox of high value: he shall heal an ox of high value for the value of an ox of average value; he shall heal an ox of average value for the value of an ox of low value; he shall heal an ox of low value for the value of a sheep; and he shall heal a sheep for the value of a meal of meat[2]." So too in the fees of the Cleanser we read: "Thou shalt cleanse a priest for a blessing; the lord of a province for the value of a camel of high value; the lord of a town for the value of a stallion; the lord of a borough for the value of a bull; the master of an house for the value of a

---

[1] *Rig-Veda, Mandala*, VII. 90. 6, VIII. 67. 1—2, VI. 47, 23—4.

[2] *Vendidâd, Fasgard*, VII. 41 (Darmesteter's translation in Sacred Books of the East).

cow three years old; the wife of the master of an house for the value of a ploughing cow; a menial for the value of a draught cow; a young child for the value of a lamb[1]." Again in the chapter on Contracts: "The third is the contract to the amount of a sheep, the fourth is the contract to the amount of an ox, the fifth is the contract to the amount of a man (human being), the sixth is the contract to the amount of a field, a field in good land, a fruitful one in good bearing[2]."

From these extracts it is plain that the ancient Persians had a system of clearly defined relations in value between all their worldly gear, whether the object was a slave or an ox, or a lamb or a field, precisely like that existing at the present moment among the hill tribes of Annam. But not simply was it between one kind of animal and another, but they had evidently strict notions as regards the inter-relations in value of different animals of the same kind; thus the ox of high value, the ox of low value, the cow of three years old, or the bull all stood to one another in a fixed relationship. We may without hesitation conclude that the same system of conventional values prevailed among the ancient Hindus. Nor can we doubt that articles of every kind, such as arrows, spears, axes, and articles of personal use and adornment all had their regularly recognized prices, and that the less valuable of them were used as small change. Gold, no doubt, occupied an important place in relation to the other forms of property in portions of fixed size or weight, as in the days of Marco Polo. In mediæval times in parts of India money consisted of pieces of iron worked into the form of large needles, and in some parts stones which we call cat's eyes, and in others pieces of gold worked to a certain weight were used for moneys, as we are told by Nicolo Conti, who travelled in India in the 15th century[3]. If iron was so employed at this late date we may well infer that bronze and afterwards iron were probably so used by the ancient Indo-Iranian people.

Among the fishermen who dwelt along the shores of the Indian Ocean, from the Persian Gulf to the southern shores of

---

[1] *Vendidâd, Fasgard*, IX. 37. [2] *Ibid.* IV. 2.
[3] Hakluyt Society, 1857, p. 35.

Hindustan, Ceylon and the Maldive islands, it would appear that the fish-hook, to them the most important of all implements, passed as currency. In the course of time it became a true money, just as did the hoe in China. It still for a time retained its ancient form, but gradually became degraded into a simple piece of double wire, as seen in Nos. 3 and 4 of our

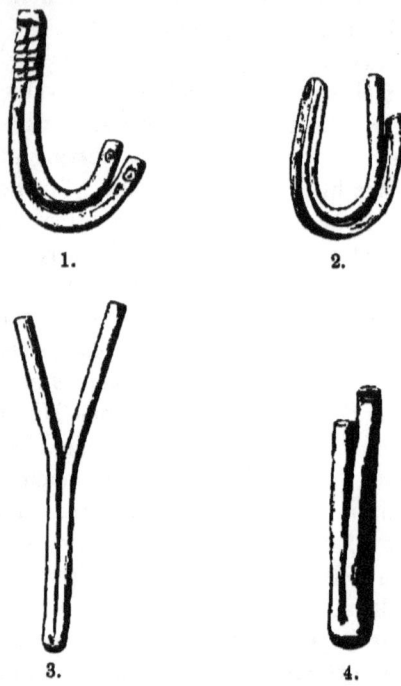

FIG. 6. Fish-hook money (*Larina*).

illustration. In its conventional form it is known as a *larin* or *lari*, a name doubtless derived from Lari on the Persian Gulf. These *larins* made both of silver and bronze were in use until the beginning of the last century, and bear legends in Arabic character. Had the process of degradation gone on without check, in course of time the double wire would probably have shrunk up into a bullet-shaped mass of metal, just as the Siamese silver coins are the outcome of a process of degradation from a piece of silver wire twisted into the form of a ring and doubled up, which probably originally formed some kind of ornament. The bullet shaped *tical* is now struck as a coin of European form. Just as perhaps the silver shells of Burmah be-

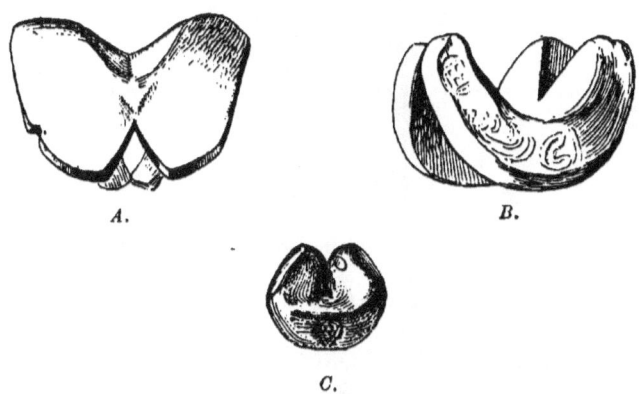

Fig. 7. Siamese silver bullet money: A. B. Early form as simple piece of wire. C. Last stage of degradation.

came the multiple unit of a large number of real cowries, so the fish-hook made of silver came into use as a multiple unit, when the bronze fish-hook had already become conventionalized into a true coin. The silver *larins* of Ceylon weigh about 170 grs. troy, and those of Southern India are said by Professor Wilson to weigh the same, although some of them weigh only 76 grs. or less than half. As the rupee weighs about 180 grs. the silver fish-hook may represent the usual unit employed for silver, strong national conservatism requiring that the silver currency should take the same form as the ancient fish-hook currency of bronze[1]. There are still in circulation in Nejd in Arabia small bars of silvered brass, which bear on the back Arabic inscriptions. It is hardly possible to doubt that in these little pieces of metal we have the last surviving descendants of

[1] For *larins* cf. Prof. Rhys Davids, "On the Ancient Coins and Measures of Ceylon" (*Numismata Orientalia*, Vol. I. 68—73). Mr Rhys Davids makes no mention of the bronze fish-hooks, but there are a number of them in the British Museum.

the old fish-hook. In the Maldive Isles a silver *larin* was worth 12,000 cowries.

FIG. 8. Silvered brass bars used as money in Nejd[1].

Advancing westward we find the Ossetes of the Caucasus at the present moment employ the cow as their unit of value, the prices of all commodities being stated as one, two, three or four cows, or even at one-tenth or one-hundredth of the value of a cow. The ox is worth two cows, and the cow is worth ten sheep. This people regulate compensation for wounds thus: they measure the length of the wound in barley corns, and for every barley corn which it measures a cow has to be paid[2]. We can have little doubt that over all Hither Asia the same method of employing the cow as the principal unit of value obtained. It is that which we found among the Greeks of the Homeric Poems, who were in full contact with Northern Asia Minor, and was almost certainly that of the Semites who dwelt in the South. Just as we find the buffalo, and the pots, bronze platters, arrows, lances and hoes standing side by side in well defined mutual relation among the Bahnars of Cochin China, so we find in Homer that whilst the cow is the principal unit, the slave is employed as an occasional higher unit, and the kettle (*lebes*), the pot (*tripous*), the axe and the half axe,

---

[1] I am indebted to the kindness of Mr A. Galetly of the Edinburgh Museum of Science and Art for the drawing from which the figure here shown is reproduced, as also for the drawing of the Calabar wire money and West African axe money figured lower down. My friend Mr J. G. Frazer (one out of countless kindnesses) called my attention to all three objects.

[2] Haxthausen, *Transkaukasia* II. p. 30 (Engl. Trans. p. 409).

hides, raw copper and pig iron stand beside the cow as multiples or sub-multiples. When Ajax and Idomeneus make a bet on the issue of the chariot race, the proposed wager is a pot or a kettle[1], whilst from another passage we learn that the usual prizes given at the funeral games of a chieftain were female slaves and pots (Tripods).

Passing from Greece into Italy we have no difficulty in proving that the cow was the regular unit of value in that peninsula and the adjacent island of Sicily. Down to 451 B.C. all fines at Rome were paid in cows and sheep. By the Tarpeian Law these were commuted for payments in copper, each cow being set at 100 asses, each sheep at 10 asses. As I shall deal with the whole question of the Roman As at considerable length later on I shall here simply note that the Italian tribes had evidently the same system of adjusting the relations between their cattle and sheep and their metals which we found among the Persians and modern Ossetes. In Sicily it is clear that the cow had played the same part as elsewhere, for we learn from Aristotle[2] that when the tyrant Dionysius burdened the Syracusans by excessive taxation, they ceased in a great degree to keep cattle, inasmuch as the unit of assessment was the cow. If then in the 4th century B.C. at Syracuse, the most advanced community in Sicily, the cow still continued to be the unit of assessment, *à fortiori*, at an earlier period that animal must have been the monetary unit of the whole island.

From the Italians we pass on to their close kinsmen the Kelts. We are told by Polybius[3] that when the Gauls entered Italy, their wealth consisted of their cattle and gold ornaments, but although an argument will be offered below to show that the cow was the monetary unit of both Gauls and Germans, we have no definite evidence respecting the barter system. But fortunately the Ancient Laws of Wales and Ireland afford us ample insight into the Keltic system. Irish tradition goes back far beyond the date at which the Brehon Laws were compiled, and from it we get a glimpse of a system almost Homeric: thus we read in the *Annals of the Four Masters*

[1] *Il.* XXIII. 485.   [2] *Oecon.* II. 21.
[3] II. 18.

under the year 106 A.D. that the tribute (*Boroimhe*, literally cow-tax) paid by the King of Leinster consisted in 150 cows, 150 swine, 150 couples of men and women in servitude, 150 girls and the king's daughter in like servitude, 150 caldrons, with two passing large ones of the breadth and depth of five fists[1]. As this tradition makes no mention of payment in metals, but only of slaves, cattle and caldrons, which doubtless stood to one another in well defined relations, we need have no hesitation in assuming that the cow formed the chief unit of the earlier, as it did of the later Kelts.

The Welsh naturally adopted the monetary system which sprang up after the reign of Constantine the Great in the Later Empire. Accordingly we find in certain of their Ancient Laws[2] tables giving in *denarii*, *solidi* or *librae* the values of various kinds of property. From these we can learn with accuracy the relations in value which existed between various kinds of property. Thus the calf from March (when the cows calved) to November was worth 6 *denarii*, to the following February 8 *den.*, till May 10 *den.*, till August of the second year 12, till November 14 *den.*, till February 15 *den.*, till February of the third year 28 *den.* The heifer is then in calf, her milk is worth 16 *den.* Thus the milch cow is worth 46 *den.*, and up to August she is worth 48 *den.*, up to November 50 *den.*, and up to May of the fourth year is worth 60 *den.* A month's milk is worth 4 *den.*; a bull calf 6 *den.*, the young ox when put to the plough is worth 28 *den.*, when he can plough, 48 *den.*, that is the same as the young milch cow of the same age; a gelding is worth 80 *den.*, a farmer's mare 60 *den.*, a trained horse is worth half a *libra*; a bow with twelve arrows is worth 7 *denarii* and an *obolus*; a queen bee (*modred af*) is worth 24 *den.*, the first swarm 16 *den.*, the second 12, the third 8; a foal is worth 18 *den.* to 24 *den.*, a two year old 48 *den.*, a three year old 96 *den.* A young male slave (*iuvenis captivus*) is worth 1 *libra*, a slave both young and of large stature (*captivus iuvenis et magnus*) is worth 1½ *libra*. It would appear that the Welsh, when taking over the Roman system, had adjusted their own highest barter-unit, the slave

---

[1] *Annals of the Four Masters*, Anno 106 A.D. (O'Donovan's ed.).
[2] *Ancient Laws of Wales*, p. 795.

(probably female as well as male), to the *libra* or pound, the highest unit in the Roman system. Of course slaves of exceptional strength or beauty would always command a higher price. But the regulations for the value of cattle are especially of interest, as shewing the extraordinary minuteness with which pastoral peoples discriminate the values of animals of different ages, and estimate the milk of a cow in proportion to her actual value. The full-grown cow is worth exactly ten times the new-born calf, an estimate which holds good just as much in 1890 as it did 1000 years ago, for it is not a mere convention but is based upon a natural law. At the present moment a calf is worth from 30 to 35 shillings, a cow from £15 to £17. 10*s*. The yearling calf was worth one-sixth of the full-grown cow, a relation which still holds good.

The Irish Kelts borrowed their silver system from Rome at a period probably before Constantine, as they seem never to have employed the *libra* and *solidus*, but simply the *uncia* (*unga*) and *scripulus* (*screapall*), adding thereto a subdivision called the *pinginn* or penny, borrowed doubtless from the Saxon invader at a later period. Thus 1 unga = 24 screapalls; 1 screapall = 3 pinginns. They equated the principal silver unit, the *uncia*, to the old chief barter-unit, the cow (*bo*). As elsewhere, however, the slave formed occasionally the highest unit, and was reckoned nominally at three cows. The slave woman (*cumhal*, *ancilla* in Latin writers) was in course of time used as a mere unit of account.

Slave woman (*cumhal*, *ancilla*)	= 3 ounces (*unga*)
Full-grown cow (*bo mor*)	= 1 ounce = 24 screapalls
Heifer now in third year (*samhaisc*)	= $\frac{1}{2}$ ounce = 12 screapalls
Heifer of second year (*colpach*)	= 6 screapalls
Yearling (*dairt*)	= 4 screapalls
A cow's milk for summer and harvest	= 6 screapalls
A sheep	= 3 screapalls
A goat's milk for summer and harvest	= $1\frac{3}{4}$ pinginn
A sheep's fleece	= $1\frac{1}{2}$ pinginn
A sheep's milk	= $\frac{1}{2}$ pinginn
A kid (*meinnan*)[1]	= $\frac{2}{3}$ pinginn.

[1] O'Donovan's Supplement to O'Reilly, s.v. *Lacht*: *Senchus Mor*, I. 287.

Here again the yearling is worth one-sixth of the cow. Gold was abundant among the ancient Irish, (almost certainly obtained in large quantities from the Wicklow mountains,) and passed from hand to hand in the form of rings, which were weighed on a system different from and probably far older than that employed for silver (see Appendix A).

Passing to the Teutonic peoples we find traces of the same ancient practice. For according to one system a *mancus* of silver (a mere unit of account) corresponded with the value of an ox. Similarly the *pound* (*libra*) was generally regarded as the silver equivalent of the worth of a man[1]. But the strongest proof is that Charlemagne in his dealings with the Saxons found it necessary to define the value of his *solidus* of 12 pence (*denarii*) by equating it to the value of an ox of a year old of either sex in the autumn season, just as it is sent to the stall. In the same law we find a list of regulation prices for other commodities, such as oats, honey, rye, similar to those already quoted from the Welsh laws[2]. The English word *fee*, which originally meant an ox, as is shown not only by the German *Vieh*, which still retains its original meaning, and by such expressions in Anglo-Saxon as *gangende feoh*, is in itself a proof that cattle served as the most generally recognized form of money. It might be expected that much the same state of things existed among the Scandinavian peoples. Their chief media of exchange were cows, and woollen cloths, slaves, and gold ornaments. By the laws of Hakon the Good penalties could be paid in cows, provided that they were not too old,

[1] Thorpe, *Laws of the Anglo-Saxons*, I. 357. Cunningham, *History of English Commerce*, I. 117.

[2] Illud notandum est quales debent solidi esse Saxonum : id est, bovem annoticum utriusque sexus, autumnali tempore, sicut in stabulum mittitur, pro uno solido: similiter et vernum tempus, quando de stabulo exiit; et deinceps, quantum aetatem auxerit, tantum in pretio crescat. De annona vero bortrinis pro solido uno scapilos quadraginta donant et de sigule viginti. Septemtrionales autem pro solidum scapilos' triginta de avena et sigule quindecim. Mel vero pro solido bortrensi, sigla una et medio donant. Septemtrionales autem duos siclos de melle pro uno solido donent. Item ordeum mundum sicut et sigule pro uno solido donent. In argento duodecim denarios solidum faciant. Et in aliis speciebus ad istum pretium omnem aestimationem compositionis sunt. *Capitulare Saxonicum*, II. Migne, XCVII. 202.

in slaves, provided they were not under fifteen years of age, in cloths, and in weapons[1].

Gold and silver were employed by the northern peoples in the form of rings.

This has led people to talk much about *ring money* as if it was a true currency, circulating like the stamped money of later times. The truer view seems to be that these rings, whether employed by the ancient Egyptians or the prehistoric inhabitants of Mycenae, the Kelts or Teutons, were nothing more than ornaments and passed in the ordinary way of barter, having a recognized distinct relation to other forms of property, such as cattle and slaves. It has been the custom in all countries for the person who desires to have an article of jewellery made to give to the goldsmith a certain weight of gold or silver, out of which the latter manufactures the desired ornament. Such is the practice at the present day in India; you give the goldsmith so many gold mohurs or sovereigns, or rupees, as the case may be; he squats down in your verandah, and with a few primitive tools quickly turns out the article you desire, which of course will weigh as many mohurs or sovereigns as you have given him (provided that you have stood by all the time, keeping a sharp look-out to prevent his abstracting any of the metal). That in like fashion gold ornaments for ordinary wearing purposes were regularly of known weights in ancient times is shown clearly by the account of the presents given to Rebekah by Abraham's servant, 'a gold earring of half a shekel weight and two bracelets for her hands of ten shekels weight' (Genesis xxiv. 22). The same word appears in Job xlii. 11: 'Then came there unto him all his brethren and all his sisters and all that had been of his acquaintance before...every man also gave him a piece of money and every one an *earring* of gold.' Consequently Rebekah's golden ring (whether it was to adorn her nose or ear) of half a shekel weighed 65 grains, being half the light shekel or ox-unit. We are not told the weight of the earrings contributed by his sympathetic kinsfolk for the afflicted patriarch, but it is evident that they were of a uniform standard. No doubt such rings had from time im-

[1] Schive and Holmboe, *Norges Mynter* (Christiania, 1865), pp. i—iii.

memorial passed in the ordinary course of barter from hand to hand. This is strongly supported by a piece of evidence produced independently of the previous suggestion by Dr Hoffmann of Kiel, who has showed[1] that *betzer* (בצר), the word used for gold in Job xxii. 24—25 (*bĕtzēr*) and in Job xxxvi. 19 (*b'tzar*), from a comparison of its cognates in Hebrew and Arabic means simply a *ring*, which through the extended meaning *ring-gold* came finally to be used as a name for the metal simply. To take another example from a very different region, the golden ornaments of the ancient Irish (of which numerous specimens exist in the Museum of the Royal Irish Academy) were made according to specified weight. Thus queen Medbh is represented as saying: 'My spear-brooch of gold, which weighs thirty ungas, and thirty half ungas, and thirty crosachs and thirty quarter [crosachs].' O'Curry, *Manners and Customs of Ancient Irish*, iii. 112. But we need not go beyond Greek soil itself for such illustrations. The well-known story of Archimedes and the weight of the golden crown, which led to the discovery of specific gravity, is sufficient to show that the practice in Greece was such as I describe.

The rings seen on Egyptian monuments (of which we give a representation in a later chapter) are of round wire; those found by Schliemann in the tombs of Mycenae[2] (Fig. 9) consist both of round wire rings like the Egyptian, and likewise of spirals of quadrangular wire. As *finger* rings (δακτύλιοι) are not mentioned in Homer, it has been assumed that the Homeric Greeks did not employ rings at all. Hence in a famous passage where the ornaments made by Hephaestus for the goddesses are described, we find mention of brooches, *bent spirals* (ἕλικες) ear-drops[3], and chains. Helbig[4] explains the *helikes* as a kind of brooch made of four spirals, such as are worn in parts of Central Europe, but it is difficult to believe that people who were using brooches with pins and necklaces would not have known and employed the far simpler ring. Again, why should

---

[1] G. Hoffmann, *Zeitschrift für Assyriologie*, Vol. II. (1887) p. 48.
[2] Schliemann, *Mycenae*, and *Tiryns*, p. 354.
[3] *Il.* XVIII. 401 πόρπας τε, γναμπτάς θ' ἕλικας, κάλυκάς τε, καὶ ὅρμους.
[4] Homer, *Epos*, 279—281 (2nd ed.).

we find two distinct words for brooches coming thus together? Is it not far more likely that in the spirals of Mycenae we have

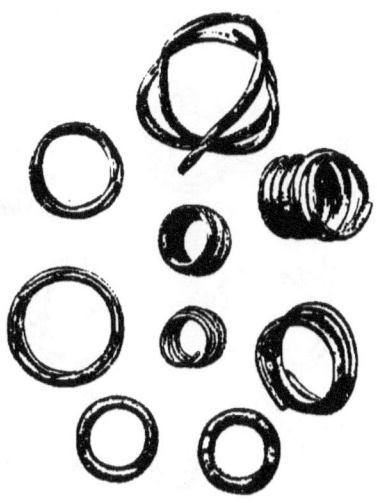

Fig. 9. Rings found in the tombs of Mycenae.

the real *bent helikes* of Homer? These spirals would serve not only for finger rings, but might be used in the hair, or more probably still were used as a means of fastening on the dress, being passed through eyelet holes or loops, on the principle of the modern key ring[1]. On comparing them with the Scandinavian spiral (Fig. 1) the reader will see that this primitive form of employing gold was widely diffused over Europe. The Scandinavians used such ornaments of *bent* wire (O.N. *baugr*, A.S. *beag* from root BUG, *to bend*) very commonly, beside oxen and other property, as media of exchange. Thus both *beag* in Anglo-Saxon, and *baugr* in Old Norse became used as general names for treasure. Thus *baugbrota* (cf. *hring brota*), literally

---

[1] Hesychius s.v. ἕλικες explains them as *earrings* (ἐνώτια), or *armlets*, *anklets* (ψέλλια), or *rings* (δακτύλιοι). Eustathius on *Iliad* XVIII. 400 explains them as ἐνώτια ἢ ψέλλια παρὰ τὸ εἰς κύκλον ἐλίσσεσθαι, "earrings or armlets (anklets), so called from being rolled up" (*helissesthai*). Cp. Ebeling, *Lexicon Homericum*, s.v. ἕλιξ.

ring-breaker, was used as an epithet of princes, meaning *distributor of treasure*[1].

The same spirals of quadrangular wire were probably employed by the Kelts, as that shown in Fig. 10, No. 3 was found

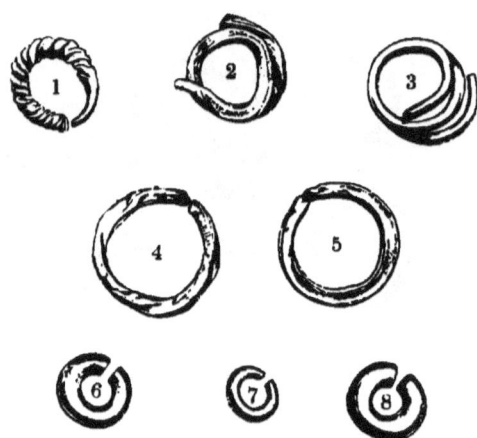

Fig. 10. Nos. 1, 2, found in Tipperary; 3, Scandinavian; 4, 5, found in Co. Mayo; 6, 7, 8, ordinary Irish type.

in Ireland; Nos. 4 and 5 are of quadrangular wire but are simple hoops, whilst in Nos. 6, 7, 8, we get the regular Irish type of a round wire not completely closed[2]. The latter probably represent a more advanced state of art, as their makers must have had considerable metallurgic skill, No. 8 being made of gold plated over a copper core.

As we shall see further on, the Egyptian rings are made on a standard almost identical with the Homeric talent, and I have shown elsewhere that the rings from Mycenae were made on almost the same standard[3]. I shall endeavour to show in an Appendix that the Irish rings also show evidence of being

[1] Keary, *Catalogue of Anglo-Saxon Coins*, I. p. vii. From *beag* Mr Max Müller derives *buy* in spite of a phonetic difficulty.

[2] Nos. 1, 2, 3, 4, 5 are in the collection of my friend Mr R. Day, F.S.A., of Cork. The others are in my own possession.

[3] *Journal of Hellenic Studies*, Vol. x. Here is the description and weight

made on a definite standard, whilst it has been long well known that the Scandinavian rings and armlets have likewise a standard of their own.

When occasion arose they cut off a piece of this bent wire (for it was really nothing more), and gave it by weight. Such a piece was called a *scillinga*, and is the direct ancestor of our own *shilling*[1]. It is not unlikely also that the ancient inhabitants of Portugal employed similar pieces of wire, as Strabo tells us that the Lusitanians have no money, but that they employ silver wire, from which they cut off a portion when necessary[2].

We now pass on to Africa, where we shall find most varied systems of currency. Thus on the West Coast of Africa the *bar* is the unit. In fact all merchandise is reckoned by the bar[3], which now at Sierra Leone means 2s. 3d. worth of any kind of the rings (which I have been enabled to figure by the kindness of Mr John Murray):

METAL	DESCRIPTION	WEIGHT	
		GRAMMES	GRAINS TROY
Silver	Plain ring	8·8	137
Gold	Spiral	8·5	132
,,	,,	9·9	153
,,	,,	10·8	167
,,	Plain ring	15·9	248
,,	,,	16·5	257
,,	,,	19·0	297
,,	,,	19·4	303
,,	Spiral	20·5	320
,,	,,	21·5	335
,,	Plain ring	22·0	340
,,	Spiral	29·3	452
,,	,,	39·0	612
,,	,,	39·5	617
,,	,,	41·5	643
,,	,,	42·2	654
,,	,,	42·3	655
,,	,,	42·8	662

[1] Cf. Keary's *Catalogue of English Coins in the British Museum*, p. 6.

[2] Strabo iii. p. 155. ἀντὶ δὲ νομίσματος οἱ λίαν ἐν βάθει φορτίων ἀμοιβῇ χρῶνται ἢ τοῦ ἀργύρου ἐλάγματος ἀποτέμνοντες διδόασιν.

[3] Gordon Laing, *Travels in Western Africa* (1825), Prefatory Note.

of commodity, although originally it meant simply an iron bar of fixed dimensions, which formed the chief article of exchange

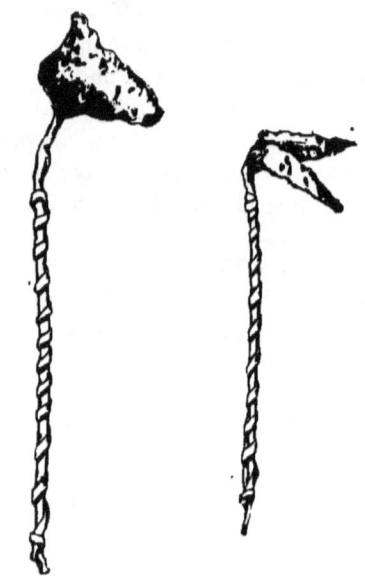

Fig. 11. Axe Money (West Africa).

between the natives and the earliest European traders. In other parts of the same region axes serve as currency; these are too small to be really employed as an implement, but are doubtless the survival of a period not long past when real axes served as money. Thus we get a complete analogy to the hoe money of the Chinese and the fish-hook currency of Ceylon and the Maldive Islands. In Calabar they formerly employed bunches of quadrangular copper-wire as currency. Each wire was about 12 inches long, and they were of course meant to be made into necklets and armlets[1].

In other parts of the West Coast, as in the Bonny River territory, iron rings very closely resembling in shape the bronze

[1] The specimen figured was brought home about 30 years ago and is now in the Edinburgh Museum of Science and Art.

fibulae found in Ireland, which probably were armlets, are employed as money. Those which I have seen seem too small to

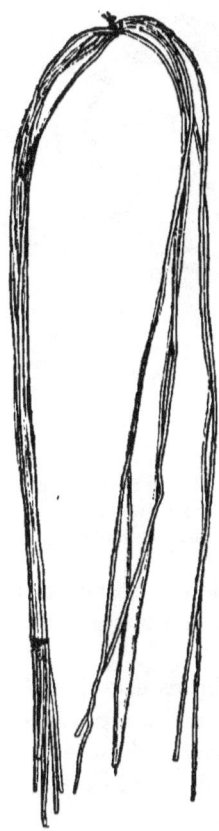

FIG. 12. Old Calabar copper-wire formerly used as money.

be used as bracelets, and are now probably a true money, retaining the old conventional shape (see Fig. 12)[1].

In the region of the Upper Congo brass rods are employed as currency for articles of small value. This wire, made at Birmingham, about the thickness of ordinary stair-rod, is sent out in coils of 60 lbs., and is then cut into pieces of a foot

[1] The specimens here figured are in the splendid collection of my friend Mr R. Day, of Cork.

long[1]. Short brass rods and armlets are also largely exported from Birmingham for the African trade.

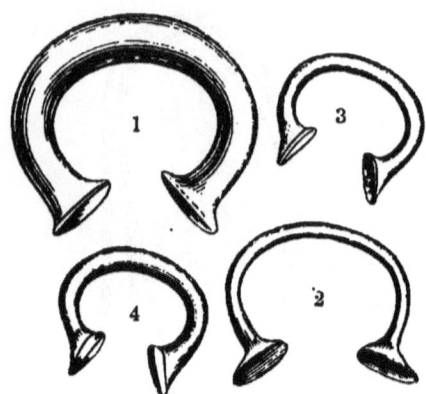

Fig. 13.

1. Bronze Irish Fibula found in Co. Cork.
2. Bronze Irish Fibula found in King's Co.
3. Iron Manilla from W. Africa.
4. Iron Manilla used as money in Bonny River Territory.

There is no absolute standard length—and thus while 36 inches is the one most commonly used, the length varies from 32 to 36 inches.

They go out in boxes containing 100, in straight lengths, and soft to admit of their being wound into armlets, &c.

The diameter of the rod varies from $\frac{3}{10}$ in. to about $\frac{3}{8}$ in.— but a rod weighing about 24 oz. to 3 ft., and $\frac{3}{8}$ in. thick, is the one most often made.

Arm rings are made from solid brass rod about $\frac{7}{10}$ in. thick and are usually 2¼ in. to 3½ in. in diameter—they are also made in large quantities from brass tubes of ½ in. to $\frac{5}{8}$ in. diameter, more frequently from $\frac{9}{10}$ in., the rings being from 2½ in. to 3½ in. in diameter, and weighing from 2½ to 4 oz. each[2].

Slaves and ivory tusks form the chief units in the same region. The slave usually is worth a tusk. In other parts

---

[1] This information I owe to Lieut. Troup.
[2] I am indebted to Messrs James Booth and Co. for this information.

pieces of precious wood of a red colour, each piece being a foot long, were employed as currency[1].

When we come to regions where the ox can live we at once find that animal occupying a foremost place. Thus when the Cape of Good Hope was first colonized, the Hottentots employed cattle and bars of iron of a given size as currency[2], and at the present moment the cow is the regular unit among the Zulus, ten cows being the ordinary price paid for a wife, although as in Homeric Greece fancy prices are paid by the chiefs for ladies of uncommon attractions. But our chief interest must centre in the peoples north of the Equator, who from time immemorial have been in contact with the ancient civilization of the Mediterranean.

Thus among the Madis of Central Africa, a pure negro tribe, cattle form the chief wealth; a rich man may have as many as 200 head, a very poor one only 3 or 4. The average number possessed by one man is from 30 to 40. They keep the milk in gourds.

"A regular system of exchange is carried on in arrows, beads, bead necklaces, teeth necklaces, brass rings for the neck and arms, and bundles of small pieces of iron in flat, round, or oval discs. All these different articles are given in exchange for cattle, corn, salt, arrows, etc. The nearest approach to money is seen in the flat, round pieces of iron which are of different sizes, from three-quarters to two feet in diameter and half an inch thick. They are much employed in exchange. This is the form in which they are kept and used as money, but they are intended to be divided into two, heated and made into hoes. They are also fashioned into other implements, such as knives, arrow-heads, etc. and into little bells hung round the waist for ornament or round wandering cows' necks. Ready-made hoes are not often used in barter. Iron as above-mentioned is preferred and is taken to the blacksmith to be

---

[1] Dapper *Description de l'Afrique* (Amsterdam, 1686) p. 367. "Le bois rouge de Majumba et la *pao* de Hiengo de Benguela tiennent aussi le lieu de monnaie: on en coupe des morceaux d'un pied de long; on leur met une certaine taxe selon laquelle le prix des vivres se règle."

[2] Peter Kolben, *Present state of the Cape of Good Hope*, p. 262.

fashioned according to the owner's requirements. Any tools may be obtained ready made from a smith, and can be used in barter when new.

"Compensation for killing a woman or any serious crime must be paid for in cattle. No cowries are used as coins in this district, no measure of weight, quantity or length is used. The payment for a wife must be made in cows of a year old, or in bulls of two or three years[1]."

But it is in Darfour and Wadai that we find the primitive system in its fullest form. Wives are bought with cows, 20 of which with a male and female slave are the usual price of a wife, hence the Darfouris prefer daughters to sons. Hence the proverb that girls fill the stable, but boys empty it, which recalls the *cow-winning maidens* of Homer (παρθένοι ἀλφεσί-βοιαι). There is absolutely no metal of any kind in Darfour, except that which is imported. Having no money, they accept certain articles as having a certain monetary value.

Facher was the first place in Darfour which had anything like a currency; it consisted of rings made of tin, which were employed in the purchase of every-day necessaries of life. These rings are called *tarneih* in Darfouris. There are two kinds, the heavy ring and the light ring; the light serves for buying the most trivial articles. For purchasing articles of value they have the *toukkiyeh*, a piece of cotton cloth six cubits long by one broad. There are two kinds of this stuff, *chykeh* and *katkât*. Four pieces of the former and 4½ pieces of the latter are worth a Spanish dollar. Buying and selling is also carried on by means of slaves: thus one says, "this horse is worth 2 or 3 *sedâcy* (a name given to a negro slave, who measures six spans from his ankle to the lower part of his ear)[2]." A *sedaciyeh* is a female negro slave of the same height. A *sedâcy* is worth 30 *toukkiyeh*, or six blue *chauter*, or 8 white *chauter* or six oxen, or 10 Spanish pillar dollars, the only coined money known in Darfour, where it is called *abou*

---

[1] R. W. Felkin, "Notes on the Madi or Moru Tribe of Central Africa," *Transactions of Royal Society of Edinburgh*, Vol. XII. p. 303 *seqq*.

[2] *Voyage au Darfour*, Mohammed Ibn Omar el Tounsy (translated by Perron), Paris, 1845, pp. 218, 315.

medfa, i.e. *cannon piece*, the pillars being taken for cannons. The inhabitants of Kobeih employ beads for money, which are called *harich*. They are green and blue and circulate in strings of 100 each. This bead takes the place of the tin ring (*tarneih*) used at Facher in the purchase of cheap commodities. The *harich* as money is employed in numbers of from 5 to 100 beads (the string), from one string to ten and indefinitely further[1].

The *toukkiyeh* is worth in the markets mentioned 8 strings of *harich*. Thus a *seddcy* is worth 240 strings. At Guerly and its environments the *falgo* or stick of salt almost as big as one's finger is employed. This salt is obtained artificially, and when liquid is poured into little moulds of baked clay. This salt is sold by the *falgo*, not by weight, and one buys by 1, 2 or 3 *falgo* according to the value of the article.

At Conca tobacco is used as money. At Kergo, Ryl, and Chaigriyeh articles of moderate value are bought with hanks of cotton thread. These threads are ten *ells* long, and there are only 20 threads in each hank. For common articles raw cotton with the pods attached is given; it is not weighed but simply estimated by guess. At Noumleh onions are employed as money for common articles, and the *rubat* or hank of thread, and *toukkiyeh* for the more valuable, whilst the *chauter* and dollar are unknown.

At Ras-el-Fyk[2] the hoe (*hachâchah*) serves as currency. It is simply a plate of iron fitted with a socket. A handle is fitted into this socket, and one has an implement suited for chopping the weeds in the corn fields. Purchases of small value are made with the hoe from 1 to 20: above that amount the *toukkiyeh* is employed and likewise the *chauter*.

At Temourkeh they use as moneys cylindrical pieces of copper (called *damleg*) for articles of some value, whilst a kind of glass bead called *chaddour* is used for small articles. Near Ganz, the eastern part of Darfour, the principal article of exchange is the *doukha* for articles of moderate value. They give it by the handful, or by the double handful up to the

[1] *Voyage au Darfour*, p. 316.
[2] *Ibid.* p. 319.

amount of half a *moda*; whilst as elsewhere articles of value are bought by the *toukkiyeh* or dollar. In a very great number of places merchandise is exchanged against oxen; thus the horse is worth 10 to 20 oxen.

Accordingly while each district of Darfour has some peculiar form of currency for small change the higher currency is the same everywhere, the piece of cloth, the ox, the slave[1].

In the region of Wadai the same shrewd Arab tells us that cattle are kept by even the most barbarous tribes[2]. Thus the Fertyt tribe, who go in a state of almost complete nudity, and thus have no need of cloth, possess large herds of cattle, which are not branded, but each owner distinguishes his cattle by giving a peculiar shape to their horns as soon as they begin to grow. In the less barbarous communities of Wadai slaves and beads are employed as currency as well as cattle. The bead used is called the *mansous*. It is of yellow amber and of different sizes. Number 1 is so called because one string (containing 100 beads) weighs one *rotl* (pound) of 12 ounces; Number 2 because two strings weigh a *rotl*; Number 3 because 3 strings make a *rotl* and so on. The first is the most costly of all beads. Often a single bead of this sort (*soumyt*) is worth two slaves; if it is abundant each bead is worth a slave.

[1] *Voyage au Darfour*, p. 321.
[2] *Voyage au Ouadai*, Mohammed Ibn Omar el Tounsy (French translation by Perron), p. 559.

CHAPTER III.

THE DISTRIBUTION OF THE OX AND THE DISTRIBUTION OF GOLD.

> And round about him lay on every side
> Great heapes of gold that never could be spent,
> Of which some were rude owre not purified
> Of Mulciber's devouring element.
> Some others were new driven and distent
> Into great Ingowes and to wedges square,
> Some in round plates with outen ornament,
> But most were stampt and in their metal bare
> The antique Shapes of Kings and Kesars straunge and rare.
>
> SPENSER, *Faerie Queen*, II. vii.

LET us now take a general survey of the results of our observations. First of all it is apparent that the doctrine of a primal convention with regard to the use of any one particular article as a medium of exchange is just as false as the old belief in an original convention at the first beginning of Language or Law. Every medium of exchange either has an actual marketable value, or represents something which either has or formerly had such a value, just as a five-pound note represents five sovereigns, and the piece of stamped walrus skin formerly employed by Russians in Alaska in paying the native trappers represented roubles or blankets[1].

To employ once more the language of geology, we have found evidence pointing to certain general laws of stratification. In Further Asia we have found a section which presents us with an almost complete series of strata, whilst in other places where we have been only able to observe two or three layers,

---

[1] Elliot's *Alaska*, p. 8. This is an interesting parallel to the ancient tradition that the Carthaginians employed leather money. (*Vide* Smith's *Dict. of Geogr.* I. 545.)

we have nevertheless found that certain strata are invariably found superimposed upon others, just as regularly as the coal seams are found lying over the carboniferous limestone. As soon as the primitive savage has conceived the idea of obtaining some article which he desires but does not possess by giving in exchange to its owner something which the latter desires, the principle of money has been conceived. Shells or necklaces of shells are found everywhere to be employed in the earliest stages. When some men began to make weapons of superior material, as for instance axes of jade instead of common stone, such weapons naturally soon became media of exchange; when the ox and the sheep, the swine and the goat are tamed, large additions are made to the circulating media of the more advanced communities; then come the metals; the older ornaments of shells and implements of stone are replaced by those of gold (and much later by silver) and by weapons of bronze as in Asia and Europe, and by those of iron in Africa. Copper and iron circulate either in the form of implements and weapons, such as the axes of West Africa, the hoes of the early Chinese and modern Bahnars, and the ancient Chinese knives, all of which remind us of the axes and half-axes in Homer; or in the form of rings and bracelets, like the manillas of West Africa and the ancient Irish fibulae; or else in the form of plates or bars of metal, ready to be employed for the manufacture of such articles, as we saw in the case of the iron bars of Laos, the iron discs of the Madis, and the brass rods of the Congo. Again we are reminded of the mass of pig-iron, which Achilles offered as a prize[1].

It is of the highest importance to observe that such pieces of copper and iron are not weighed, but are appraised by measurement. We shall find that it is only at a period long subsequent to the weighing of gold that the inferior metals are estimated by weight. The custom of capturing wives which prevails among the lowest savages is succeeded by the custom of purchasing wives. The woman is only a chattel on the same footing as the cow or the sheep, and she is accordingly appraised in terms of the ordinary media of exchange employed in her community, whether it be in cows,

[1] *Il.* XXIII. 826.

horses, beads, skins or blankets. Presently male captives are found useful both to tend flocks and, as in the East and in the modern Soudan, to guard the harem. With the discovery of gold, ornaments made at first out of the rough nuggets supersede other ornaments, and presently either such ornaments or portions of gold in plates or lumps are added to the list of media, and the same follows with the discovery of silver. Such ornaments or pieces of gold and silver are estimated in terms of cattle, and the standard unit of the bars or ingots naturally is adjusted to the unit by which it is appraised. Thus we found the Homeric talent, the silver bar of Annam, the Irish *unga* all equated to the cow, and the Welsh *libra*, Anglo-Saxon *libra*, similarly equated to the slave. With the discovery of the art of weaving, cloths of a definite size everywhere become a medium, as the silk cloth of ancient China, the woollen cloths of the old Norsemen, the *toukkiyeh* of the Soudan, and the blanket of North America. This fact once more recalls Homer and makes us believe that the robes and blankets and coverlets which Priam brought along with the talents of gold to be the ransom of Hector's body all had a definite place in the Homeric monetary system[1].

We have seen the Siamese piece of twisted silver wire passing into a coin of European style, and we shall find that the Chinese bronze knife has finally ended by becoming a *cash*, just as we have already found the Homeric talent of gold appearing, in weight at least, as the gold stater of historical times. Thus in every point the analogy between what we find in the Homeric Poems and in modern barbarous communities seems complete. We may therefore with some confidence assume that we are at liberty to fill up the gaps in the strata of Greek monetary history which lie between Homer and the beginning of coined money on the analogy of the corresponding strata in other regions. This assumption, resting on a broad basis of induction and confirmed, as we shall see, by a good deal of evidence special to Greece and Italy, will be found to explain the origin, not only of weight standards in those countries, but

[1] *Il.* xxiv. 230—2.

also of the Greek *obol* and Roman *as*, as well as of the types on the oldest coins, such as the cow's head of Samos, the tunny fish of Olbia and Cyzicus, the axe of Tenedos, the tortoise of Aegina, the shield of Bocotia, and the silphium of Cyrene.

Let us now turn to the races who both in modern and in ancient times have dwelt around the basin of the Mediterranean and Black Sea, whether in Asia Minor, Central Asia, Europe or Africa. In what did their wealth consist? When we first meet in history the various branches of the Aryan, Semitic, and Hamitic races, they are all alike possessed of flocks and herds. To deal first with the Aryans; we have already had ample evidence that such was the case with the early Greeks. The ox plays a foremost part, and they likewise possessed sheep, goats and swine, whilst slaves formed also an important commodity. Further east again, in the Zend-Avesta the cow is found playing the principal part in every phase of the primitive life there unfolded, both as the chief article of value and in reference to their religious ceremonies. Still further to the east we find from the Rig-Veda that among the ancient Hindus the same important *rôle* was assigned to the cow. Turning now to Mesopotamia we find that in the time of Abraham the keeping of herds and flocks was the chief pursuit of the Semites. Passing on to Egypt, the hoary mother of civilization, we find evidence that although "every shepherd was an abomination to the Egyptians," yet the worship of their great divinity Apis (Hapi) under the form of a bull and the worship of the sacred ram indicate that at a period preceding the invasion of the Hyksos the Egyptians regarded the ox and the sheep with love and veneration. Whether the Egyptians came from Asia into the valley of the Nile, or whether they came from some region of Africa more to the south, one thing at least is certain, and that is that in either case they came from a country eminently fitted for the rearing and keeping of cattle. The functions of the ox became limited under altered conditions, and their ancient esteem for the cow as one of their chief means of subsistence survived only in religious observances. So too in modern India the reverence for the sacred cow amongst a people who regard as an abomination the eating of beef is a survival from the time when in a more northern clime cattle formed the principal wealth of their forefathers.

In the Soudan, as we have seen to this day, slaves and oxen are the chief kinds of property. Crossing back to Europe we find the Italian tribes represented in the earliest records as a cattle-keeping people. The story of their invasion of Italy took the form of their driving before them a steer and following obediently to whatever new home it might lead them[1].

The same holds of the more northern peoples. When the Gauls entered the plains of Northern Italy they drove before them vast herds of cattle. Caesar found the Britons keeping large numbers of cattle, and especially those in the interior of the island subsisting almost entirely on their produce[2]. Strabo writing about A.D. 1, mentions hides as among the articles exported from Britain to the Continent[3].

The linguistic argument fully supports the literary evidence. All the Aryan or Indo-European peoples possess a common name for the cow. The Sanskrit *gaus*, Greek βοῦς, Lat. *bos*, Irish *bo*, German *kuh*, Eng. *cow*, taken together indicate that before the dispersion of the various stocks (whether the original home of the Aryans was in Northern Europe, as Latham first suggested, or in the Hindu Kush, as Prof. Max Müller maintains) they all possessed the cow. This is further supported by the name for the bull which is found amongst various stocks, the Greek ταῦρος, Lat. *taurus*, Irish *tarb*, and the name of the *ox*, which corresponds to the Sanskrit *uksha*, and finally the name of *steer*[4]. Here then we have undoubted evidence of the universal possession of cattle by the Aryans at a very early period.

Archaeology lends its support likewise. We have already found in the case of the Greeks the cow used as a unit of currency side by side with gold. This leads us to the question of the precious metals, which in course of time have come to be almost the sole medium of exchange. In the case of the Greeks we saw reason to believe that the barter-unit was older than

[1] Timaeus 12.
[2] B. G. v. 12.   [3] 199.
[4] Schrader. *Prehistoric Antiquities of the Aryan Peoples*, p. 260.

the metallic. Is this the case universally? The evidence, I think, which I shall adduce will lead us to this belief.

First of all it is certain that man must have been acquainted with the ox long before he ever gathered a grain of gold from the brook. When primæval man first stood on the plains of Europe and Asia vast herds of wild cattle met his eye on every side. The process of domestication was long and slow, but yet in all the ancient refuse heaps of Scandinavia and Germany, whilst the remains of the ox are found in plenty there is yet no trace of gold.

At this point it will be well to remind the reader that the area occupied by the cattle-keeping races whom we have enumerated was continuous. There was no insuperable barrier between Indian and Persian, Persian and Mede, Mede and the dweller in Mesopotamia, or again, between Persian and Armenian, Armenian and the Scythian who lived in his ox-waggon on the plains of what is now Southern Russia: the Scythian was in contact with the tribes of the Balkan Peninsula, who in turn were in contact with the Greeks and the dwellers along the valley of the Danube, who in their turn joined hands with the peoples of Italy, Helvetia and Gaul. Hence the value of cattle would be more or less constant from one end of this entire region to the other. The purchasing power of the cow might be greater in some parts than in others, just as with ourselves a sovereign has the same value from Land's End to John o'Groats, although the purchasing power of the sovereign as regards the necessaries of life may differ widely in different places within the limits of Great Britain.

It is only when some impassable natural barrier intervenes that there will be a difference in the value of the unit of barter. Thus, in the case of Britain we cannot suppose that the value of oxen was necessarily the same there as it was on the Continent. If it was it would be merely a coincidence. The difficulty of transporting live cattle in such ships as the Gauls or Britons possessed would have been too great to permit of such a free circulation of the unit as would have kept its value exactly even on both sides of the Straits. In fact it was only with the invention of steam that facilities for transmarine cattle-trading came in which could tend to level the value on both sides of an arm of the sea. In the earlier half of this century cattle were extraordinarily cheap in Ireland in proportion to the prices which they fetched in England, but yet the difficulty and expense entailed in sending them across in sailing ships effectually prevented the export. When the first steamers began to convey cattle from Ireland to England the profits were enormous, although the freight of a single cow cost, I believe, several pounds. Steam-power has done much to equalize prices, but still there is a considerable difference in the value of cattle on both sides of the Irish Sea. But where no impassable barrier of sea or forest intervened, we may fairly assume the ox carried much the same value from Northern India to the Atlantic Ocean.

We have already proved in the case of most of the peoples with which we have to deal that the ox was the unit of value. We have likewise found that these primitive peoples, whilst employing a cow or ox of a certain age as their standard of value, had adjusted accurately to this unit their other possessions: for instance, the heifer of the second year bore a distinct value relatively to the cow of the third year, so likewise the calf of the first year and the milk of a cow for a certain period. These thus acted as submultiples of the standard unit, and as they were the same in kind and only differed in degree, the various sub-units of the cow remained in constant proportion to the chief unit and to one another. On the other hand, when there was a distinction in kind between animals, as between oxen and sheep, the relative value would probably differ according to the scarcity or abundance of either kind of animal, which difference would probably arise from a difference in the nature of the pastures and climate. Thus we have found in some places ten sheep regarded as the equivalent of an ox, in others again eight. The same holds good of goats. In the case of these smaller animals we have seen the same fixed scale of values according to age, and the same method of rating the value of the milk of an ewe or the goat as we find in the case of the cow. Amongst people who possessed horses, camels and asses, the same principle holds good, horses and camels on account of their great value being treated as higher units for occasional use, just as the elephant is regarded at present in parts of Further India. The slave, as we have before remarked, played an important part as a higher unit or multiple of the ox, the average slave having a fixed value, whilst of course in the case of female captives of unusual beauty a fancy price would be paid. As climate and pasture would not affect the keeping of slaves, and as human beings were fairly universally spread over the area of the ox, the probabilities are that it was almost as easy proportionally to get slaves as oxen, and to keep the one as to keep the other from being stolen. Thus there would be more or less of a constant ratio between slaves and oxen. There would be a tendency likewise to regulate the number of slaves by the amount of work to be done, and as this work in the pastoral stage is almost entirely that of the neatherd, the shepherd, the swineherd and the goatherd, the number of *male* slaves at least would be to a certain extent conditioned by the extent of the flocks and herds. Such we may infer from the picture of the household of Ulysses in the Odyssey was the practice in early Greece. The faithful swineherd Eumaeus, and his fellow the good neatherd, with the rascally goatherd Melanthius, and their underlings, seem, with the addition perhaps of a few house slaves who would assist in tilling the chieftain's demesne (*temenos*), to have comprised all the men-servants. The master of the house worked hard himself in his field and at various handicrafts, as we find Ulysses boasting of his expertness both as a ploughman and mower; he was also a skilled carpenter, having with his own hands built the chamber of Penelope and constructed a cunningly wrought bedstead[1]. Hence the amount of help to be required from *male* slaves, exclusive of their duties as herdsmen, would be but insignificant. When we come to deal with the question of *female* slaves, the conditions of their number seem at first sight entirely different. The question of polygamy here comes in, and we must bear in mind that they were acquired not merely as servants to perform menial duties, but likewise to be

---

[1] *Odyssey*, XXIII. 198.

wives and concubines. It is evident then that the number of such attendants will depend on the inclination and wealth of the house-master. But here again the problem is simplified, for inasmuch as his wealth consisted in cattle, a man's power to purchase handmaidens depended on the amount of his kine. Thus at the present day the number of women owned by a Zulu depends entirely on the number of cattle he possesses. Hence there was likely to be a fairly universal ratio in value between female slaves and oxen, over such a region as we have sketched above. The facility too in transporting human chattels from one place to another would be an important element in keeping the price almost the same over all parts of the area. It is a very ancient principle with the slave captor and slave dealer to sell their captives far away from their original home. Among our Anglo-Saxon forefathers the slave from beyond the sea was always worth more than a captive from close at hand[1]. The explanation of this fact was suggested by Dr Cunningham, and the proof of it was found by Mr Frazer in Further India; for there the slave brought from a great distance is always more valuable than one who comes only a short way from his native land, as the possibility of the former's running away and succeeding in escaping is so much less than that of the latter. This too seems to be the true explanation of the fact that in Homer we regularly find persons sold into slavery beyond the sea. Achilles sold the son of Priam to Euneos the son of Jason of Lesbos[2], the nurse Eurycleia had been brought from the mainland, Eumaeus the swineherd had been sold to Laertes by the Phoenicians who had captured him with his nurse in his distant home[3]. This constant tendency to sell in one country the captives taken from another would do much to equalize prices everywhere, and the price being paid in oxen the ratio in value between oxen and *female* as well as *male* slaves would tend to be constant.

We have now reviewed the ordinary kinds of wealth amongst primitive pastoral people, but we have touched but lightly as yet on the subject of the metals.

[1] Cunningham, *Hist. of English Commerce*, I. p. 117.
[2] *Il.* XXI. 41.   [3] *Od.* XV. 460.

We saw above that the two earliest kinds of currency consisted either of some article of absolute necessity, such as the skins of animals in the colder climates, or of some form of personal ornament, which being both universally esteemed as well as durable and portable will be readily accepted by all members of the community. It is of pre-eminent importance that it be universally esteemed. Travellers who have ignored this principle have found out its truth to their cost in Central Africa in modern times. As the chief currency consists of glass and porcelain beads, which the traveller must carry with him or starve, the European is too apt to assume that provided the beads are bright and gaudy in colour all sorts will be taken with like readiness by the natives. Sir Richard Burton in a valuable appendix to his *Lake Regions of Central Africa* warns travellers against this dangerous 'error. The African has his own firmly rooted canons of aesthetics, and will take as payment only those sorts of beads which he considers suitable and becoming. Again, some explorers brought supplies of cheap Birmingham trinkets, thinking that they would captivate the negro eye, but they proved a complete commercial failure, for the natives much prefer trinkets and jewellery of their own manufacture, and which are more in keeping with their standard of good taste. Again, the Arabs of the Soudan will not take gold as payment, in consequence of which our army in the late expedition had to take with them large and inconvenient supplies of silver dollars, coined for the purpose. The Maria Theresa dollar is the recognised currency in that region, not because of any notions as regards currency properly speaking, but because the Arab's taste lies in silver ornaments for himself, his weapons and his horse. He values then the silver because of its utility as an ornament, whilst gold he cannot employ to the same advantage.

I have thus digressed in order that it may be clearly seen that mankind were not seized with the *sacra fames auri* from the very first moment when the eye of some wild hunter or nomad first lighted on a gold nugget as it glistened under the sunlight in the stream.

A considerable period may have elapsed after mankind

became acquainted with gold or silver before man cast away his necklets or bracelets of shells such as have been found along with the most ancient remains of the human race yet discovered in Europe, and put on his person in their stead similar ornaments beaten out of the gold from the brook. It is perfectly reasonable to assume that the primitive Aryan or primitive Semite, who wore ornaments of shells, used these as instruments of barter, or even currency, in the same way as we have found the peoples of Asia and Africa using their strings of cowries, the aborigines of North America their wampum belts, and the Fijians their whales' teeth.

In what particular region mankind first employed the precious metals to adorn his person, it is of course impossible for us to say. But beyond all doubt already in Egypt at the very dawn of history gold was playing an important part. The question of the relative dates at which the metals were first employed by man is one of great interest and importance in studying the history of human development. Of the four chief metals, gold, silver, copper and iron, we have no difficulty in deciding that iron is most certainly the latest to come into use. It is only within historical time that implements and weapons of iron have superseded those of copper and bronze, at least within the area occupied by the great civilized races. The reason for this is obvious: iron is not found native, but must be obtained by a difficult process of smelting, and even when obtained requires great skill to make it available for use. The Greeks of the Homeric Poems were still in the later bronze age, although iron was known and employed for weapons and implements. But as we have no immediate need to discuss the date of the introduction of iron, we may pass on to the three remaining metals.

It is obvious that if a metal is found naturally in such a condition that it can be immediately wrought into various forms for ornament or utility, such a metal is likely to have been employed at a much earlier period than one which is rarely if ever found in a native condition. Now silver is a metal which is rarely found pure, and considerable metallurgical skill is needed to render it fit for use. On the other hand gold and copper are both found in a pure state. We may then on this ground alone infer that mankind was acquainted with gold and copper before they as yet had learned the art of working silver ore. It next comes to be a question of the priority of gold or copper. The probabilities will undoubtedly be in favour of that metal which is most universally found native, and which is the most likely by its hue to attract the eye, and which is the most easily worked. On all these counts gold can claim priority over copper. Still copper is found native in various countries, Hungary, Saxony, Sweden, Norway, Spain and Cornwall.

It is of course quite possible that in a region where gold is not native and copper is, the latter may have been the first metal known to the aboriginal inhabitants. This can be well illustrated from the case of iron and copper in Central Africa. The negroes never had a copper or bronze age, but passed directly into the iron age, for the very sufficient reason that no native copper was found in their country, and consequently they had no metal suited for implements until they had learned to smelt iron. Gold of course on the other hand was known to them from the most remote period. Finally, from a famous modern occurrence we may come to the general conclusion that wherever gold is a natural product of the soil there it has been the first metal to come under the observation of man. The great gold-field of California was first discovered on a memorable Sunday morning, when the eye of a lounger who was smoking his pipe by the side of Captain Sutter's millrace happened to light on some glittering body in the sandy bottom of the stream. This was the first scrap of gold found in California, and whilst that fertile land has produced many natural treasures besides gold within the scarcely more than forty years which have since elapsed, its gold it will be observed was the earliest of its metals, both from the nature of its deposit and from the brilliancy of its colour, to attract the attention of man. In certain parts of Southern Europe, notably parts of Southern Italy and Southern Greece, where copper is found but not gold, copper perhaps may have been known before gold, and certainly before silver. It will be important to bear this in mind with reference to a stage in our future arguments.

That silver came under men's notice at a later time than either gold or copper can be put beyond doubt by historical evidence. In the Rig Veda, where gold (*heranya*) is already well known and likewise copper (for there can be no doubt that the *ayas* of the Veda, Lat. *aes*, means copper), silver is entirely unknown; the word *rayatam*, which in later Sanskrit means silver, does indeed occur, but only as an adjective applied to a horse and meaning *bright*. Again, we know as a matter of fact that it was only at a comparatively late period that the famous silver mines of Laurium in Attica were developed. At least Plutarch (*Solon*, ch. 16) tells us that, owing to the scarcity of silver coin, Solon reduced the amount of the fines levied and also of the rewards for killing a wolf or wolf-cub, the former to five drachms, the latter to one drachm, the rewards representing the value of a cow and a sheep respectively. If they had already learned to work that "well of silver, the treasure-house of their land," in the time of Solon (596 B.C.), there certainly could have been no such dearth of silver. Finally let us take a comparatively modern case, that of the Aztecs of Mexico. When the Spanish conquerors reckoned up their great tale of treasures found in the royal palace, whilst the gold amounted to the large sum of *pesos de oro* 162000 lbs., the silver and silver vessels only weighed the small sum of 500 marks[1]. Yet this was in the country that is now known as the richest silver-producing region that the world has ever seen.

We thus find a people in a highly advanced state of civilization, who had invented a calendar, had devised a system of picture-writing, who had actually a currency in gold-dust, as we have found, and who were skilled and artistic craftsmen in gold, and yet who were scarcely able to make the slightest use of the silver, with which almost every crevice in their native hills was charged.

We may thus with safety rest in the conclusion that silver only comes into use at a stage always and probably much later than gold.

We have been thus led to the conclusion that gold is

[1] Prescott, *Mexico*, p. 234.

known to man at a far earlier stage than silver; furthermore that copper is also prior in discovery and use to silver owing to its natural form of deposit, and that, although in a region where gold does not exist, copper may have been the first of the metals to come under human notice, yet wherever gold-bearing strata are found, there is a great probability that gold was the first metal known. Schrader (*op. cit.* p. 174) has discussed the evidence from the Linguistic Palaeontological point of view, and whilst much of what he says is interesting, there are some points in his conclusions which shake one's faith in the infallibility of the Linguistic method for determining disputed points in archaeology. Gold he considers was known to the Egyptians from the remotest times, and so also to the Semites of Asia. As gold is found in abundance in the tombs of Mycenae (circ. B.C. 1400) he considers that just about that time the Greeks had acquired a knowledge of gold from the Phoenicians. The Greek *Chrysos* (χρυσος), *gold*, is derived, according to many scholars, from the Phoenician equivalent for *charutz*, the Hebrew name for the same metal.

There is plainly no relationship between the Egyptian name *Nub* and the Semitic appellation. The question, however, may arise as to whether, even granting that *chrysos* is derived from *chârûz*, it follows that the Greeks had no knowledge of gold prior to their contact with the Phoenicians. It is the skilful manufacture of a metal into beautiful and useful articles which gives it its real value. Hence arises the high esteem in which the cunning workman is held in early times. In Homer he is ranked along with the *prophet*, a sufficient proof in itself of the great importance attached to his functions. Again, in the Homeric Poems all articles of gold and silver of especially fine workmanship, if they are not the work of the divine smith Hephaestus himself, are the productions of the Sidonian craftsmen. The priest Maron gave Odysseus, amongst other presents, seven talents of well-wrought gold. Whether this took the form simply of rings we cannot tell, but plainly the value of the gift is enhanced by the epithet. From these considerations it seems not unreasonable to suppose that the Greeks, although possessing a name of their own for gold, may have adopted a Phoenician name, because they obtained the fine-wrought ornaments of that metal which they prized so highly from the Semite traders.

If any one thinks that this is a mere suggestion unsupported by analogy, my answer is not far to seek. The Albanian word for gold is φλjορι[1], so called because the first coined gold moneys of the middle ages with which they became acquainted were those of Florence. Now I think Dr Schrader will hardly maintain that the Albanians were unacquainted with gold as a metal until sometime in the mediaeval period they first obtained it from the Florentines. What took place in the case of the Albanians may have taken place again and again at earlier periods. A rude nation already acquainted with a certain metal receives by trade from a more advanced people the same metal wrought into various shapes and forms for personal decoration or use, and along with the superior articles it takes over the name by which the makers of those objects of metal described them.

These considerations well serve to show how unsafe is the basis afforded by Linguistic Palaeontology alone on which to build any theory of ethnical development. Let us now take another case where Schrader and his followers dogmatize without the slightest suspicion that the facts of recorded history may step in and rudely upset their conclusions. Schrader[2] holds that the Kelts were not acquainted with gold until their invasion of Italy in the beginning of the 4th cent. B.C. His argument is that the Celtic word for gold (Irish *or*, Cymric *awr*) is a loan-word from the Latin *aurum*. As the Sabine form of the latter is *ausum*, and the change of *s* to *r* did not take place in Latin until the fifth century B.C., and as the change of primitive *s* into *r* does not take place in the Keltic languages, he infers that it was only after the change in the form of the word had taken place in Latin that the Gauls became acquainted with the metal. Yet who will, on reflection, maintain that the Gauls had not already learned the use of gold from the Etruscans with whom they had been in con-

[1] Schrader, p. 255.
[2] Schrader, *op. cit.* p. 255.

tact long before they ever reached the Allia or sacked Rome? The Italian dialects were still employing the form of the word with *s*. Why should the Gauls have taken the form of the word with which they must have come least in contact in their invasion of Italy in preference to that used amongst the other Italians? Finally comes the irresistible evidence of Polybius that when the Gauls invaded Italy their only possessions consisted of their cattle and an abundance of gold ornaments, both of which could be easily transported from place to place[1].

Again, we can argue forcibly that it is contrary to all experience for primitive peoples to suddenly exhibit so strong a predilection for metals, or objects of which they have not had previous knowledge, as the Gauls showed in their rapacious demands that the ransom of Rome should be in gold. The legend that Brennus threw his sword into the scales, and ordered them to make up its weight in addition to the stipulated sum, shows, if it is true, that the Gauls were well acquainted with the art of weighing, which would be only gained from a long knowledge of the precious metals. The solution of the difficulty involved in the Keltic *or* can be readily found. The Iberians in Spain had long been skilled in the working and use of the precious metals. Tradition told how Colaeus of Samos, the first of the Greeks who ever sailed to Spain, brought back a fabulous amount of precious metal, and that the Phoenicians when they first traded in that region found silver so plentiful that in their greed for gain, when the ship could hold no more, they replaced their anchors by others made of that metal. The Phocaeans had traded with Iberia and Gaul from the end of the 7th century, Massalia had been founded by this bold people about 600 B.C. Are we to suppose that in all those centuries when the Kelts are in constant contact with the Iberians, and when already all Keltike, Helvetia, Northern Italy and even perhaps 'the remote Britanni,' were in constant touch with the traders of Massalia, the Kelts waited to learn the use of gold and silver until B.C. 400? The Basque name for gold is *urrea*. It is quite possible that the Keltic

[1] Polybius II. 19.

name was obtained from the Iberians, whom they found already in possession of Western Europe. But there is another alternative which is probably to be preferred. As we found the Albanians calling gold by a name derived from the gold coins of Florence, so the Kelts may have adopted the Latin names for gold used by their Roman conquerors. This is made almost certain by the fact that *aura*, in old Norse, derived from Latin *aurum*, became the regular word for treasure, although no one will deny that the Teutonic peoples had already *gold* and its cognates as terms of their own for the metal. Everyone is familiar with the influence exercised by the Roman coinage even in the countries of the East, where Rome met with a civilization hoary in age before Romulus founded Rome, and from which Rome herself had ultimately derived the art of coining. Yet by the time of Christ the Roman *denarius*, the *penny* of our Authorized Version, had already asserted itself in the Greek-speaking provinces of the East, and became in later days, when the rule of Rome and Constantinople fell before the Arab conquerors, under the form of *dinar*, the standard coin of the great Mahomedan Empires. Did then in like fashion the Roman form of the name for *gold*, which in all probability varied but little from the cognate Gaulish word, supplant at a comparatively early period that native form?

The same argument may be urged in reference to the silver. The Irish form is *airgid*, according to some a loan-word, being simply the Latin *argentum*. We have already seen that it is not possible that the Kelts, in constant contact with the Iberians who were so rich in silver, could have remained in ignorance of that metal. The Gaulish form of the name for silver was plainly in Roman times almost the same as the Latin, as is shown by *Argentoratum*, the ancient name of Strasburg. It is plain then that before the Roman Conquest the Gauls had a town called by the name for *silver*, whilst the Irish form has no nasal, the Gaulish coincides completely with the Latin. Is it not possible, that in this case too a native Keltic name, a close cognate of Latin *argentum*, whose lineal descendant is seen in the Irish form, may have been assimilated to the Latin form? But there is plenty of evidence from other quarters to show that the mere existence of a foreign name for a particular object in any language is no proof that the object in question came into use for the first time along with the borrowing of the name. When the Franks conquered that portion of the Roman empire to which they gave their name, they must have had Teutonic words of their own for *silver* and *gold*, closely related to our own forms of the words. Yet whilst many Teutonic words lingered and became absorbed into what became in process of time the French language, their names for the metals disappeared and the Latin derivatives remained in possession.

Again, we get another instance of such borrowing in the case of our own *penny*, old English *pendinga, penning*, German *Pfennig*. The philologists seem agreed in recognizing this as a loan-word from the Latin *pecunia*. Yet money was familiar to the northern peoples long before they ever came into contact with even the advanced posts of the Empire. The use of rings and spirals of gold as a form of currency in Scandinavia is well known; our word *shilling* seems to mean no more than portions of such a coil of gold or silver wire cut off, to be used as small change. But as the first coined money with which they became familiar was the currency of Rome, they seem to have taken the generic Roman name for money as their own expression for the Roman silver coins with which they became familiar, just as the Latin *aurum* under the form of *aura* (*eyrir*) became in old Norse the general term for coined money or treasure in money.

We may ask why did the Kelts especially choose the Roman form of the name for gold, if they were then for the first time getting a name for the substance then (according to Schrader) first known to them? Before they ever reached Latium they had been in contact with peoples in Northern Italy who undoubtedly were well acquainted with gold. The Etruscans were a wealthy people, who coined gold pieces before Rome had struck coins of any kind[1]. The Umbrians on the east side, the ancient Italic race who had in the days before the Etruscan Conquest held all Northern Italy up to the Alps, which was hence known to the earliest Greek geographers by

[1] W. Deecke, *Etrusk. Forschungen*, p. 5.

the name of Ombriké[1], were, beyond all doubt, acquainted with the use of gold, and had a name for it probably the same as the Sabine *ausum*. Why then did the Gauls remain entirely ignorant of gold and of a name for it when they had been in constant contact with those peoples who had most undoubtedly abundance of the metal and names of their own for it? Until some sufficient answer is given to the objections here raised, we must on every logical and scientific ground refuse our assent to an argument, the sole basis of which is philological. It may not be inappropriate also here to remark that it is most desirable in all historical enquiries to rely as little as possible on Etymology. From the days when the Stoics laid such importance on arguments based on the *originatio verborum* down to the present time reasonings based on such foundations have been as a rule founded on the sand. Comparative Grammar as yet can hardly be described as a science. New principles and laws are brought to light each year, and, although of course the solid *residuum* of what may now be regarded as more or less positive knowledge is slowly growing in bulk, those laws which were the shibboleth of Philologists a decade ago, are now rudely hurled from their preeminence. The only sound scientific method in historical research is to employ linguistic science as merely ancillary to our enquiries.

We have now seen the importance of the ox over the whole area of Europe, Asia and Northern Africa, in which those ancient peoples dwelt of whom history has preserved for us some knowledge. We have likewise found that over the same area gold was known and played an important part from a very remote antiquity. This proof has depended of course almost entirely on the literary remains and archaeological evidence. Political Economists, when discoursing on the oft-vexed question of monetary standards, lay down as one of the reasons why gold has been found so convenient, that it is universally found. Whether that fact is of much importance in modern times, when the facilities of communication are so great, may perhaps be doubted (especially when we see some of the largest stocks of

[1] Herod. IV. 49.

gold existing in countries like England and France, where there has been no production of gold for many years), but most certainly in early times it was of great importance, as we shall see, that the supplies of gold were not all concentrated in one or two places, but that at many points in all the different countries which came within the area of the ancient world, nature had had her treasure-houses.

To begin in the East, we shall first find that in all Central Asia there are rich auriferous deposits in many places. The stories told of the gold-digging ants and of the Griffins and Arimaspians are familiar to all readers of Herodotus. That historian (III. 102—5) gives an explanation of how the Indians are so rich in gold. To the north of India lies a region desert and waste by reason of sand. Close to this desert dwells an Indian tribe, who border on the city of Kaspaturos, and the land of Paktuiké, dwelling to the north of the other Indians, who live in the same manner as the Bactrians, and are the most valiant of the Indians. These men go on expeditions in search of gold. In this desert and in the sand are ants, which are in size smaller than dogs, but larger than foxes. As these ants make their habitations under ground they carry up the sand just as the ants in Greece do, and they are very like the latter in form. But the sand which is carried up is of gold. The Indians then make expeditions in quest of this sand, each man having yoked three camels. He then relates how the Indians time their arrival at the ant region so as to reach the ant-diggings at the hottest time of the day, which in that region is the early morning. The ants are then not to be seen for they have returned into their burrows to avoid the heat of the sun. The Indians hastily fill the sacks they have brought with the precious sand, and depart with all speed, as the ants from their keen sense of smell quickly detect their presence, and at once give chase. Their speed is such that though the camels are as swift as horses, the Indians would never manage to return in safety, unless they succeed in getting a good start whilst the ants are still assembling from their various habitations.

This story has been very ingeniously explained in modern times by Lassen (*Alt-Ind. Leben*) and others. Lassen pointed out that a kind of gold brought from a people of Northern India was called *pipilika* 'ant' (*Mahābhārata* 2, 1860) and that it was probable that the story referred to a kind of marmot which to this very day lives in large communities on the sandy plateaus of Thibet. On the other hand more recent explorations in Thibet show us that there are still communities of gold-diggers, who in the rigour of the Himalayan winter clothe themselves in skins and furs, which are drawn up right over their ears in such a fashion that they present at first sight the appearance of large shaggy dogs[1]. Whichever explanation may be right, it may be inferred that from a very early time the region north of the Panjab afforded vast supplies of gold. The remark of Herodotus (III. 105) that it was from this source that the Indians obtained their wealth, and that there was not much gold mined in their own land, is probably correct. It is beyond all doubt that the gold of Thibet at all times found its way largely into what is now the Panjab. We need have little hesitation in believing that from a very remote epoch the rude tribes of the Himalaya must have been acquainted with the gold-dust, which lay in rich deposits in the various mountain streams.

To come towards the west, the great wealth of the Persian kings seems to have been derived from the basin of the Oxus, which was famous in antiquity for its golden sands. Thus in the *Book of Marvels* (a work ascribed to Aristotle and largely composed of extracts from his writings) it is stated that the river Oxus in Bactria carries down nuggets of gold many in number[2]. But the region from which Herodotus thought that in his time came the greatest supply of gold was the Oural-Altai region of Central Asia. The Greek Colonies on the northern coast of the Black Sea, the most important of which was Olbia at the mouth of the river Borysthenes, had a large and lucrative trade with the Scythians, who inhabited the wide plains of that bleak region. The Scythians were rich in gold which they obtained from the still remoter country

---

[1] *Ausland*, 1873, No. 39.
[2] Arist. Θαυμ. 833 b. 14, φασὶ δὲ ἐν τοῖς Βάκτροις τὸν ῎Ωξον ποταμὸν καταφέρειν βωλία χρυσίου πλήθει πολλά.

of the Issedones, that people who, though righteous in all other respects, had the singular fashion of devouring their dead fathers. The Issedones again obtained by barter the gold from the Arimaspians, a race who had but one eye, and were hardly human[1]. They in turn, so report went, obtained the precious article not by traffic, but by theft from the gold-guarding griffins, who occupied the land where the gold was found. At least Herodotus says, "How the gold is produced I cannot truly tell, but the story is that the Arimaspians, people with one eye, carry it off from the Grypes[2]." He describes elsewhere (IV. 17) this region, which lay beyond the Scythians, where the cold was so great that the ground was frozen hard for eight months of the year, and that it was even cold in the summer season, that the air was so full of feathers that no one could see, by which, as Herodotus very properly explains, the thick falling feathery flakes of snow were meant, and that the cattle could not grow horns. All this seems to point beyond all doubt to the Ural and Altai ranges. Unquestionably there was a well-established trade route extending from the Black Sea through the country inhabited by the Scythians proper, which Herodotus describes as consisting of plains of rich soil, a true description of the fertile steppes of Southern Russia. Then beyond this lay a large area of rugged, stony land, inhabited by a people called Argippaei, who, males and females alike, were born bald. Their territory formed the lower part of a range of lofty mountains. They were a peaceful and a harmless race, dwelling in tents of white felt in the winter. It was easy to learn about them and their country from the Scythian traders who held intercourse with them, as likewise from the Greeks from the factories of the Borysthenes, and from the other Greek trading ports on the Euxine. No man could say of a truth what lay to the north of the "Baldheads," as on that side rose the lofty,

[1] Herod. IV. 13.
[2] Herod. III. 116, λέγεται δὲ ὑπὲκ τῶν γρυπῶν ἁρπάζειν Ἀριμασπους ἄνδρας μουνοφθάλμους.

For the gold-fields of India, cf. Dr Valentine Ball's excellent chapter (IV.) in his *Geology of India*.

impassable range of mountains, but Herodotus had heard (but did not believe) that according to the "Baldheads" a race of men having the feet of goats dwelt there[1], a legend which may be plausibly rationalized into a simple statement that a race of mountain-folk, sure-footed as the wild goat, inhabited the mountains. But on their east the existence of the Issedones was an established fact.

It is plain then that from a date lost in the distance of time the gold of the Ural-Altaic region had been worked and exported, and that consequently it was known and prized by all the tribes who came within the influence of this wide district. The Scythians in the fifth century before Christ were engaged in regular trade with this region, and possessed abundant store of the prized substance. This is shown by Herodotus in a very remarkable passage wherein he describes the burial of a Scythian king. After recounting the ceremonials he thus proceeds: "In the open space round the body of the king they bury one of his concubines, first killing her by strangling, and also his cupbearer, his cook, his groom, his lacquey, his messenger, some of his horses, firstlings of all his other possessions and some golden cups; for they use neither silver nor copper[2]." From this passage we learn the interesting fact that the Scythians, although possessing great quantities of gold and being able to work it into articles of use, were yet ignorant of silver and copper, which nevertheless, as we know now, exist in large deposits in the Ural region. This is one of several cases which we shall have to notice which go far to prove that the knowledge and working of gold preceded not only that of silver, but also that of copper.

The remoteness of the age at which some branch of the Turko-Tartar family who dwelt in the Altai region, first discovered the treasures which Nature had stored up there, is evidenced, as Schrader (following Klaproth) rightly points out (p. 253), by the fact that among all the branches of that widespread family of languages, from the Osmanli Turks on the Dardanelles to the remote Samoyedes on the banks of the

---

[1] Herod. IV. 25.
[2] Herod. IV. 71, $ἀργύρῳ\ δὲ\ οὐδὲν\ οὐδὲ\ χαλκῷ\ χρέωνται$.

Lena, the same word for gold is found in slightly varying forms, *altun, altyn, iltyn*, etc., which can hardly be etymologically separated from *Altai*, the locality from which it first became known in far-off days. In the ancient graves of the Tschudi in the Altaic districts, have been found abundance of gold and silver utensils which according to Sjögren (Schrader 136), exhibit the representation of the Griffin of Greek fable.

Before passing further west into Europe we shall complete our survey of the gold-fields of Western Asia. One of the most beautiful of Greek stories hangs around the eastern end of the Black Sea, where lay the land of Colchis, the goal which Jason and his fellow Argonauts sought in their quest of the Golden Fleece. In the Homeric poems the voyage of the ship Argo is referred to as an event which had taken place in a past generation. In the time of the geographer Strabo (B.C. 63–A.D. 21) gold was still found in Colchis in a district occupied by a tribe called Soanes, scarcely less famous for their personal uncleanliness than their neighbours the Phtheirophagoi (Lice-eaters) who bore this appellation from the filthiness of their habits. "It is said that in their country the mountain torrents bring down gold, and that the barbarians catch it in troughs perforated with holes, and in skins with the fleece left on, from which circumstance they say arose the fable of the Golden Fleece[1]."

Strabo's explanation, which seems from his words to have been the current one in his day, is extremely plausible, and it appears highly probable that from the first dawn of history the torrent-swept treasures of the Colchian land were well known to the dwellers in both Asia Minor and Europe. But this was not the only place in Asia Minor where gold was found. We shall have occasion again and again to refer to the Electrum of Sardis, obtained from the sand of the river Pactolus which flowed down from Mount Tmolus. Scholars are familiar with the account which Herodotus gives of these gold deposits, but probably the most convenient thing for our present purpose will be to quote Strabo's enumeration of the

---

[1] Strabo, xi. p. 499, παρὰ τούτοις δὲ λέγεται καὶ χρυσὸν καταφέρειν τοὺς χειμάρρους, ὑποδέχεσθαι δ' αὐτὸν τοὺς βαρβάρους φάτναις κατατετρημέναις καὶ μαλλωταῖς δοραῖς· ἀφ' οὗ δὴ μεμυθεῦσθαι καὶ τὸ χρυσόμαλλον δέρος.

kings and potentates of antiquity in Asia and Europe who were famous for their wealth, as he has added in each case the source from which their wealth was obtained. The current account as given by Callisthenes and others was, "that the wealth of Tantalus and the Pelopidae was derived from the mines of Phrygia and Sipylus, whilst the wealth of Cadmus came from the mines of Thrace and Mount Pangaeum, but that of Priam from the gold-mines at Astyra in the vicinity of Abydus, of which even now there are still scanty remnants. But the quantity of earth cast up is vast, and the diggings are proofs of the ancient mining operations. But the wealth of Midas came from the mines round Mount Bermion, whilst that of Gyges and Alyattes and Croesus came from the mines in Lydia. But in the district between Atarneus and Pergamus there is a deserted city, with places containing worked-out mines[1]." This passage gives a good picture of the gold-fields which in ancient days were worked round the shores of the Aegean.

In the time of Strabo some of them were already worked out and gave but a scanty yield, for he says, "above the territory of the people of Abydus lies in the Troad Astyra, which now belongs to the people of Abydus, a ruined city, but aforetime it was independent, possessing gold-mines, now affording but a scanty yield, as they are exhausted, just like the mines on Mount Tmolus in the neighbourhood of the Pactolus." The latter district was still productive in the days of Herodotus, who declared that the land of Lydia had few marvels to chronicle except the gold-dust that is borne down from Tmolus[2]. Strabo too, elsewhere[3], when describing the river system of this part of Asia Minor says, " the Pactolus flows from Tmolus, carrying down that ancient gold-dust from which they say that the famous wealth of Croesus and of his ancestors became renowned. But now the gold-dust has failed, as has been stated."

It is interesting to observe that according to tradition the wealth of Midas, the king of Phrygia, who is perhaps more

[1] Strabo, xiv. p. 680.
[2] Herod. I. 93, πάρεξ τοῦ ἐκ τοῦ Τμώλου καταφερομένου ψήγματος.
[3] xiii. p. 625 sq.

famous for his ass's ears than his riches, came from the Bermion Mount in that part of Macedonia, which was occupied in historical times by the powerful tribe of the Bryges. This in itself is an interesting indication of the intimate connection and close communication between the countries and peoples on both sides of the Dardanelles from the earliest epoch. There were on either side lands gifted by nature with stores of wealth, as well as possessing the portals of either continent. Hence the Hellespont and Bosphorus have ever been the seat of rich cities, and have ever been regarded amongst the greatest of prizes in the struggles of the nations.

It is possible that the ancient legend connecting the wealth of Priam of Troy with the mines of Astyra, still worked in Strabo's days, may serve to explain the real cause of that invasion of the Achaeans, which in all probability did occur, although on what form or at what time we know not, and around which there grew in the mouths of the rhapsodists the tale of Troy Divine. In all our enumeration of gold-mines we do not find a single one allotted to Greece Proper. The wealth of Cadmus, the old Phoenician founder of Thebes, who was said to have introduced the art of writing into Hellas, came, according to Strabo's tradition, from Thrace and the mines of Pangaeum. As Cadmus is the typical wealthy potentate of Northern Greece, so the line of Pelops are the typical wealthy potentates of Peloponnesus. Their wealth, like that of Cadmus, is adventitious, for it is the product of the mines of Phrygia and Mount Sipylus. This is quite consistent with the statement of Thucydides that "those Peloponnesians who have received the clearest accounts by tradition from the men of former time declare that Pelops first by means of the mass of wealth with which he came from Asia to men who were poor, having acquired for himself power although he was a new-comer, gave occasion for the land to be called after him."

Of the three cities which are called rich in gold by Homer, two are in Hellas proper, namely Mycenae in Peloponnesus, and the Minyan Orchomenus in Boeotia. Gold has been found in abundance in the prehistoric tombs at Mycenae, thus confirming the ancient tradition. This gold, beyond doubt, was imported from outside Greece, and we may without hesitation accept the view of the Greeks themselves that it came from Asia Minor. The story of the wealth of Cadmus, who came to Bocotia as Pelops did to Peloponnesus is equally in harmony with the Homeric tradition of a great wealthy city in Bocotia. Dr Schliemann excavated the remains of Orchomenus, as he did those of Mycenae, and of the ancient city at Hissarlik, but his labours unfortunately gave no confirmation of the accounts of the ancient wealth of Orchomenus. The reason probably was that he came many centuries too late, as the great prehistoric tomb known as the Treasure-house of the Minyans had long since been repeatedly plundered and ransacked; not even one bronze plate of those that once had probably lined its walls was left. Still less likely was it that any vestige of gold would have escaped the rapacity of the spoiler.

The wealth of Northern Greece, then, by the earliest tradition is connected with the rich gold regions of Thrace, which, if we accept the same tradition, must have been worked from the remotest age. The connection of the Cadmus legend with this region points clearly to very early Phoenician trade in the days when as yet the Phoenicians had undisputed mastery over the Aegean Sea and the Hellenes had not begun to develop maritime enterprize.

As a matter of fact the name of the island of Thasos, which lay off the Thracian shore, was directly ascribed to a Phoenician settler. In the time of Herodotus the Thasians had a large revenue both from the mines on the mainland and from those in their own island. For he tells us that "from the gold-mines of Scapte Hyle they had a revenue on the average of eighty talents, and from those in Thasos itself a lesser one, but yet so good that the Thasians enjoyed exemption from taxation on produce and had a yearly revenue from the mainland and the mines together of two hundred talents on the average, but when the revenue was at its maximum, it was three hundred talents. And I myself likewise saw these mines, and by far the most wonderful were those which the Phoenicians who had colonized the island along with Thasos had opened up, it was this Phoenician *leader* Thasos who gave his name to the island. These

Phoenician mines lie in the part of Thasos between the district of Aenyra and Coenyra; a great mountain has been upturned in the search[1]." But the most famous mines on the mainland of Thrace were those of Mount Pangaeum, Crenides, and Datum. Strabo gives a succinct account of this wealthy district: "There are other cities round the gulf of the Strymon, as for instance Myrcinus, Argilus, Drabescus, Datum. The last-named has very excellent and fruitful land and shipbuilding-yards, and mines of gold, from which comes the proverb a *Datum of riches*, just like *loads of wealth*." And in another passage he says that, "there are very numerous gold-mines at Crenides[2]. The city of Philippi is now seated close to the Pangaeum Mount. And the Pangaeum Mount too has mines of gold and silver, and so has the region both on the other side of and on this side the Strymon as far as Paeonia. And they say likewise that those who plough the Paeonian land find some morsels of gold."

It was in a struggle with a Thracian tribe, the Edonians, for the possession of the mines at Datum that Sophanes, the son of Eutychides of Decelea, who had distinguished himself above all other Athenians at the battle of Plataea, was killed[3]. The possession of Thasos and the coast of Thrace was not the least important means by which Athens held her supremacy in Greece, and when Philip (360—336 B.C.) finally got supreme control over all this region, and built his new capital of Philippi, his path of conquest was henceforward made easy by the golden Philippi, the *regale nomisma* of Horace,

                Diffidit urbium
Portas uir Macedo, et subruit aemulos
Reges muneribus.
                          (*Carm.* III. 16. 13.)

Passing on now to Southern Asia we find that there gold was found in Carmania (the modern Kerman) on the Persian Gulf. Strabo states on the authority of Onesicritus that in Carmania a river carries down gold-dust, and that there is like-wise a mine of dug gold and of silver and of copper[4].

[1] Herod. vi. 46 *sq.*
[2] Herod. ix. 75.
[3] Strabo, 331.
[4] Strabo, 618. 29. Didot.

That there was gold in Arabia is placed beyond doubt by various notices in antiquity. "He shall live and unto him shall be given of the gold of Sheba (Saba¹)," says the Psalmist (Ps. lxxii. 13), showing that the inhabitants of Palestine regarded that country as a source from which the gold-supply came.

Strabo and Diodorus give somewhat similar accounts of the gold found along the Red Sea littoral. The former, describing the land of the Nomads who live entirely by their camels, which they employ for warfare and for travelling, and on whose milk and flesh they subsist, says: "a river flows through their land which carries down gold-dust, but they have not skill to work it up. Now they are called Debae[2]; some of them are nomads, others are tillers of the soil. But I do not mention the numerous names of the tribes on account of their uncertainty and outlandish pronunciation. Next to them come more civilized men, who inhabit a more genial soil. For it is well supplied with both river and rain water. And dug gold is produced in their land, not from dust but from nuggets of gold, which do not need much refining. The smallest nuggets are of the size of olive-stones (?) (πυρήν), the medium-sized are as big as medlars, and the largest are of the size of chestnuts (?) (κάρυον). Having perforated these they pass a thread of flax through them in alternation with transparent stones and make themselves chains, and put them round their necks and wrists. And they offer their gold for sale to their neighbours likewise at a cheap rate, giving thrice as much gold as they get copper in exchange and twice as much gold as they get silver in exchange, for they have not the skill to work the gold, and the metals which they receive in exchange are rare in their country and more necessary for life[3]."

This is a most interesting and important passage, as it brings us face to face with primitive peoples in the very earliest stage of the use of metals. The Nomads do not possess skill

---

[1] Cf. Isaiah xlv. 14.

[2] The Debae of Agatharchides and Artemidorus are held by almost all scholars to be the people of Ptolemy's Θῆβαι πόλις, i.e. Dhahabân, from *Dhahab*, gold, with term. -ân.

[3] Strabo, 661. 45. Didot.

enough to work the gold-dust of their river, although evidently aware of its existence. Their neighbours being more favoured by the nature of their gold deposit are able to use the metal in the way in which we may with safety conclude that mankind everywhere first employed it. Accustomed to use ornaments of shells made into rude beads, they had no difficulty in adapting for like use the small lumps of native gold. They readily pierced the soft metal and making the nuggets into beads used them to form their necklets and armlets. But although this people had made some progress in the working of gold, they were incapable of working copper and silver. We shall have to return to this passage hereafter. Let us now hear Diodorus in reference to the same region.

He speaks of it in two separate places in his Collections, first in his Second Book, when giving a brief general statement of Arabia and its natural products, and again in the Third Book, when he is giving a more detailed account of the tribes who dwelt along the shores of the Red Sea or, as he called it, the Arabian Gulf.

The first passage runs thus (he has just been describing certain quarries): "There are mines in Arabia likewise of the gold that is termed 'fireless.' It is not refined down from gold-dust as in other countries, but it is obtained straightway on being dug up in size like unto chestnuts, and so fiery in colour that the most precious stones when set in it by the craftsmen make the most lovely of ornaments. And so great abundance of all sorts of cattle is found in the country that many tribes having chosen a pastoral life are able to get a comfortable subsistence, and being completely furnished with the plenteousness derived from their herds, they even have no need of corn in addition[1]." In his second reference, after describing the hill district, where lay the Mount Chabinus, densely clad with forests of all kinds of trees he says: "The land which comes next to the mountain region those Arabs called Debae inhabit. Now these people are camel-keepers and make use of this animal for all the most important affairs in life. For from them, they fight against their enemies and conveying their

[1] Diodorus Sic. II. 50. 1 sq.

wares on the backs of these effect successfully all their business, and they subsist by drinking their milk, and they range over the whole region on their fleet camels. Now about midway in their land flows a river which brings down so much shining gold-dust that the alluvial mud deposited at its mouth positively glitters. Now the natives are completely unskilled in the working of the gold, but they are hospitable to strangers, not to all comers, but to those alone who come from Boeotia[1] and Peloponnesus because of a certain ancient affinity of Heracles with their nation, a tradition of which in legendary fashion they relate they have received from their forefathers. The next region is settled by the Alilaean and Gasandan Arabs, not being torrid, like those near it, inasmuch as it is often overcast with soft dense clouds, and from these arise snowstorms and seasonable rains which make the summer season temperate. And the land is capable of producing everything and surpasses in excellence, yet it does not meet with proper attention, owing to the ignorance of the folk. And finding gold in the natural cavities in the earth they collect it in quantities, not that which is obtained by fusion from gold-dust, but that which is native and from the circumstance called 'fireless.' And as to size the smallest piece found is similar to an olive-stone, whilst the largest is not much less than a walnut. And they wear it round their wrists and necks when it is perforated, the nuggets alternating with transparent stones. But since this kind of metal is plentiful with them, but copper and iron are scarce, they barter these wares with the traders at an equal rate[2]." Strabo probably got his information from Artemidorus, who is his chief authority for everything connected with the Red Sea. Diodorus, whose authority is Agatharchides, substantially agrees with Strabo in all the main facts, such as the name of the tribe who cannot work up the gold-dust, whilst he adds the names of the Alilaeans and Gasandans, which are not given by Strabo[3].

[1] This story about their connection with Boeotia doubtless arose from the confusion between Δέβαι and Θῆβαι.
[2] Diod. Sic. III. 45. 4.
[3] His description of the size of the largest nuggets of gold varies slightly; in his second reference he compares them to "royal nuts" (κάρυα βασιλικά), which

From Arabia we naturally pass on to Egypt. We have already seen that the archaeologists assign reasons for supposing that the Egyptians were acquainted with gold from the remotest ages. The Egyptian word for gold is *nub*, from which the name Nubia, *i.e. El Dorado*, is commonly derived. Having fresh in our minds the interesting fact noticed above (p. 69) that the universal word for gold in use amongst the Turko-Tartaric races is probably derived from the Altai, the source from which they first got the metal, we are tempted to reverse the ordinary doctrine, and to derive the Egyptian name for gold from that of the region whence they first obtained it. The principle of naming products after the region or place from which they have been first brought is too well known to need illustration. Instances are familiar in all languages: *Cappadocae*, the Latin name for lettuce; *Persica* from which has come our *peach*, through the French; Indian corn, india-rubber, etc. are sufficient examples. The negroes of Eastern Africa call a certain kind of cloth *Merikano, i.e.* American. Perhaps, then, the name *nub* is rather a word of this class, and Nubia is not like Gold Coast, which belongs to the category of names formed by epithets applied in consequence of some article already well known having been found there.

Strabo (p. 821), describing Meroe, that large and fertile island formed by the Nile, says: "the island has many great mountains, and some of its inhabitants are shepherds, some hunters, and some husbandmen. And there are likewise copper-diggings and iron-works, and gold-mines, and varieties of valuable marbles. It is shut off from Libya by great sands,

are generally admitted to be walnuts, though walnuts are sometimes also called "Persian nuts" (κάρυα Περσικά), the latter name reminding us of the derivation of *walnut* itself; in the first passage he likens them in size to chestnuts (κάρυα καστανάικά or καστανάια), the name being said to be derived from Castanaea, a city of Pontus. It would seem from this then that Diodorus got his accounts from two slightly different sources. Strabo has been so cautious as not to give us any specific epithet for the large nut, which we may accordingly regard as we please either as a chestnut or a walnut. There can be no doubt about the fruit to which Strabo compares the medium-sized nuggets. The *mespilon*, Latin *mespilum* (from which comes the French *nèfle*), is undoubtedly the medlar, whilst perhaps the most likely meaning for the smallest of the three fruits is *olive-stone*.

from Arabia by unbroken heights, and from the upper region from the south by the junctions of the rivers, Astaboras, Astapus, and Astasobus. On the north the Nile flows all the way to Egypt in that tortuous fashion which I have described." This island virtually coincides with the modern province of Atbar. It is probably to this same region that Diodorus refers in his famous description of the Egyptian gold-mining. Although the passage is one of considerable length, it is of such interest and importance that it is perhaps advisable to give it in full: "On the confines of Egypt, Arabia which marches with it, and Ethiopia is a spot possessed of many great mines of gold, where the gold is got together with much suffering and expense. Since the earth is black and has lodes and veins of quartz of surpassing whiteness, and which excel in brilliancy all those natural objects which are noted for their lustre, those who are in charge of the mining works by the numbers of the labourers prepare the gold. For the kings of Egypt collect together and consign to the gold-mines those who have been condemned for crime, and who have been made captive in war, and furthermore those who have been ruined by false slanders, and who owing to an outburst of anger have been cast into prison, sometimes only themselves, but sometimes likewise with all their kindred, at one and the same time both exacting punishment from those who have been condemned, and obtaining great revenues by means of those who are engaged in the labour. Those who have been consigned to the mines, being many in number and all bound with fetters, toil at their tasks continuously both by day and all night long, getting no rest, and jealously kept from all escape. For guards composed of foreign soldiers, and who speak languages which differ from theirs, are set over them, so that no one is able by association or any kindly intercourse to corrupt any one of the warders. The hardest of the earth which contains the gold they burn with a good deal of fire, and make soft, and work it with their hands, but the soft rock and that which can easily yield to stone chisels or iron is worked down by thousands of hapless beings. And the craftsman who distinguishes the stone takes the lead in the whole process, and he gives instructions to the workmen. And of those who have been appointed to this misery those who surpass in bodily strength cut with iron pickaxes the glittering rock, not by bringing skill to bear upon their tasks, but by mere brute force, and they hew out galleries, not in a straight line, but according to the vein of the glittering rock. They then living in darkness owing to the bends and twists in the pits carry about lamps fitted on their foreheads, and changing in many ways the posture of their bodies according to the peculiarity of the rock throw down on the floor the fragments that are being hewn, and this they do unceasingly under the severity and stripes of an overseer. But the boys who have not yet reached manhood going in through the shafts into the excavations in the rock, laboriously cast up the rock that is being thrown down bit by bit, and convey it to the place outside the mouth of the shaft into the light. But the men who are more than thirty years old take a fixed measure of the quarried stone, and pound it in stone mortars with iron pestles until they reduce it to the size of a vetch. From these the women and older men receive the stone now reduced to pieces the size of a vetch, and as there is a considerable number of mills there in a row, they cast the stone upon them, they stand beside them at the handle in threes or twos, they grind until they have reduced the measure given them to the fineness of wheaten flour. And since they are all regardless of their persons, and have not a garment to cover their nakedness, no one who saw them could refrain from pitying the hapless creatures owing to their excessive misery. For there is absolutely no consideration nor relaxation for sick, or maimed, for aged man, or weak woman, but all are forced to toil on at their tasks until, worn out by their miseries, they die amid their toils. Wherefore the unhappy beings regard the future as more to be dreaded than the present owing to the excess of punishment, and expect death as more to be longed for than life.

"But finally the craftsmen get the ground-up stone, and complete the process. For they rub the ground-up quartz on a broad board placed on a slight incline, pouring water on it. Then the earthy part of it, melting away by the action of the liquid, flows down along the sloping board, but the part that contains the gold adheres to the board owing to its weight. Repeating this process frequently at first with their hands they gently rub it, but after this pressing it lightly with delicate sponges they take up by these means the soft and earthy part until the gold-dust is left in a state of purity.

"Finally other craftsmen, taking over the collected gold by measure and weight, put it into earthenware pots, and in proportion to the amount they put in a piece of lead and lumps of salt and furthermore a small quantity of tin, and they add barley bran. Then having made a well-fitted cover and having laboriously smeared it over with mud, they bake it in kilns for five days and as many nights continuously. Then after letting it cool, they find none of the other things in the vessels, but get the gold in a pure state with but a slight reduction in quantity. With so many and so great sufferings is the production of gold at the frontiers of Egypt completed. For Nature herself makes it plain, I think, that gold is produced with toil, is guarded with difficulty, is most eagerly sought for, and is enjoyed with mixed pleasure and pain. The discovery of these mines is of very ancient date, inasmuch as it was made known by the ancient kings[1]."

Such then is the vivid picture drawn by the humane Diodorus of the horrible torments of the unhappy bondsmen who worked these famous mines, sufferings only to be paralleled by the miseries endured by the miners in Spain under Roman rule, by the Indians in the mines of Peru under the yoke of the Spaniard, and by the helpless sufferers under Muscovite cruelty who at this hour endure a living death in the mines of Siberia.

For our immediate purpose it is interesting to notice that the Egyptians from a far back time obtained an abundant supply of gold from the confines of their own territory, and doubtless drew a further supply from those rich gold districts along the Red Sea of which we have just spoken.

Whilst in the latter case we had a most instructive instance of the first attempts to utilize the metals made by men, so in

[1] Diodorus, III. 12—14.

the case of Egypt we find an example of the most elaborate and scientific process of gold-mining known to the ancients. For we shall find that the process employed in Spain by the Romans for refining the crude gold was not nearly so elaborate as that employed by the Egyptians.

It is of course quite possible that supplies of gold either in the form of dust or of rings may have reached Egypt from the interior of Africa, but of that we have not as far as I am aware any historical record. For the negroes who are depicted in Egyptian paintings bringing tribute of gold rings might have brought them from Nubia or from a region on the coast of the Mediterranean further west. It is indeed a fact of great interest that down to the present day gold in the shape of rings or links is brought to Massowah on the Red Sea from Sennaar (Nubia). This is the best of the three qualities which reach Massowah; the second quality is Abyssinian gold, "in grains or beads," and the third is also Abyssinian gold "in ingots." Thus two most ancient ways of using gold are employed in this region still, for the gold in grains or beads reminds us at once of the story of its being employed by the Debae to form necklaces[1].

Once more let us advance westward, and notice the last gold-field on the continent of Africa. That gold was obtained by the Carthaginians from a district in North Africa is put beyond doubt by a passage of Herodotus (IV. 195), who, after describing a certain people called the Gyzantes, who coloured themselves red with raddle, and ate apes, says that "the Carthaginians declare that opposite this people lies an island named Cyraunis, two hundred stades long (25 miles) but narrow in breadth, with a crossing from the mainland; the island is full of olives and vines, and there is a lake in it from which the native maidens by means of birds' feathers smeared with pitch take up gold dust out of the silt." Whatever may be the exact spot meant on the coast of the Libyan nomads we may at least conclude that there is a distinct indication that the Carthaginians were well acquainted with gold deposits in this quarter. Whether or not the Carthaginians and in later times the Romans may have obtained by caravans across the desert supplies of gold from the great gold-bearing

[1] Mansfield Parkyns, *Life in Abyssinia*, Vol. I. p. 405 (London, 1853).

regions of West Africa, we have no means of judging, but it is on the whole probable that they did. The voyage of Hanno, the Carthaginian admiral, along the western side of Africa can hardly have failed to make known to them the existence of rich gold fields, even if they had been previously ignorant of them; but it is still more likely that it was the knowledge of such an Eldorado far away beyond the great Sahara that induced them to send out the expedition.

It has often happened in the history of both ancient and modern commerce that the products of a certain region are known long before travellers or merchants from civilized lands have ever reached the country that produces them. Thus the merchants of Marseilles were probably familiar with the tin brought from Devon and Cornwall across Gaul before the famous Pytheas ever coasted round Spain and Gaul and visited our shores. Again, in modern times, it is only within the last thirty years that the source of that most familiar of drugs, Turkey rhubarb, has been discovered.

By whatever means they may have learned its existence the following passage of Herodotus (IV. 196) puts it beyond all doubt that the Carthaginians in the fifth century B.C. traded by sea for gold to the west coast of Africa, and that consequently the savages of that region must have been long acquainted with the metal: "The Carthaginians," he says, "also relate the following: there is a country in Libya and a nation beyond the Pillars of Heracles, which they are wont to visit, where they no sooner arrive than forthwith they unlade their wares, and having disposed them after an orderly fashion along the beach, leave them and returning aboard their ships, raise a great smoke. The natives, when they see the smoke, come down to the shore, and laying out to view so much gold as they think the worth of the wares, withdraw to a distance. The Carthaginians upon this come ashore and look; if they think the gold enough, they take it and go their way, but if it does not seem to them sufficient, they go aboard once more and wait patiently. Then the others approach and add to their gold, till the Carthaginians are content. Neither party deals unfairly with the other, for they themselves never touch the gold until it comes up to the worth of the goods, nor do the natives ever carry off the goods till the gold is taken away[1]."

Let us now retrace our steps to Europe and take up our investigation at the point from which we diverged into Asia. We found Thrace and Thasos to have been for many ages an inexhaustible source of gold. We must now pass on from the Balkan peninsula to the Italian.

Although according to Helbig (*Die Italiker in der Poebene*, p. 21) no traces of gold have as yet been found in the lake-dwellings of Northern Italy, which were erected and occupied by the Umbrians, who occupied all that region until conquered by the Etruscans[2], we cannot take this negative evidence as at all conclusive proof that the inhabitants of these dwellings were utterly ignorant of gold and its use. Helbig has shown that the inhabitants of the lake-dwellings were in the bronze age at the time of the Etruscan conquest, which can be hardly placed later than B.C. 1100. Bronze implements are found in the remains. But as a matter of fact ornaments of gold are not generally found in the ruins of the habitations of the living, but rather in the tombs of the dead. That certainly has been the case at Mycenae, at Spata, on Mount Hymettus in Attica, in the island of Thera, and at Ialysus in Rhodes. Contrast the wealth of gold ornaments found in the tombs at Mycenae with the complete absence of that metal in the palace at Tiryns. Of course it may be urged on the other side that at Hissarlik amid the ruins of a burnt city great treasure in gold and silver has been found, and we must undoubtedly admit that in certain cases such as that of a city suddenly destroyed by a fire before there was time either for the owners to remove or the enemy to pillage the valuables therein, there is the possibility of finding such remains. If we were to apply this negative method consistently we must conclude that Orchomenus, which Homer called "rich in gold," was inhabited by men who were not yet acquainted with that metal, and we should I believe be constrained to arrive at the same conclusion in the case of Nineveh

---

[1] For similar ways of trading in Africa in modern times see Rawlinson's note *ad locum*.

[2] Herod. IV. 49.

and Babylon. At least Sir Henry Layard discovered scarcely a fragment of any articles of gold in the course of his excavations on the site of those two cities, which nevertheless we have the strongest grounds for believing were amongst the wealthiest of those of ancient days. In dealing with the question of Northern Italy we cannot separate it from the contiguous region of Switzerland or Helvetia. Dr Keller, in his well-known work on the Lake-dwellings (p. 459), gives instances where gold has been found in lake-dwellings amongst remains that indicated the owners to have been in the bronze period. Of course it may be said and said with truth that the lake-dwellings of Switzerland continued to be occupied down to a time posterior to those found in the Aemilia. But when we find that a gold ornament has been found in a dwelling of neolithic age, we have a positive proof not simply of the knowledge, but probably of the skill requisite to manufacture the metal. If any upholder of the negative method urges that gold has been found very sparingly in these lacustrine dwellings, let him remember that the existence of one single object of gold in these remains is sufficient to demolish all his argument. The objects found in the lakes are chiefly débris, the offal of the house, bones of animals, which had formed the food of the former owners, broken and disused implements, and such like. Ornaments of gold were not likely to have been flung into the bottom of the lake for the purpose of getting rid of them. Such precious articles were probably handed down with great care from generation to generation, and possibly in later days gold that once graced the neck or arms of prehistoric men and women has reappeared time after time in the form of coins, first the rude imitations of the staters of Philip of Macedon, again under the form of Roman *aurei*, and perhaps even bore the impress of some mediaeval monarch at a later time. There have been issues of coins both in ancient and modern times of which not a single specimen is at present known; yet if any one were to argue from this against the truth of the documentary evidence, the spade of a peasant by turning up a single coin might on the moment wreck all his logic. The sum of positive knowledge which we obtain from this discussion is therefore that some people who inhabited Switzerland in what is called the neolithic age (a vague and often misleading phrase) were acquainted with the use of gold ornaments. Could we but fix the inferior limits of this neolithic age, we should at least obtain an approximate date before which gold was already known. But it is most probable that stone, bronze and even iron long continued to be used side by side in the same areas. The man who had no articles to barter for bronze continued to use stone implements of his own manufacture, whilst his more fortunate coeval used weapons made of the superior but more costly material.

Granting now that bronze implements made their way from the Mediterranean into the middle and north of Europe, brought most likely by traders from the more civilized shores of the Aegean, let us ask ourselves how did the men of the neolithic stage obtain them. Did the kindly Phoenician trader generously bestow as free gifts these articles on the barbarians of the West? Does the trader of today among the isles of Melanesia lavish for mere thanks his wares upon the natives who gather round him on the beach? In Homer those Phoenician shipmen are described by an epithet, which by the mildest interpretation means *knaves*. The men who brought bronze got some valuable objects in exchange for it. Such objects must be portable: slaves, gold, silver, copper, tin, skins and furs would probably form the main objects of barter. If we make use of the philological method of Schrader and his school, there can be no doubt that copper was known to the Italians before ever a Phoenician keel grated against their shores, for the Latin *aes* is as we said a true Aryan word. There is no suspicion of borrowing here from the Semitic as there is in the case of the Greek *chalkos*. In such a case as this the philological argument has some distinct force; for whilst, as I argued, it is easy to realize a state of things under which a native name for a particular substance already known may give place to a foreign one, on the other hand it is difficult to see how a people who are receiving such a substance for the first time from foreigners, and who would therefore naturally apply to it a term obtained from the foreigners' language, could afterwards replace this name by one which is found applied to the same substance by a cognate people dwelling thousands of miles away from them. The Italians therefore probably had copper from a very early age. But we have already seen good reason for believing that a knowledge of gold precedes that of copper whenever both are found in the same area. We saw that the Scythians, who got copious store of gold from the Ural-Altai region, made no use of copper in the fifth century before our era, although copper is found abundantly in the same area. From this we may infer with some probability that the Italian stock were acquainted with gold sooner than with copper. We may apply the same argument to gold in Italy as we did to *copper*. *Aurum* (older *ausum*), the Latin word for gold, is plainly not borrowed, as is perhaps the Greek *chrysos*, from the Semites. Hence it cannot be maintained that it was only with the Phoenicians that the knowledge of gold reached Italy.

It now only remains for us to see if the Italians had the means within their reach of discovering gold. No one I suppose will dispute that the Italian stock entered the peninsula from the north, driving before them older occupants. They must then have either entered Italy by the head of the Adriatic, coming round from the valleys of the Balkan peninsula, or through the Alpine passes. If they came from the first quarter it is impossible to suppose that a people in close contact with the tribes who occupied the Balkan peninsula, and who as we have seen above must have been acquainted with gold from a remote time, could have remained without a knowledge of the metal. On the other hand it will be seen from the following evidence that there was every opportunity for the discovery of gold in the Alpine valleys. Strabo gives various notices of the gold workings of this region. "Polybius states that in his own day in the vicinity of Aquileia, in the territory of the Taurisci of Noricum, was found a gold mine so productive that on clearing away the surface dirt to a depth of two feet gold which could be dug was straightway found, and that the pit did not exceed fifteen feet, and that part of the gold was pure on the spot, being the size of a bean or a lupin, only one-eighth being lost in refining, whilst some of it required a process of smelting which, though more elaborate, was still very remunerative. When the Italians worked them along with the barbarians for a space of two months, straightway gold coin went down one-third in value throughout the whole of Italy; but when the Taurisci became aware of this they expelled their partners and held the monopoly. But now all the gold mines are in the hands of the Romans. And there too, just as in Iberia, the rivers in addition to the dug gold produce gold dust, but not in such quantities[1]."

In another passage, speaking of the town of Noreia in Noricum, he says "this district possesses productive gold-washings and iron-works[2]."

Moving on again westwards, we easily find strong evidence of active gold-mining in the Alpine regions. All the granite strata on the southern side of the High Alps from the Simplon to Mont Blanc are auriferous. Not only have extensive mining operations been carried on at different points down almost to the present day, but the mines were beyond all doubt vigorously worked, not merely in Roman but in pre-Roman days. In the district of La Bessa, at the foot of Mont Grand on the right bank of the Cervo between Biella and Ivrea, are still to be seen very extensive traces of gold washings and gold diggings[3]. These are no other than the once famous mines of Victumulae alluded to by Strabo when, in speaking of this region, he says that "there is not now as much attention bestowed on the mines as there used to be, because the mines in the country of the transalpine Kelts and in Spain are more profitable, but formerly they were well worked, since at Vercelli there was a gold-digging. Vercelli is a village near Ictumulae which is itself a village, and both of them are in the vicinity of Placentia[4]." So important were these mines that Pliny[5] says there existed a Censorian law relating to them, by

---

[1] Strabo, 173. 34—49, Didot.   [2] Ibid. 178 Didot.
[3] Th. Mommsen (*Nordetruskische Alfabete*, p. 250, *seqq.*) gives an admirable summary of the metallurgical history of this region.
[4] Strabo, 218.
[5] Pliny, xxxiii. 4. § 78, extat lex censoria Victumularum aurifodinae, qua in Vercellensi agro cavebatur, ne plus quinque м hominum in opere publicani haberent.

which it was provided that the capitalists who farmed the mines were not to employ more than 5000 workmen.

There are also traces of ancient gold-washings on the Cervo, on the Evenson, a small stream which comes down from Monte Rosa, and which falls into the Doria at Bardo, and likewise on the Doria itself from Bardo down to its junction with the Po. This latter region was anciently the territory of the powerful and wealthy tribe of the Salassi. The traces I speak of are beyond doubt the remains of the gold-workings described by Strabo. "The territory of the Salassi contains gold mines, which the Salassi, when aforetime they were strong, kept possession of, just as they had likewise the control of the passes (*i.e.* the Great and Little St Bernard). The river Durias (Doria) gave them very great assistance in their gold washing, and on this account dividing over many places the water into many side-channels they used to empty completely the main bed of the river.

"This was of service to them in their quest of gold, but it did harm to the cultivators of the plains below, who were being deprived of the means of irrigation, since the river was not able to water their land from the others having possession of the stream in its upper course. From this cause there were incessant wars between the two peoples. But when the Romans got the mastery the Salassi were expelled from the gold-mines and from their territory, but still being in possession of the mountain, they used to sell the water to the farmers who had hired the gold-mines, and with whom there were constant quarrels because of the grasping conduct of the contractors[1]." This passage shows plainly that for a very long period before the Roman Conquest the Salassi had not merely worked the gold of their mountains, but had attained to very considerable engineering skill in so doing. Further, in this region have been found gold coins bearing the inscriptions *Prikou*, etc. in one of the North Etruscan alphabets. These coins were most probably struck by the Salassi, who were probably not Kelts, but a remnant of the ancient Rhaetian stock[2].

[1] Strabo, 205.
[2] Th. Mommsen, *Die nordetruskischen Alfabete*, p. 223; Pauli, *Altitalische Forschungen*, p. 6.

Passing northwards by the Pennine Alps, the regular road in ancient days from Italy into Switzerland, into the valley of the Rhone, the so-called *Vallis Poenina*, the modern Canton of Valais, we come to the Helvetii, whom Posidonius of Apamea, the famous Stoic philosopher who travelled in Western Europe about 100—90 B.C., describes as "wealthy in gold." This gold was probably derived from the same Alpine region. The Helvetii struck both silver coins in imitation of the silver coins of Massalia with the Lion type, and gold ones after the type of Philip's staters. We may now pass on to Gaul Proper, many peoples of which were famous for their wealth, especially the Arverni, who have left their name in Auvergne, and the Tectosages, whose chief town was Tolosa (Toulouse). The former, whose original home was on the upper waters of the Loire, probably had no gold in their native mountains (for if they had, Strabo would hardly have failed to mention it), but in the second century B.C. they became the most powerful state of Central and Southern Gaul, for "they extended their dominion even as far as Narbo (Narbonne) and the borders of the territory of Massalia (Marseilles), and they likewise had the control of all the tribes as far as the Pyrenees, and as far as the Ocean and the Rhine. And it is said that Lucrius, the father of Bituitus, who fought against Maximus and Domitius (121 B.C.), came to such a pitch of wealth and luxury that on one occasion, making a display of his riches to his friends, he drove on a waggon through a plain sowing broadcast gold and silver coin, while his friends followed him gathering it up[1]." It was the Arverni who first[2] struck gold coins in imitation of the gold staters of Philip II., a fact explained by the passage just quoted, which shows that their empire extended up to the frontiers of the great Greek emporium of Massalia, by which they would be brought into immediate contact with all kinds of Greek currency; furthermore their conquests put them in possession of those districts where we have direct evidence of the existence of gold fields[3].

[1] Strabo, 191. [2] Hucher, *L'Art Gaulois*, 19.

[3] We must then in all probability place the first striking of the Gaulish limitations of the Philippus about 150 B.C., rather than as is usually stated about 250 B.C.

Again Strabo says: "The Tectosages adjoin the Pyrenees, and to a slight extent they likewise touch upon the northern side of the Cevennes (Κέμμενα), and they occupy a land rich in gold[1]." It is no doubt with reference to the same region that Strabo, whilst describing the Spanish gold-mines, remarks incidentally that "the Gauls advance the claims of the mines in their country, both those in the Cevenne mountain and at the foot of the Pyrenees, themselves[2]." Beyond doubt from those mines came "the gold of Tolosa," those vast treasures which were plundered by the Roman General Caepio. They were said to have amounted to fifteen thousand talents of unwrought gold and silver. There was a current story that, for laying sacrilegious hands on the consecrated treasure, misfortune dogged the steps of Caepio and his family, he himself dying in exile and his daughters, after lives of degradation, coming to a shameful end. This was the account given by one Timagenes, who also stated that the treasure of Toulouse was part of the spoil taken by the Gauls from the temple of Delphi in 279 B.C., the Tectosages as he alleged having formed part of the invading host. This story doubtless is due to the circumstance that one of the three tribes of Gauls who settled in Asia Minor (the "foolish Galatians" of St Paul's Epistle) was called by the same name as the Tectosages of Gaul (the other two being called Trocmi and Tolistobōgii). The treasures were partly stored in shrines or sacred enclosures, partly deposited in the sacred lakes. There can be little doubt that Posidonius was right (as Strabo also thought) in considering them ancient native offerings, not spoils of war. He put forward the good argument that at the time of the attack on Delphi the temple there was bare of treasure, as it had been plundered by the Phocians in the Sacred War some seventy years before, that any treasure that remained was distributed among many, and that it was not likely that any of the Gauls returned to their own land, since after their retreat from Greece they broke up and were scattered into various regions. This is confirmed by what Diodorus tells us in a remarkable chapter: "The Kelts of the interior have a singular peculiarity with respect to the sacred

[1] Strabo, 187.   [2] Strabo, 146.

enclosures of the gods. For in the temples and sacred enclosures consecrated in their country gold is deposited in quantities, and not one of the natives touches it owing to superstition, although the Kelts are excessively avaricious[1]." This passage seems to explain thoroughly the real nature of the treasures of Tolosa; they were doubtless ancient votive offerings under a taboo, not, as Timagenes imagined, some of the treasure of Delphi, dedicated to appease the wrath of Apollo, with additions from the private resources of the Tectosages themselves. In the same chapter Diodorus says that "there is no silver at all found in Gaul, but gold in abundance, of which the natives get supplied without mining or hardship. The currents of the rivers, which are tortuous in their course, beat against the banks formed by the adjacent mountains, and bursting away considerable hills, fill them with gold dust. This the persons who are engaged in the workings collect, and they grind or break up the lumps which contain the gold dust. Then having washed away the earthy part with water, they transfer the gold to furnaces for smelting. In this fashion heaping up quantities of gold, not only the women but likewise the men employ it for adornment. For they wear bracelets round their wrists and arms, and thick torques of solid gold round their necks and rings of remarkable size, and moreover breastplates of gold." The statement regarding silver is not accurate, as the more careful and trustworthy Strabo mentions silver mines in various places in Gaul. Finally, in the land of the Tarbelli, an Iberian tribe of Aquitania, who dwelt in the extreme south-west corner of Aquitania on the shore of the Bay of Biscay, there were extremely productive gold-mines. "For in spots dug only to a shallow depth are found plates of gold that sometimes require little refining, and the rest consists of dust and nuggets which involve but little working[2]."

I have purposely gone somewhat minutely into the goldfields of ancient Gaul, and the story of the sacred treasures. For I think that no one who considers carefully the statements of Posidonius, Strabo, and Diodorus, can help regarding as wholly inaccurate the conclusion of Schrader, based on the Irish

[1] Diodorus, v. 27.   [2] Strabo, 190.

word *or*, that the Keltic peoples were not acquainted with gold until the fourth century B.C. The sacred treasures point to a ceremonial consecration of gold extending back through untold ages.

It must also be borne in mind that in the treasure of Tolosa there was a good proportion of silver which probably came from the silver mines mentioned by Strabo[1] as existing in the land of the Ruteni and Gabales (Γαβάλεις), two peoples of Aquitania, whose names are represented by the modern *Rovergue* and *Gevaudan*. As the working of silver is so much later than that of gold, it is impossible to believe that if the Gauls in Italy only learnt the use of gold in the 4th century B.C. we should find consecrated treasures of silver, evidently of ancient date, at Tolosa in the time of Servilius Caepio. It is also important to observe that it is among the Iberians of Aquitania, not the Kelts, that we find silver mines being worked. The former people were entirely free from Roman influence, and we shall see shortly that there is the strongest evidence for believing that the Iberians south of the Pyrenees were acquainted not merely with gold but with silver, centuries before ever Brennus stood in the Roman Forum. But before we cross the Pyrenees, we shall conclude our survey of the ancient gold fields of Europe in the north-west by glancing briefly at Britain. When Julius Caesar invaded the island he found the natives using

A. B.
Fig. 14. Ancient British Coins.
A. Coin of Iceni.   B. Common type with plain obverse[2].

gold not simply as ornaments, but in the shape of coins, for he says, "They have great numbers of cattle, they use for money

---

[1] Strabo, 191.

[2] Both are from coins in my own possession; A found near Mildenhall (Suffolk) in 1884, cf. Dr Evans, *Ancient British Coins*, Pl. XXIII. 4; B at Potton in Bedfordshire, 1888; cf. *op. cit.* Pl. B. 8.

either bronze, or coins of gold, or rods of iron of a fixed standard of weight. Tin is produced there in the inland, iron in the coast districts, but the supply of the latter is scanty; the copper which they use is imported[1]." Caesar's statement is fully confirmed by the existence of ancient British coins, chiefly in gold and copper; although silver coins are likewise found, they are for the most part imitations of the types of Roman denarii, whilst the gold are the descendants of the Philippus, from which the Gauls got their chief gold type. All the Britains did not employ coins, but only the Belgic tribes in the south and east, who had crossed over at a comparatively late period. About a century before our era a king of the Suessiones (*Soissons*) by name Divitiacus ruled over all Northern France and a large part of Britain[2]. Coins similar in type and weight are found on both sides of the Channel, indeed the French numismatists claim them as struck in Gaul, whilst their English brethren have maintained that they are of British origin. Those found in Kent are regarded by Dr Evans, in his *Coins of the Ancient Britons*, as the prototypes of the whole British series. Hence we may infer that the Belgic invaders brought the Philippus type of coin into Britain, as it is most probable that the time when the same coins were in circulation on both sides of the Straits of Dover corresponds with the period when Divitiacus held sway on both sides of the sea[3]. Strabo substantiates Caesar's account; "It (Britain) produces wheat and cattle, and gold and silver and iron. These are exported from it, also hides and slaves and good hunting dogs. But the Kelts employ even for their wars these, and their own native dogs[4]."

[1] Caesar, *B. G.* v. 12, pecorum magnus numerus. Utuntur aut aere aut nummis aureis aut taleis ferreis ad certum pondus examinatis pro nummo. Nascitur ibi plumbum album in mediterraneis regionibus, maritimis ferrum, sed eius exigua est copia, aere utuntur importato.

[2] Caesar, *B. G.* ii. 4.

[3] W. Ridgeway, "The Greek Trade Routes to Britain" (*Folklore*, March 1890, p. 23).

[4] Strabo, 199, leaves out tin here although he mentions it when quoting from Poseidonius. The reason is that after the tin-mines of Northern Spain had been developed by Publius Crassus, Caesar's lieutenant, the British tin trade ceased.

There can therefore be no doubt that gold was found in Britain although we are not told in what particular part. Gold is still found in Wales and in several parts of Scotland, although not in sufficient quantity to be worth working. Two observations remain to be made on the statements of Caesar and Strabo. Caesar tells us definitely that whilst they used copper as money, they had to import that metal. He omits all mention of silver, whilst Strabo, writing half-a-century later, speaks of it as a British product. I have remarked already that the silver coins of the Britons are all late, and exhibit as a rule Roman influence. It would therefore seem as if the working of silver had developed some time after Caesar's invasions. Thus once more we have an instance of gold in full use long before silver. But what is still more important, though the Britons are in the bronze period and are actually using copper money, they have to import that metal, although copper is actually found native in Cornwall. It still remained undiscovered in Strabo's time to judge by his silence, but as he is equally silent about tin, which was known long before, we cannot press the argument *ex silentio*. However, it is of great importance to find a people who possess gold and copper in a native state, already working the gold long before they have even discovered the copper. This is completely in harmony with what we have already seen in the case of the Scythians and Arabs of the Red Sea coasts. At a later stage we shall have to notice the rods or bars of iron used as currency by the Britons in connection with a similar practice elsewhere.

The writers of the classical age have left us no information respecting Ireland save that the people practised polyandry, and ate each other[1]. Nevertheless there is abundant evidence to show that there were large deposits of gold on the east side of Ireland, in the Wicklow Mountains, and that the natives from a very early period wrought it into ornaments of various kinds. The vast quantity of gold ornaments to be seen in the Museum of the Royal Irish Academy is a proof of its abundance.

We shall now return to Aquitania and the Bay of Biscay, from which we digressed to Britain, and coming into Northern

[1] Strabo, page 201.

Spain enter that region which was to the Greek of the sixth
and fifth centuries B.C. what the Spanish Main was to the
Europeans of the sixteenth and seventeenth centuries. It
seems beyond doubt that when the Phoenicians first reached the
Spanish coasts the natives were fully acquainted with both gold
and silver. Tradition told how the Phoenicians found the native Iberians feeding their horses from mangers made of silver,
and that after having filled every available portion of their
ship with freight of treasure, they replaced their anchors by
others made of silver. Colaeus of Samos in the eighth century B.C. had been the first of all Greeks to reach Tartessus,
the Tarshish of Holy writ, having been carried away by a storm
when on a voyage to Egypt, and driven right through the
Straits of Gibraltar, "under some guiding providence," says
Herodotus[1]: "for this trading town was in those days a virgin
port" (i.e. unfrequented by merchants). "The Samians in consequence made a profit by their return freight, a profit greater
than any Greeks had ever made before, except Sostratus, son
of Laodamas, of Egina, with whom no one else can compare."
From the tenth part of their gains, amounting to six talents,
the Samians made a brazen vessel. At a later period the
Phocaeans made great profit by trade with Iberia, which at
that time meant East Spain as opposed to Tartessus, as well as
with the Tartessians. The king of this people, by name
Arganthonius, who reigned over them for eighty years, and
attained to the patriarchal age of one hundred and twenty,
became such a friend of the Phocaeans that he invited them
to settle in his land, perhaps through motives of policy, wishing
to have their support against the Phoenicians of Gadeira, or
Gades (*Cadiz*), the most ancient of all the daughter cities of
Tyre. When he did not succeed in persuading the Phocaeans,
afterwards having learned from them of the great growth of the
power of the Medes, he gave them treasure to enable them to
fortify their city with the strong wall by means of which they
were to withstand Harpagus, the general of Cyrus, until they
launched their ships, and embarked their wives and children,

---

[1] IV. 151.

with that firm resolution to be free, which has made their name memorable through the ages[1].

The evidence of these passages is sufficient to show that already in the seventh century B.C., not simply the gold, but likewise the silver, of the Spanish peninsula was known to and wrought by the Iberians, the oldest race of whom written history affords any traces in the west of Europe.

We shall now deal with the actual localities and mines described for us by the ancient writers. Strabo once more is our chief helper: he seems as usual for all statements about the mines of the west to have drawn his information chiefly from Posidonius, although he likewise makes use of Polybius and others. "Posidonius averred that in the country of the Artabri, who are the most remote people in Lusitania towards the north and west [occupying the present province of Galicia], the earth crops out in silver, tin and white gold (for the gold is mixed with silver), and that the rivers carry down this earth, and that the women scrape it up with hoes and wash it in sieves into a box[2]." Here we have a description of the method employed by the natives in the remote regions of the north-west of Spain about 100 B.C., before Roman influences had time to affect them, and we may not unreasonably infer from it that the same process was universal amongst the Iberians and Celtiberians of Spain.

In his general description of Spain Strabo declares that nowhere in the world down to his day was such plenty of gold, silver, copper and iron to be found as in Turdetania, the district named after the Turdetani, one of the two great tribes into which the Turti were divided [from the name of Turti it is probable that Tartessus, the Greek name for this region, as also for the Baetis (*Guadalquivir*), and also the Phoenician *Tarshish* were formed]. "Not merely is the gold got by mining but it is swept up. The rivers and torrents carry down the golden sand, which in many localities is likewise to be found in places where there is no water, but there it is invisible, but in those that water flows over the gold dust gleams out. And flushing with water that has to be fetched the arid spots, they make the gold dust glitter, and by digging wells and by

[1] Herodotus, I. 163—4.   [2] Strabo, 147.

devising other means they get out the gold by washing the
sand, and what are called gold-washings are now more numerous
than the gold diggings. But they say that in the gold dust
are found nuggets sometimes even half a pound in weight
(βώλους ἡμιλιτριαίας) which they term *palae*, which need but
little refining, and they say likewise that when stones are split
little nuggets like teats are discovered, and when the gold is
refined and purified with a kind of earth which contains alum
and vitriol, the residuum is electrum. When this residuum,
which consists of a mixture of gold and silver, is again refined,
the silver is burnt away and the gold remains. But the gold is
very fusible, and on this account it is melted with chaff rather
than with coal, because the flame being gentle acts moderately
upon a metal which is yielding and easily fused, whereas the
charcoal causes excessive waste by melting it too much by its
violence, and detracting from it. In the river-beds the sand
is swept up and then washed in troughs beside the river; or
else a well is dug, and the earth that is brought up out of it is
washed. They make the furnaces for the silver high, that the
smoke from the ore may be carried up into the air: for it is
noisome and pestilential[1]." Then he adds that "some of the
copper works are called gold mines, from which people infer
that gold was formerly dug from them. Posidonius, when prais-
ing the number and excellence of the mines, refrains from none
of his wonted rhetoric, but warms up with hyperboles, for he
says he cannot doubt the truth of the story that once on a
time when the woods caught fire, the earth having been melted,
inasmuch as it was permeated with silver and gold, boiled out
on to the surface over the whole mountain, and that a whole
hill was a mass of money heaped up by the bounteous hand of
fortune. And to speak generally (he says) any one who saw
these regions would say that they were Nature's perennial store
chambers or Sovereignty's inexhaustible treasure house. For
not merely the surface but the under-soil is rich (πλουσία—
ὑπόπλουτος), and with those people it is not Hades who dwells
in the region beneath the earth, but Pluto (Πλούτων). So
spake he in a fine figure as though he himself too were drawing

[1] Strabo, 146.

from a mine his diction in copious store. There was a saying of Phalereus in reference to the eagerness of the miners of Laurium in Attica, that they dug as continuously and earnestly as if they expected to drag up Pluto himself. This saying Posidonius quotes anent the energy and vigour of those who worked the Spanish mines, for they cut deep and winding galleries, and by means of 'Egyptian pumps' combated the springs which burst into the workings[1]."

So rich were the silver mines of New Carthage (*Cartagena*) that in the time of Polybius (140 B.C.) 40,000 men were employed in working them for the Roman State, and the daily out-put was reckoned at 25,000 drachms, or roughly speaking about 3,000 ounces Troy.

Diodorus Siculus[2] gives an account of mines and mining in Spain, which, as it is clearly derived from the same passage of Posidonius as the account of Strabo, is worth quoting, especially as it gives probably *in extenso* what Strabo has summarized. For although it more particularly refers to the discovery of silver mines, yet it is very relevant to our subject, since silver invariably is later in point of discovery than gold; thus if we can fix at an early period an inferior limit for the knowledge of silver in Spain, we may with confidence fix the inferior limit for the knowledge of gold at a still earlier epoch. Diodorus has been describing the range of the Pyrenees, which like all the early geographers he represents as running north and south, and thus proceeds: "Since there are on them (the Pyrenees) many forests dense with trees, they say that in ancient times the whole mountain region was completely burned by some shepherds having cast away a firebrand. Then since the fire kept burning on for many days continuously, the surface of the earth was burned and the mountains from the circumstance were called Pyrenaean (Πυρηναῖα, *scorched*), and the surface of the burnt region flowed with much silver, and since the natural ore had been smelted, there ensued many lava-like streams of pure silver. But inasmuch as the natives did not understand the use of it, the Phoenicians trading with them, and having learned about the occurrence, bought the silver for some small

[1] Strabo, 146 *sq.* [2] Diodorus, v. 35.

return in other wares; accordingly the Phoenicians by conveying it to Greece and to Asia and all the rest of the world acquired great wealth. And so covetous were the merchants that though their ships were fully freighted, when much silver still remained over they cut out the lead that was in their anchors and replaced it with silver. The Phoenicians by means of such trade increased greatly and sent out many colonies, some to Sicily and the adjacent islands, others to Libya, others again to Sardinia and Spain. But many years afterwards the Spaniards, having become acquainted with the peculiarities of silver, started remarkable mines. Wherefore as they prepared very excellent silver in very great quantities they used to get great revenues." Diodorus then gives a detailed account of the working of the shafts and winding galleries which followed the course of the veins of gold and silver, the difficulties caused by the bursting in of springs and subterranean streams, and the ways in which the miners overcame this latter obstruction by means of the Egyptian pumps. But Diodorus, as a patriotic Sicilian, takes care to tell his reader that this pump was invented by Archimedes, the famous mathematician of Syracuse, when, in the course of his travels, he paid a visit to Egypt. Finally, he gives a short but graphic picture of the sufferings of the wretched slaves who were bought wholesale by the mine owners and endured incredible miseries until death, the only friend they had to look to, came to end their sufferings. Strabo, the stoic, is silent on this point, which here, as in Egypt, so strongly moved the heart of Diodorus.

The story of the discovery of silver by the burning of the woods at first savours of the mythical, but there is really good reason for believing that there is in it a solid nucleus of truth. Tin was unknown in Sumatra until in 1710[1] it was discovered by the accidental burning down of a house (an incident which recalls Charles Lamb's delightful account of the discovery of Roast Pig). It is highly probable that it was owing to some such accident that men first became acquainted with silver, as that metal is rarely if ever found native. It may well be therefore that mankind has learned the art of smelting metalliferous

[1] Marsden's *History of Sumatra*, p. 172.

ore from observing the results of some such conflagration as that described by Posidonius.

Finally, we shall turn to Pliny the Elder for a moment. That industrious collector has given us a minute account of the various methods of mining carried on in Spain in his time, but as that is beside our present purpose I shall only quote a short passage, in which we get some interesting technical expressions relating to gold-mining. After detailing the method of washing soil containing gold by bringing streams of water to bear on it, just as we found the Salassi doing in the valley of the Doria, by which process he says 20,000 lbs. of gold were annually obtained in Asturia, Gallaecia, and Lusitania, he proceeds: "Gold obtained by shafting (*arrugia*) does not require refining, but is straightway pure. Nuggets of it are found in this way; likewise in pits nuggets are found exceeding ten pounds each. The Spaniards call them *palacrae*, others *palacranae*. The same people term the gold dust *balux*[1]." Here then we have an interesting group of technical terms, *arrugia*, *palacra* or *palacrana* and *balux*. The latter forms at once remind us of Strabo's *palae* ($\pi\acute{a}\lambda\alpha\iota$), and we can have little doubt that *palacra* and *pala* are simply dialectic variants, just as *palacrana* evidently was considered by Pliny to be a bye-form of *palacra*. Corssen has sought to find a Latin etymology for *arrugia*, connecting it with *runco*, *ruga*, but it is hardly possible to regard it as otherwise than Spanish, especially as this appears to be the only place where it is found. *Balux* (also *baluca*) is undoubtedly a native Iberian term. On Schrader's principles we might at once argue that as the technical words for gold-mining and for the different kinds of gold are native Spanish words, it is beyond doubt that the Spaniards were acquainted with gold and knew the art of working it before any foreign traders brought that metal to them. Without dogmatizing in this fashion and keeping to

[1] Pliny, *H. N.* xxxiii. 4, 21 aurum arrugia quaesitum non coquitur sed statim suum est; inueniuntur ita massae; necnon in puteis denas excedentes libras; palacras Hispani, alii palacranas, iidem quod minutum est balucem uocant.

May the French *paille* (in the phrase *pailles d'or*), Ital. *paluola*, Span. *palazuola*, all used technically of gold, be derived from *pala*, the old technical term, rather than from *palea*, chaff?

our more cautious principles we may say that the evidence of those words is strongly in favour of such a conclusion, unless a Semitic origin be sought for those terms, which is highly improbable. For we know beyond doubt that the Spanish mines were worked for centuries before ever a Roman soldier passed the Ebro. Unless then the technical terms were introduced by the Greeks (which they were not, as Strabo considers *pala* a native word) or by the Phoenicians, they are ancient Iberic terms connected with gold from its first discovery. We saw that in the Red Sea the first form in which gold was utilized by the Arabs was that of nuggets used as rude beads. The *palae* of the Iberians may represent the same period of development as well as the same kind of gold. From the traditions given us by the ancient writers there can be little doubt that the art of mining silver was of extremely ancient date in Spain. The founding of Gadeira (Cadiz) is placed at 1100 B.C. and the tradition of Posidonius regards the Phoenician colonies in the west as long posterior to their trading for silver with the rude natives. If this tradition could be relied on, silver must have been known to the Spaniards in the twelfth century B.C. And there is no reason to doubt the story. At Mycenae gold and silver were found along with Baltic amber. The two former prove that amongst the civilized races around the Aegean the precious metals were abundantly used, the latter that the trade routes across Europe from the Baltic and North Sea to the Adriatic were already in use. Accordingly there is no improbability in the supposition that in the twelfth century B.C. the shipmen of Tyre traded for silver to North Eastern Spain as well as to Northern Italy for amber. If the knowledge of silver came so early in Spain, much earlier must that of gold have been.

Let us now take a general survey of the region over which we have travelled. In the far east we had both the literary evidence of the Rig Veda and the evidence of the traditions and legends handed down by the historians to show that well back in the second millennium B.C. the gold deposits of Thibet were known and worked. Silver is as yet unknown to the people of the Rig Veda. Again in the region of the Altai and Oural mountains, the tale of the "Arimaspian pursued by a griffin" pointed to great antiquity for gold-mining in this district; the barbarous Massagetae[1], who occupied the modern Mongolia and Sangaria, were rich in gold; and to the west the Scythians, who used neither silver nor copper, had abundant store of gold. These tribes stretched right across Russia until they touched on the west the Getae and the other tribes of the great Thracian stock. Gold must early have been known throughout all Thrace. Greek tradition and history unite in demonstrating the great antiquity of the first Phoenician gold-seeking in Thasos and on the mainland. The evidence in Greece itself puts it beyond doubt that gold was in use 1500 years B.C. The Balkan Peninsula was occupied on the north-west by Illyrian tribes, some of whom, like the Dardani, dwelt interspersed among the Thracian clans. The Illyrians inhabited all the northern end of the Adriatic, and originally much of the east side of all Italy, although under the pressure of the Umbrians and Kelts they had been almost completely crushed out of the Italian Peninsula, only maintaining themselves in the extreme southeast where the Messapians remained independent of both Italian and Greek alike. The Keltic tribes were their neighbours in Noricum, where they had succeeded the ancient Rhaetian stock, the survivors of which, like the Salassi, had managed to maintain themselves in the fastnesses of the Alps. We found strong evidence that these Rhaetians must long have known the art of working gold, for they had devised elaborate pieces of engineering work for the purpose of developing their gold fields; added to this was the fact that gold as an ornament seems to have been used by the inhabitants of the Swiss lake dwellings in the neolithic age. The Kelts must have been in contact with this people for a considerable time before they ever invaded Italy; again in Spain we found every token of great antiquity in the working of gold and silver. Again, before they invaded Italy, the Kelts must have been long in contact with the Iberians of what in later days was Aquitania, for the Keltic conquest of Northern Spain can hardly be placed later than in the fifth century B.C., and it is most probable that that con-

[1] Herod. iv. 11.

quest only took place after long and stubborn struggles. The Kelts too in Southern Gaul must have come in contact with the Ligyes (or Ligurians), whose territory at one time extended from the Iberus (Ebro) along the coast of the Mediterranean to the frontiers of Etruria. The Ligurians had been in touch with the Iberians on their western border; in fact the two races had blended to a considerable degree, and since they had also had communication with Etruscans, Phoenicians and Greeks (with the last from at least 600 B.C., when Massilia was founded in their country), it is impossible to suppose that this people could have remained ignorant of the use of gold. The Kelts thus at every point along their southern front, as they advanced, must have been for centuries in full knowledge of gold before they ever entered Rome. Add to this the fact that when they entered Italy they appear to have brought nothing but their gold ornaments and their cattle, and that in Gaul it had been the habit to dedicate great piles of the precious metal in the sacred precincts of their divinities.

CHAPTER IV.

PRIMAEVAL TRADE ROUTES.

THERE can be little doubt that from the extreme West of Europe to Northern India, or rather to China and the Pacific shore, there was complete intercourse in the way of trade, from the most remote epochs. In the lake dwellings of Switzerland are found implements of Jade, a stone which is not found at any spot in Europe; in fact the nearest point from which the material was fetched must have been Eastern Turkestan on the borders of China[1]. If in neolithic days such communication

[1] How trade was carried on in early days may be well illustrated from Torres Straits of to-day. (Haddon, "The Western Tribe of Torres Straits," *Journal of Anthrop. Inst.* XIX. p. 347.)

Dance masks made of turtle shell (340) occasionally used as money.

If a Muralug man wanted a canoe he would communicate with a friend at Moa, who would speak to a friend of his at Badu; possibly the Muralug man might himself go to Badu, or treat with a friend there. The Badu man would cross to Mabuiag to make arrangements, and a Mabuiag man would proceed to Saibai.

If there was no canoe available at the latter place word would be sent on, along the coast, that a canoe was to be cut out and sent down.

The canoe would then retrace the course of the verbal order and ultimately find its way to Muralug. The annual payment for a canoe was say three *dibi dibi* or goods of about equal value. There were three annual instalments.

There is no money in the Straits; but certain articles have acquired a generally recognized exchange value, a value which is intrinsic, and not irrespective of the rarity of the material or the workmanship put into it. These objects cannot be regarded as money; they are the round shell ornaments (*dibi dibi*, shell armlet, *wai wai*, dugong, harpoon, *wap*, and canoe). A good *wai wai* is the most valuable possession; the exchange of a *wai wai* was a canoe, or harpoon. Ten or twelve *dibi dibi* was considered of equal value to any of the above. A wife was the highest unit of exchange, being valued at a canoe, or a *wap* or *wai wai*. "The intermediaries (in the purchase of a canoe) are paid

existed between Further Asia and Western Europe, it is not unreasonable to suppose that when gold, an article existing in almost every country across the two continents, came into use, a like facility of intercourse must have existed. In one of the passages of Herodotus which I have given above we had explicit information respecting a trade route extending from the Greek factories on the northern shores of the Black Sea through the medium of the Scythians right away to the remote region of the Altai. On the other hand there is good evidence for the existence of a great trade route from the Black Sea westward up the valley of the Danube, and so reaching the head of the Adriatic; and again, there is equally good reason for believing that from the mouth of the Po there ran a similar route across Northern Italy through Liguria and Narbonese Gaul and into Spain. In reference to the first of these routes we may quote a tradition preserved in the Book of Wonderful Stories before alluded to. It is there stated that once on a time travellers who had voyaged up the Danube finally by a branch of that river which flowed into the Adriatic made their way into that Sea. It is there alleged[1] that "there is a mountain called Delphium between Mentorice and Istriana, which has a lofty peak. Whenever the Mentores who dwell on the Adriatic mount this crest, they see, as it appears, the ships which are sailing into the Pontus (Black Sea). And there is likewise a certain spot in the intervening region in which, when a common mart is held, Lesbian, Chian and Thasian wares are set out for sale by the merchants who come up from the Black Sea, and Corcyraean wine jars by those who come up from the Adriatic. They say likewise that the Ister, taking its rise in what are called the Hercynian forests, divides in twain, and disembogues by one branch into the Black Sea, and by the other into the Adriatic. And we have seen a proof of this not only in modern times, but likewise still more so in antiquity, as to how the regions there are easy of navigation (reading

for their services 'by charging on,' the amount depending on individual cupidity, or they may be recompensed for their trouble by presents from the purchaser" (p. 341).

[1] [Aristotle,] *De Miris Auscult.* 104—5 (839ᵃ 84 *seqq.*).

εὔπλωτα). For the story goes that Jason sailed in by the Cyanean Rocks, but sailed out from the Black Sea by the Ister."

The story of the meeting between the traders from the Black Sea and Adriatic has every mark of probability, whilst we are possibly justified in regarding the legend of Jason as evidence that for long ages the Greeks knew that up the valley of the Danube traders from the Pontus made their way. Doubtless too it was with a view to tapping the trade of this very route that the trading factories like Istropolis were founded on the Danube.

The branch of the Danube flowing into the Adriatic can only mean that travellers from the Danube by passing up one of its tributaries would reach a point from which it was but a short journey to the Adriatic shore. But a famous story in Herodotus will yield us more efficient aid. To the Greeks of the fifth century B.C. the extreme north was represented by the land of those happy beings the Hyperboreans, just as the furthest south was represented by the sources of the Nile. Thus Pindar sings: "Countless broad paths of glorious exploits have been cut out one after another beyond Nile's fountains and through the land of the Hyperboreans[1]."

Some of the oldest legends of the young world's prime cluster around this shadowy region. Herakles had wandered there in quest of the hind of the golden horns, consecrated to Artemis Orthosia by Taygeta[2]; "In quest of her he likewise beheld that land behind the chilling north wind; there he stood and marvelled at the trees." The judge at the Olympic festival placed round the locks of the victor "the dark green adornment of the olive, which in days of yore Amphitryon's son had brought from the shady sources of the Ister, a most glorious memorial of the contests at Olympia, when he had won over by word the Hyperborean folk that are the henchmen of Apollo[3]." The hero Perseus too had reached that land where no ordinary mortal could find his way. "Neither in ships nor yet on foot wouldst

---

[1] Pind. *Isth.* v. 22 sq. μυρίαι δ' ἔργων καλῶν τέτμηνθ' ἑκατόμπεδοι ἐν σχερῷ κέλευθοι | καὶ πέραν Νείλοιο παγᾶν καὶ δι' Ὑπερβορέους.

[2] *Ol.* III. 31 sq.

[3] *Ol.* III. 13 sqq.

thou find out the marvellous ways to the assembly of the Hyperboreans, but once on a time did the chieftain Perseus enter their houses and feast, having come upon them as they were sacrificing glorious hecatombs of asses to the god. Now Apollo takes continuous and especial delight in their banquets and hymns of praise, and he laughs as he beholds the rampant lewdness of the beasts[1]."

Herodotus felt puzzled where to place the Hyperboreans; "For concerning Hyperborean men neither the Scythians say anything to the point nor any other of those that dwell in this region, save the Issedones. But as I think, not even do they say anything to the point; for in that case the Scythians too would have told it, as they tell about the one-eyed people" (the Arimaspians[2]). "But a certain Aristeas, the son of Caÿstrobius, a man of Proconnesus, alleged in a poem that under the influence of divine afflatus he had reached the Issedones, and that beyond them dwelt the Arimaspians who have but one eye, and that beyond these are the gold-guarding griffins, and beyond these the Hyperboreans, stretching to the sea[3]." But where Pindar and Herodotus hesitated, the priest of Apollo at Delos stepped in with an explicit statement of that "marvellous road" which Pindar said no one could find by sea or land. Accordingly Herodotus has to resort to the men of Delos for his information about the Hyperboreans: "Much the longest account of them is given by men of Delos, who have alleged that sacred objects bound up in wheaten straw are brought from the Hyperboreans to the Scythians, and that the Scythians receive them and pass them on to their neighbours upon the west, who continue to pass them on until at last they reach the Adriatic, and from thence they are sent on southwards. First of the Greeks do the men of Dodona receive them, and from them they travel down to the Melian Gulf and cross over to Euboea, and city sends them on to city as far as Carystus. The Carystians take them over to Tenos without stopping at Andros; and the Tenians convey them to Delos." Then he adds a further story that on the first occasion the Hyperboreans sent two maidens, Hype-

---

[1] Pind. *Pyth.* x. 29 *sqq*.
[2] Herod. IV. 13.
[3] Herod. IV. 32.

roché and Laodicé, with five male protectors, but as they died at Delos, and returned home no more, they for this reason "bring to their borders the sacred objects packed up in wheaten straw and lay a solemn injunction on their neighbours, bidding them send them forward to another nation, and the men say that being forwarded in this fashion they arrive at Delos[1]."

From the various passages quoted we may draw the probable conclusion that there was a well-defined trade route existing for untold ages between the heart of Asia, the valley of the Danube and the head of the Adriatic. The nameless poets who framed the legends of Herakles and his wanderings would certainly make the hero travel by the routes where both in their own time and from tradition they knew of the existence of highways from nation to nation. Thus in his journey to the Hyperboreans Herakles is represented as having visited the shady forests of the Danube, which points to the same road as that assigned to the Hyperborean maidens by the Delian tale. Finally it may not be farfetched to conjecture that the sacrifice of hecatombs of asses may be taken as evidence that the Hyperborean legend points to a people of Central Asia, which is the natural habitat of the wild ass. However, as it seems that there was an annual sacrifice of asses to Apollo at Delphi[2], we must be careful not to lay much stress on this argument, although it is quite possible that a vague knowledge of a far-off region where asses abounded and were sacrificed may have given the Greeks the idea that the Hyperboreans were worshippers of their own god Apollo, at whose altar like offerings were made.

Having seen some reasons for believing that before the beginning of history there was a well-defined route from Central and perhaps Further Asia across Southern Russia to the valley of the Danube, and then by one of the valleys of its tributaries to within a short distance of the Adriatic, whence after crossing the watershed it reached the head of that sea, we are now in a position to enquire whether we have similar evidence for the further continuance towards the west of this highroad of nations.

[1] Herod. IV. 33.
[2] Boeckh, *Corp. Inscr. Graec.* Vol. I. p. 807.

We have had occasion already to remark that the legends of the Voyage of the Argo in quest of the Golden Fleece, and the journeyings of Herakles and such-like stories, really represent the earliest knowledge of the regions which lay far away to the east and north-west. There is no tale of the hero Herakles more famous than that of his travelling to the very marge of Ocean, where in the Pillars of Hercules he left an imperishable record of his wayfaring for the men of aftertime. His object, so goes the story, was the capture of the famous kine of the giant Geryon who dwelt in the island of Erythia, in after years the site of Gaddir, or Gadeira as the Greeks called it, the Gades of the Romans, and the modern Cadiz. Many vague stories relating to the early ethnology of Western Europe and Northern Africa cycle round this expedition[1]. But for our present purpose it is only the fabled route by which he went with which we are concerned. As might naturally be expected that part of Italy with which the Greeks seem first to have become acquainted was the district lying in the Adriatic around the mouths of the Po (Eridanus). The reason why they came thither is not far to seek. They doubtless simply followed the example of the Phoenicians who probably had long traded thither to obtain both the highly prized golden amber from the Baltic, and the red amber of Liguria, called from that region Lingurium, or *ligurion*, a name for which the Greeks found a strange etymology which connected it with the lynx[2]. According to Herodotus, "the Phocaeans were the first of the Greeks who made long voyages and discovered Adria, Tyrsenia (Etruria), Iberia and Tartessus" (I. 163). The trade routes to the amber coasts of the north have long been well known; they passed over the Alps, crossed the Danube at Passau, Linz or Presburg, and proceeded then either to Samland or to the vicinity of Jutland[3]. As these northern routes crossed that which came up the valley

[1] Cf. Sallust, *Jug.* 18.

[2] They derived it from λύγξ and οὖρον. The difference in colour between the Baltic and Ligurian amber found an easy explanation, the latter was regarded as the solidified urine of the female lynx, the former of the male animal. Pliny, *H. N.* xxxvii. 2, § 34.

[3] Cf. Boyd Dawkins, *Early Man in Britain*, 466. Von Sadowski, *Die Handelstrassen der Griechen und Römer*, p. 15.

of the Danube, we see that by this route there was complete communication between the Black Sea and the Adriatic. In later times we know that active trade was carried on with all Northern Italy from Marseilles along by the Ligurian shore, for the coinage of Massalia, and the barbarous imitations of it struck

FIG. 15. Barbarous imitation of Drachm of Massalia.

by the peoples of what was afterwards known as Cisalpine Gaul, formed the currency of that region until the Roman Conquest. But once more the Book of Wonderful Stories comes to our aid: "They say that from Italy into Keltiké, and the land of the Keltoligyes and Iberians, there is a certain road called that of Herakles, by which if any journey, whether Greek or native, he is protected by those who dwell along it, that he may suffer no wrong. For those in whose vicinity the wrong is done have to pay the penalty." Here we have a clear instance of a well-defined caravan route, connected by Greek tradition with the name of Herakles, which was placed under a kind of taboo, so that all travellers could use it with impunity. We may then conclude that as from Central Asia there was unbroken communication with Northern Italy, so likewise from Northern Italy there was from remote ages a definite trade route into Gaul and Spain, and that these routes were in turns connected with the great routes which lead from the Mediterranean to the Baltic and North Sea.

CHAPTER V.

THE ART OF WEIGHING WAS FIRST EMPLOYED FOR GOLD.

WE have seen in the preceding pages that from the Atlantic seaboard right across into Further Asia the ox was universally spread, and from a period long before the daybreak of history already formed the chief element of property amongst the various races of mankind which occupied that wide region. We have likewise seen that gold was very equally distributed over the same area, being ready to hand in the still unexhausted deposits in the sands of rivers. And lastly we have seen that from the most remote times there was complete communication for purposes of trade between the various stocks. For whilst peoples in the pastoral and nomad stage do not dwell together in large communities they nevertheless are within touch of one another. No better illustration of this can be found than the relations between Abraham and Lot as set forth in Genesis (xiii. 5 *sqq.*): "And Lot also, which went with Abram, had flocks, and herds, and tents. And the land was not able to bear them, that they might dwell together: for their substance was great, so that they could not dwell together. And there was a strife between the herdmen of Abram's cattle and the herdmen of Lot's cattle: and the Canaanite and the Perizzite dwelled then in the land. And Abram said unto Lot, Let there be no strife, I pray thee, between me and thee, and between my herdmen and thy herdmen; for we be brethren. Is not the whole land before thee? separate thyself, I pray thee, from me: if thou wilt take the left hand, then I will go to the right: or if thou depart to the right hand, then I will go to the left. And Lot lifted up his eyes, and beheld all the plain of Jordan, that it was well watered every where, before the Lord destroyed Sodom and Gomorrah, even as the garden of the Lord, like the land of Egypt, as thou comest unto Zoar. Then Lot chose him all the plain of Jordan; and Lot journeyed east: and they separated themselves the one from the other. Abram dwelled in the land of Canaan, and Lot dwelled in the cities of the plain, and pitched his tent toward Sodom." But although, from the necessity of finding sufficient pasturage for their flocks and herds, they had parted from one another, they remained within touch. For we find that no sooner had Lot and his possessions been carried away by Chedorlaomer and his confederates, after the overthrow of the kings of Sodom and Gomorrah, than Abraham at once hears of his mishap and hastens to his rescue (xiv. 13 *sqq*.).

The picture here given may be taken as holding good for a large part of Asia and Europe. There is a great intermingling of various races and untrammeled intercourse between the various communities. Thus we find that Abraham was able to journey from Haran into Egypt with his flocks and herds and suffered harm or hindrance of no man. Nay, a still stronger proof of the safety and freedom of intercourse is that when Abraham entered Egypt, although afraid that if it were known that Sarah was his wife the Egyptians might murder him, yet he had no fear that they would take her away by force if she was supposed to be his sister. Thus, when his princes told Pharaoh that the Hebrew woman was fair to look on, though the king commanded her to be taken into his house, he did not act with high-handed violence against the stranger, but "he entreated Abram well for her sake: and he had sheep, and oxen, and he asses, and menservants, and maidservants, and she asses, and camels." And when Pharaoh discovered that she was really Abraham's wife, although on account of Abraham's mendacity the Lord had "plagued Pharaoh and his house with great plagues because of Abraham's wife," he did not, as he might very justly have done, take a summary vengeance upon him, "he commanded his men concerning him: and they sent him away, and his wife, and all that he had." (Gen. xii. 12—20.)

Such then being the general distribution of cattle and sheep, and such again the distribution of gold, we can have little hesitation in coming to the conclusion that the ox, which we have evidence to show was the chief unit of value in all those countries, had the same value throughout, and in like manner that gold would have almost the same value over all the area in which we have shown that it was so impartially apportioned out by Nature. From this it follows that if the unit of gold was fixed upon the older unit, the ox, the same quantity of gold would be found serving as the metallic unit throughout the same wide area.

If then it can be proved that throughout the area in which those weight standards arose from which all the known systems of the ancient, mediaeval, and modern world were derived, the same gold-unit is found everywhere, and that wherever evidence is to hand, this unit is regarded as equal in value to a cow or ox, the truth of our hypothesis will have been demonstrated. For it would be impossible that such an occurrence should be a mere coincidence if found repeated in different areas. Furthermore, if it can be shown that in cases at a comparatively late historical period peoples who were borrowing a ready-made metallic system from more civilized neighbours, have found it impossible to do so without adjusting or equating such metallic standard to their own unit of barter, we may infer *a fortiori* that it would have been impossible for any people to have framed a metallic unit for the first time for themselves without any reference to the unit of barter. But as we have already proved that the unit of barter is in every case earlier in existence than even the very knowledge of the precious metals, it follows irresistibly that the metallic unit is based on the unit of barter. We have also given reasons for believing that gold was the first of the metals known to primitive man, but as yet we have not proved that the metals are the first objects to be weighed. If this can be proved, and if furthermore it can be proved that before silver or copper or iron were yet weighed, gold has been weighed by that standard, which we find universal in later times, we have still more closely narrowed down our argument and put it beyond all reasonable doubt that weighing was first invented for traffic in gold, and since the weight-unit of gold is found regularly to be the value of a cow or ox, the conclusion must follow that the unit of weight is ultimately derived from the value in gold of a cow.

If we begin in modern times and reflect on the articles which are usually sold by weight, we find at once that the more valuable and less bulky the commodity, the more regularly is it sold and bought by the medium of the scales and weights; furthermore, on enquiry we find that many kinds of goods which are now sold by weight were formerly sold simply by bulk or measure. At the present moment corn is generally sold by weight (though sometimes still by measure), although the nomenclature connected with its buying and selling shows beyond doubt that formerly it was sold entirely by dry measure. The English coomb, the Irish barrel, the bushel and the peck are indubitable evidence. The selling of live cattle by weight has only lately been adopted in some markets in this country; but go back to a more remote period, and you will find that even dead cattle were not sold by weight. Thus we see that it is only in a comparatively late epoch that two of the chief commodities on which human life depends for subsistence have been trafficked in by weight. Nothing now remains but man's clothing, weapons, ornaments, fuel and furniture.

The more primitive the condition of life, the more scanty and rude is the household furniture, and as even in modern times timber is not sold by weight, beyond all doubt the same must hold good in a still stronger degree of a time when wood could be had for the mere trouble of sallying forth with an axe and cutting it. The same argument applies cogently to the question of fuel. For even though coal is now sold by weight, both coal and coke are still sold in some places at least in name by the chaldron, a fact that indicates that it was only when facilities increased for weighing large and bulky commodities that such a practice came into vogue. Similarly, although firewood is now sold by weight on the Continent, beyond all doubt at a previous period it was uniformly sold by bulk, as peat or turf is now sold in Cambridgeshire, in Scotland, and in Ireland.

Weapons and ornaments and utensils now only remain. To take the last-named first, at no period have vessels of earthenware been sold by weight. On the other hand those of metal, especially when made of copper and iron, are usually sold in this fashion, although vessels of iron and tin are commonly sold by bulk, or according to their capacity, thereby following, as we shall shortly see, a most ancient precedent. The value of ornaments largely consists in the artistic skill displayed in their manufacture, hence weight is not employed in estimating their value except when the material is gold or silver, and therefore possesses a certain intrinsic value apart from the mere workmanship. We may therefore infer that in early times no decorative articles save those in metal were valued by weight. Next comes the question of weapons, one of the most important sides of ancient life. Of course gold and silver are unfit for weapons and implements, save in the case of the gods, as for instance the chariot of Hera, with its wheel-naves of silver and its tires of gold[1]. The spear-head and sword-blade must be made from tougher and cheaper metals. Hence copper or bronze (copper alloyed with tin) in the earlier periods which succeeded the stone age, and iron at a later time, have mainly provided mankind with weapons of offence and defence. But precious as copper and bronze and iron were to the primitive man, we do not find them sold by weight: a simple process was employed; the crude metal was made into pieces or bars of certain dimensions, so many finger-breadths or thumb-breadths long, so many broad, so many thick, just as wooden planks are now sold with us, when the value of a piece of timber is estimated by its being so many feet of inch board, or half-inch board, and of a fixed width. Lastly we come to the question of clothing. Skins of course were sold by bulk, the hide of an ox or a sheepskin having generally a fixed and constant value. Even when sheep came to be shorn, the fleece was set at an average value. But beyond all doubt among the peoples who dwelt around the Mediterranean the practice of weighing wool was of a most respectable antiquity. Such, too, was the practice all through the middle ages in

[1] *Il.* v. 720 *seqq.*

England and on the Continent. We have abundant specimens still left of the weights carried by the wool merchants, slung over the back of a pack-horse.

Having said so much by way of preliminary, we can now adduce testimony in support of our thesis. Once more let us start with the Homeric Poems. The weighing of gold is already in vogue, but the highest unit known is the small talent, the value of an ox, weighing 130-5 grs, or 10-15 grs more than a sovereign. Silver is not yet estimated by weight, although large and handsome vessels of that metal are described and have their value appraised. But it is not by their *weight* that their value is estimated, but by their *capacity*. Thus as first prize for the footrace Achilles gave "a wine-mixer of silver, wrought, and it held six measures, but it surpassed by far in beauty all others upon earth, since cunning craftsmen, the Sidonians, had carefully worked it, and Phoenician men brought it over the misty deep." (*Iliad*, XXIII. 741 *sqq.*) Here we have a vessel wrought in silver evidently of considerable size, but it is simply by its content that its size and value are expressed. Among the lists of prizes in the same book we find the size of vessels made of copper or bronze similarly indicated. Thus the first prize for the chariot race consisted of a woman skilled in goodly tasks; and a tripod with ears, which held two and twenty measures; whilst the third prize was a *lebes* or kettle which had never yet been blackened by the fire, still with all the glitter of newness, which held four measures. So, too, in the case of iron. As the prize for the Hurling of the Quoit, Achilles set down a mass of pig iron, which he had taken from Eetion. It is a piece of metal as yet unwrought, so that here if anywhere its size and value ought to be reckoned by weight, since no account has to be taken of workmanship. But Achilles, instead of saying that it weighs so many talents or minae, describes its value in a far more primitive fashion. "Even if his fat lands be very far remote, it will last him five revolving seasons. For not through want of iron will his shepherd or ploughman go to the town, but it (the mass) will supply him[1]."

[1] *Il.* XXIII. 826 *seqq.*

Thus of the four chief metals mentioned in the Homeric Poems, gold alone is subjected to weight. But the scales are used for another purpose still. In the Twelfth Book of the *Iliad* there is a curious simile wherein a fight between the Trojans and Achaeans is likened to the weighing of wool: "So they held on as an honest, hardworking woman holds the scales, who holding a weight and wool apart lifts them up, making them equal, in order that she may win a humble pittance for her children: thus their fight and war hung evenly until what time Zeus gave masterful glory to Hector, Priam's son[1]."

Without doubt one of the first uses to which the art of weighing was applied was that of testing the amount of wool given to female slaves[2], or in this case perhaps to a freed woman, to make sure that they would return all the wool when spun into yarn, and not purloin any portion for themselves. Thus in the older Latin writers we constantly find allusions to the *pensum* (*pendo* = to weigh), the portion of wool *weighed* out to the slave. It is quite possible that in the sale of wool the more ancient conventional fashion of estimating the fleece as worth so much in other familiar commodities long continued for mercantile purposes, the weighing of the wool in small portions being only used as a check on the dishonesty of the spinners. At all events we have found wool estimated by the fleece in mediaeval Ireland, at a time when weights are in common use for the metals.

Such then is the condition of things in the Homeric Poems. Gold is transferred by weight and by weight wool is apportioned out for spinning.

[1] *Il.* XII. 433—7,

ἀλλ' ἔχον, ὥς τε τάλαντα γυνὴ χερνῆτις ἀληθής,
ἥ τε σταθμὸν ἔχουσα καὶ εἴριον ἀμφὶς ἀνέλκει
ἰσάζουσ' ἵνα παισὶν ἀεικέα μισθὸν ἄρηται.
ὣς μὲν τῶν ἐπὶ ἶσα μάχη τέταται πτόλεμός τε κ.τ.λ.

Dr Leaf, in his introduction to Book XII., when calling attention to various marks of lateness in this book, says: "It has further been remarked with some truth that the numerous similes, though beautiful in themselves, are often disproportionately elaborated and lead up to points which are almost in the nature of an anti-climax." But the use of the word ἀληθής in an entirely un-Homeric sense seems to make it almost certain that these lines are of late date.

[2] Cf. Plautus, *Merc.* II. 3. 63. Virg. *Georg.* I. 390, carpentes pensa puellae.

Let us now turn to the Old Testament and find what are the objects which are dealt in by weight. All transactions in money are thus carried on, as for instance the purchase by Abraham of the Cave of Machpelah from Ephron the Hittite when "Abraham weighed to Ephron the silver, which he had named in the audience of the sons of Heth, four hundred shekels of silver, current *money* with the merchant" (Gen. xxiii. 16). So likewise in Achan's confession: "I saw among the spoils a goodly Babylonish garment, and two hundred shekels of silver, and a wedge of gold of fifty shekels weight" (Joshua vii. 21). And so too in the Book of Judges (viii. 26) the weight of the rings taken from the Midianites and given to Gideon was "a thousand and seven hundred shekels of gold; beside ornaments, and collars, and purple raiment that was on the kings of Midian, and beside the chains that were about their camels' necks." And again David bought the threshing-floor of Ornan the Jebusite for six hundred shekels of gold by weight (1 Chron. xxi. 25), although the same purchase is described in 2 Samuel (xxiv. 24) as being effected for fifty shekels of silver. In Solomon's time gold has become exceedingly abundant, and we find it reckoned by talents and minae (pounds). For "king Solomon made a navy of ships in Ezion-geber, which is beside Eloth, on the shore of the Red sea, in the land of Edom. And Hiram sent in the navy his servants, shipmen that had knowledge of the sea, with the servants of Solomon. And they came to Ophir, and fetched from thence gold, four hundred and twenty talents, and brought it to king Solomon" (1 Kings ix. 26—8). And after the story of the Queen of Sheba's visit and her gift to the king of "an hundred and twenty talents of gold, and of spices very great store, and precious stones," we read that "the weight of gold that came to Solomon in one year was six hundred threescore and six talents of gold, beside that he had of the merchantmen, and of the traffick of the spice merchants, and of all the kings of Arabia, and of the governors of the country. And king Solomon made two hundred targets of beaten gold: six hundred shekels of gold went to one target." Spices such as myrrh, cinnamon, calamus and cassia (Exod. xxx. 23) were sold by weight, being as costly as gold. The familiar description of Goliath of Gath, the weight of whose coat of mail "was five thousand shekels of brass," and whose "spear's head weighed six hundred shekels of iron," will serve to show that articles in the inferior metals were at that time estimated according to weight by the Hebrews and their neighbours, the Philistines. Of the weighing of wool we find no instance, but it is quite possible that it was from the practice of weighing wool that Absalom when he "polled his head, (for it was at every year's end that he polled it: because the hair was heavy on him, therefore he polled it:) he weighed the hair of his head at two hundred shekels after the king's weight" (2 Sam. xiv. 26). But it is perhaps more probable that the habit of weighing a child's hair against gold or silver to fulfil a vow (which was almost certainly Absalom's motive) may have suggested the employment of the scales for wool[1].

[1] Mr J. G. Frazer gives me the following interesting note:

As to the cutting off a child's hair and weighing it against gold or silver, the facts are these.

(1) Among the Harari in Eastern Africa when a child is a few months old, its hair is cut off and weighed against silver or gold money; the money is then divided among the female relations of the mother.

Paulitschke, *Beiträge zur Ethnographie und Anthropologie der Somâl, Galla und Harari* (Leipzig, 1886), p. 70.

(2) Mohammed's daughter Fâtima gave in alms the weight of her child's hair in silver.

W. Robertson Smith, *Kinship and Marriage in early Arabia*, p. 153.

(3) Among the Mohammedans of the Punjaub a boy's hair is shaved off on the 7th or 3rd day after birth, or sometimes immediately after birth. Rich people give alms of silver coins equal in weight to the hair.

Punjab Notes and Queries, I., No. 66.

(4) When the Hindus of Bombay dedicate a child to any god or purpose, they shave its head and weigh the hair against gold or silver.

Id. II. No. 11.

(5) In the inland districts of Padang (Sumatra) three days after birth the child's hair is cut off and weighed. Double the weight of hair in money is given to the priest.

Pistorius. *Studien over de inlandsche Huisponding in de Padangsche Bovenlanden*, p. 56; Van Hasselt, *Volksbeschrijving van Midden-Sumatra*, p. 268.

(6) There is the Egyptian custom, for which we have the evidence of Herodotus, II. 65, and Diodorus, I. 8.

Finally, once in the prophet Ezekiel do we find food weighed, but evidently under special circumstances: "And thy meat which thou shalt eat shall be by weight, twenty shekels a day: from time to time shalt thou eat it. Thou shalt drink also water by measure, the sixth part of an hin: from time to time shalt thou drink" (iv. 10, 11). In any case we should expect to find traces of later usage in the writers of the age of the prophets, but from the directions regarding the amount of water, it is evident that we cannot take this passage as a proof of the ordinary practice of the time.

Unfortunately our oldest records of Roman life and habits go back but a short way before the Christian era, and hence we cannot get much direct information as regards the first objects which were sold by weight. We have already seen that in the time of Plautus (*flor.* 200 B.C.) the habit prevailed of weighing wool out to the women slaves.

However, from the legal formula used in the solemn process of conveyance of real property (*res mancipi*) *per aes et libram*, we may perhaps infer that the scales were used for none but precious articles such as copper, silver and gold. That they were used for those metals there can be little doubt. On the other hand, as we find all kinds of corn sold at a later period by dry measure, such as the *modius* or bushel, we may with certainty conclude that such too had been the practice of the earlier period.

From the literary remains then of the Greeks, Hebrews and Latins, it is beyond all doubt that in the early stages of society nothing is weighed but the metals and wool (for the apportioning of tasks). In this the records of all three nations agree, whilst from Homer we learn that the Greeks were using gold by weight, when as yet neither silver, copper nor iron was sold or appraised by that process.

To proceed then to a people compared to whom the Greek and Hebrews in point of antiquity of civilization are but the upstarts of yesterday. The Egyptians seem to have used weight exclusively for the metals; the *Kat* and its tenfold the *Uten* seem always used in connection with metals, whilst corn is always connected with measures of capacity. The following instances taken from the list of prices of commodities given by Brugsch (*History of Egypt under the Pharaohs*, II. p. 199, English Transl.) will suffice for our purpose: a slave cost 3 *tens* 1 *Kat* of silver; a goat cost 2 *tens* of copper; 1 *hotep* of wheat cost 2 *tens* of copper; 1 *tena* of corn of Upper Egypt cost 5—7 *tens* of copper; 1 *hotep* of spelt cost 2 *tens* of copper; 1 *hin* of honey 8 *Kats* of copper. Even drugs were not weighed by the Egyptians in the time of Rameses II. The physicians prescribed by measure, as we learn from the Medical papyrus Ebers[1].

Passing then to the far East, we naturally are curious to learn whether the oldest literary monument of any branch of the Aryan race, the Rig-Veda, throws any light on our question. We get there but meagre help: but yet, scanty as it is, it is of great importance. As we saw above the Indians of the Vedic age were still ignorant of the use of silver, although possessing both gold and copper. Now, whilst we have no evidence bearing upon the latter metal, there are two very remarkable and important words used in connection with gold which beyond doubt refer to the weighing of that metal. In the *Mandala* (VIII. 67, 1—2; 687, 1—2) a hymn commences: "O India, bring us rice-cake, a thousand Soma-drinks, and an hundred cows, O hero, bring us apparel, cows, horses, jewels along with a *mana* of gold." Again, "Ten horses, ten caskets, ten garments, ten *pindas* of gold I received from Divodāsa. Ten chariots equipped with side-horses, and an hundred cows gave Açvatha to the Atharvans and the Payu". (*Mandala*, VI. 49, 23—4). As we shall have occasion later on to deal with the terms *manâ* and *hiranya-pinda* at greater length, it will suffice our present purpose to point out that we have a distinct mention of a weight of gold in the expression *manâ hiranyayâ*. In only these two passages have we any allusion to weighing, and in both it is in direct connection with gold. The Aryans of the Veda are beyond all doubt in a far less civilized state than the Egyptians, Hebrews, Greeks or Romans of the historical period. Hence we may

[1] F. L. Griffith, "Metrology of the Medical Papyrus Ebers," *Proceed. of Soc. Bibl. Arch.* June 1891.

without danger infer that they did not use weight for any cereals they may have cultivated. Therefore we may, with a good deal of probability, conclude that we have got a people who had already a knowledge of the art of weighing before they were acquainted with either silver or iron, and that this people used the scales for gold and nothing else. This, taken in connection with the fact that in Homer, although silver is known, the weighing of metals is confined to gold, leads us irresistibly to conclude that gold was the first of all substances to be weighed, or, to put it in a different way, the art of weighing was invented for gold.

CHAPTER VI.

The Gold Unit everywhere the value of a Cow.

We have now proved four things : (1) the general distribution of the ox throughout our area, (2) its universal employment as the unit of value throughout the same region, (3) the equable distribution of gold throughout the same countries, and (4) that gold is the first of all commodities to be weighed. Our next step will be to show that gold was weighed universally by the same standard, and that this standard unit in all cases where we can find record was regarded as the equivalent of the ox or the cow.

We have already seen that the gold talent of the Homeric Poems, which was in use among the Greeks before the art of stamping money had yet become known, weighed about 130 grains troy (8·4 grammes). In historical times gold was always weighed on what was called the *Euboic* (or Euboic-Attic) standard. Thus when Thasos began to strike gold coins in 411 B.C. after her revolt from Athens they weighed 135 grs. Unless this had been the time-honoured unit employed for gold in that island so famous for its mines the Thasians would hardly have employed it. Certainly they would not adopt it simply because it was the standard of the hated Athenians, especially as they had a different standard for silver.

The gold coins of Athens struck a few years later are on the same standard of 135 grs, and when Rhodes at the beginning of the fourth century B.C. began to coin gold, she used the same unit, although she employed for silver the unit of 240 grs. Cyzicus also, although coining her well-known electrum

Cyzicenes on the Phoenician standard, used the unit of 130 grs for pure gold.

FIG. 16. GOLD STATER OF PHILIP OF MACEDON.

This standard, as we shall presently see, virtually remained unchanged for gold down to the latest days of Greek independence. It likewise prevailed in Macedonia and Thrace. For when Philip II. coined the gold from the mines of Crenides into staters on the so-called Attic standard of 135 grains, he did nothing else than employ for the first gold coinage of his country the unit which had there, as in Greece Proper, prevailed for many ages for the weighing of gold. For since gold was first coined in that region about 350 B.C., and yet silver coins had been current in Thrace and Macedon since about 500 B.C., it would be absurd to suppose that there was no unit by which gold in ingots or rings could be appraised.

I have shown elsewhere that the rings found by Dr Schliemann at Mycenae were probably made on a standard of 135 grains troy. It is natural to suppose that if within the area of Greece Proper gold rings were fixed according to a definite standard, and that standard the Homeric talent, the Macedonians and Thracians would possess a similar unit in the fifth century B.C. But there is a small piece of literary evidence to show that the Macedonians were acquainted with the gold unit, which we already know as the Homeric ox unit. Eustathius tells us that "three gold staters formed the Macedonian talent[1]." Whether Mommsen is right in thinking that this name was given to the talent in Egypt in consequence of its having been introduced by the Lagidae (themselves Macedonians) or not, it equally indicates that from of old such a talent, confined in use to gold, and the threefold of the Homeric ox-unit,

---

[1] Hultsch, *Metrol. Scrip.* 209, τὸ Μακεδονικὸν τάλαντον τρεῖς ἦσαν χρύσινοι.

had existed in Macedonia. Hence Philip II. did not require to go to Athens to seek for a standard for his new gold coinage.

Fig. 17. Persian Daric.

Passing into Asia we find there the shekel as the Daric(Δαρεικός), the normal weight of which is 130 grains troy. This standard prevailed all through the Persian empire, thus extending into the countries now represented by Afghanistan and Northern India. Numismatists have pointed out the fact that Philip coined his staters some five grains heavier than the rival gold currency of the Persian empire, as if to enhance the estimation of his new coinage. This explanation is perhaps over subtle;

Fig. 18. Gold Stater of Diodotus, King of Bactria.

at all events it is interesting to find the successors of Alexander the Great in the Far East, the kings of Bactria, coining their staters not on the standard of 135 grains, but rather on that of 130, in other words following the native standard which the Daric simply represented as a coin. Thus Dr Gardner[1] in his Table of Normal Weights makes the Bactrian stater of what he calls the Attic standard weigh 132 grains and the drachm 66 grains, and it is also admitted that from the time of Eucratides the Greek kings of Bactria adopted a native standard.

This new standard seems to be identical with that called by metrologists the Persian, on which [silver] coins were struck in

---

[1] *Catalogue of Greek Kings of Bactria*, p. lxix.

all parts of the Persian empire, notably the Sigli stamped with the figure of the Persian king, which must have freely circulated in the northern parts of India that paid tribute to the king. Whether the reason given for the use of this standard is right or not, we may see hereafter, when a different explanation will be offered to the reader. That great Indian archaeologist, General Cunningham[1], goes further, and maintains "that the earliest Greek coins of India, those of Sophytes, are struck, not on the Attic standard, but on a native standard which is based on the *rati* or grain of *abrus precatorius*." Whatever may be the ultimate decision of this dispute, it is enough for our purpose that whilst undoubtedly a native silver standard sooner or later replaced the Attic, so likewise the Attic standard, if used for gold, did not remain at its full weight of 135 grains, but rather approximated to that of the native standard of the Daric (130 grains). It is almost certainly a native standard which appears as the weight of the *gold piece* (*suvarna*) in the tables of weights given in the Hindu treatise called *Lilavati*, written in the seventh century A.D., before the Muhammadan conquest of India, and which we shall notice presently at greater length. This *suvarna* is the only unit for gold mentioned in the tables, and its weight can be demonstrated to be about 140 grs troy. That the gold unit only varied 10 grains in the course of 10 centuries is very remarkable.

Let us now return to the ancient peoples of Further Asia Minor and Northern Africa. The Phoenicians and their neighbours in historical times seem to have used the double of the unit of 130 grains. It is quite possible that this doubling of the unit can be explained by a simple principle, which will likewise fit in with the threefold of the same unit, which we have just now had to deal with under its name of Macedonian Talent. But how far this double unit prevailed in earlier times among the Semites it is not easy to tell. However, the evidence to be derived from the Old Testament is in favour of the priority of the unit of 130 grains. But this is not all our evidence. The Egyptian hieroglyphic inscriptions give us con-

---

[1] *Catalogue of Greek Kings of Bactria*, p. lxvii.

128   THE GOLD UNIT EVERYWHERE

siderable information regarding the currency not simply of Egypt itself but likewise of neighbouring countries. For when Egypt was at the zenith of her glory great conquerors like Thothmes III. and Rameses II. (the Sesostris of Herodotus) carried their arms into all the surrounding lands and reduced them to the position of tributary vassals. Many of the tablets which recount their exploits contain the tale of the spoil, and describe it as consisting amongst other things of gold rings.

The wall paintings which still survive the inroads of time, and the still ruder hands of Arabs or tourist, constantly exhibit representations of the payment of tribute. Again and again we see the tribute money in the form of rings being weighed in scales, "on which solid images of animals in stone or brass in the shape of recumbent oxen took the place of our weights[1]."

FIG. 19. EGYPTIAN WALL PAINTING SHOWING THE WEIGHING OF GOLD RINGS[2].

Erman gives several representations of such weighing scenes (pp. 611—12), and infers from the fact that the weigh-master and his scales are always present at such payments, that the scales were the ordinary medium of such payments. Mere pictures however do not tell us anything about the weight of the rings therein pourtrayed. Fortunately however we have examples

[1] Brugsch, *Op. cit.* I. 386.   [2] Lepsius, *Denkmäler*, 331.

of such rings. Brandis[1], who was the first to seek for the unit on which these rings were fashioned, thought that they followed the heavy shekel (260 grs.), the double of our common unit. On the other hand F. Lenormant[2] thinks that they are really based on the light shekel, or rather on a lighter variety of the light shekel, of about 127 grains, and he is followed in this by Hultsch[3]. For our purpose it matters not whether the rings were made on the simple unit or its double, for there are not really two separate standards but simply one and the same. It is hardly likely that the Pharaohs would have done otherwise than the kings of Persia at a later time, who made their subject countries pay their tribute in the recognized currency of the kingdom, the gold being reckoned (as Herodotus says) by the Euboic talent, the silver by the Babylonian talent. There can then be but little doubt that these gold rings give us either actually the old Egyptian standard, or a standard so closely related to it that there was to all intents and purposes no material distinction between them.

Schliemann noticed a resemblance between some of the rings found at Mycenae and those represented in Egyptian paintings. It is not preposterous to suppose that the rings of Mycenae represent a kind of ring both in form and weight which was employed by the peoples of Asia Minor and Egypt, as well as in Greece. The contact between Egypt and Asia Minor is so close, communication so free, that it would be in itself most unlikely that any wide divergence of currency would exist in earlier times, whilst on the other hand her relations with the people of Ethiopia and Libya were likewise so close that they forbid any other conclusion. This is proved by the statement of Horapollo that the *Monad* ($\mu o \nu \acute{a} \varsigma$), which the Egyptians held to be the basis of all numeration, was equal to two drachms, that is, to 135 grs.[4]

Passing westward let us try and learn something from the early coinage of Italy. Unfortunately, with the exception of

---

[1] *Münz- Mass- und Gewichtswesen in Vorderasien*, p. 80 seqq.
[2] Lenormant, *La Monnaie dans l'Antiquité*, I. 103 seqq.
[3] *Metrol.*[2], p. 375.
[4] Horapollo, I. 11, Παρ' Αἰγυπτίοις μονάς ἐστιν αἱ δύο δραχμαί.

the Greek cities of Magna Graecia, all Italian mintages are of a comparatively late date. The Etruscans were probably the first of the non-Hellenic inhabitants to coin money, but unhappily their gold coins are of rather uncertain date. However, it is worth noticing that these coins are probably thirds, sixths and twelfths of the unit 130·5 grains, the weights respectively being 44 grs., 22 grs., 11 grs. This view borrows considerable additional probability from the fact that the silver coins with plain reverses, which very possibly belong to the same age as the earlier gold, are struck on the standard of 135 grains. Whilst in the latter case the Etruscans can be said to have struck their coins on the Euboic-Syracusan, or Attic-Syracusan, or Euboic-Attic standard which was in use at Syracuse, it cannot be so alleged with respect to their gold. For not only are the subdivisions of the unit unknown to the Attic or Syracusan gold, but the coins bear numerals, $\Lambda = 50$, $\Lambda XX = 25$, $XII< = 12\frac{1}{2}$, $X = 10$, which are found respectively on the coins of 44, 22, 11 and 9 grains, while on others again which weigh 18 grains we find the numeral $\Lambda = 5$ grains[1]. Here then we have clear indications of a native Etruscan gold currency, existing prior to Greek influence and able to hold its own when the art of coining, and the very coin types themselves, were borrowed from the Greeks.

The Carthaginians were the close allies of the Etruscans in the struggle for the maritime supremacy of the Western Mediterranean against the Greeks, especially the bold Phocaeans, who gained over the fleet of both peoples a "Cadmean victory" at Alalia in Corsica (537 B.C.).

The first Carthaginian coinage was issued in the Sicilian cities, especially Panormus, at a comparatively late date, certainly not earlier than 410 B.C. As this coinage was entirely under Greek influences of comparatively late date, we cannot of course get any direct evidence from it as regards the original Phoenician standard. Carthage herself did not issue coins until about a century later, B.C. 310[2]. Hence we have no data of an early date. The gold coins struck in Sicily are

---

[1] Deecke, *Etrusk. Forsch.* II. p. 1. Head, *Op. cit.* p. 12.
[2] Head, *Op. cit.* p. 747.

didrachms of about 120 grains troy, with various subdivisions. This is usually described as the Phoenician standard, or rather the Phoenician gold standard of 260 grains considerably reduced. But the full unit of 240 is never found in the coins, and although we get coins of 2½ drachms (= 147 grains), it is more natural to regard the didrachm of about 120 grains as the real unit, in other words the slightly lowered common unit, which we already found fixed at about 127 grains in the Egyptian rings. In Sicily and Magna Graecia we are fairly certain that the unit was in early times that of 130 grains. But whether this was native or brought in by the Greek colonists, it is impossible to prove. All that we know for certain is that there was in Sicily and Magna Graecia, a small talent used only for gold; which was equivalent to three Attic gold staters, or in other words the threefold of our Homeric ox-unit. Thus an ancient writer says "the Sicilian talent had a very small weight; the ancient one, as Aristotle says, 24 nummi, the later 12 nummi. But the nummus weighs three half obols[1]." From this it is plain that the ancient form of this talent weighed 36 obols, that is, six drachms, or three staters.

Lastly, let us glance at those peoples who lay between Northern Italy and the Bay of Biscay. Although we have no direct evidence as to the unit by which the Gauls reckoned that gold of which, as we saw above, they had great store, before they came under the influence of either Phoenician, Greek, or Italian, we can perhaps make a justifiable inference from the fact that when the Gauls proceeded to strike gold coins in imitation of the gold stater of Philip of Macedon, they did not, as might have been expected, follow also the weight unit (135 grs.) of that coin. For as a matter of fact scarcely any of the Gaulish imitations exceed 120 grains troy[2]. It would appear then that the Gauls had already at that time a gold unit in use, somewhat lighter than the usual weight of our "ox-unit," although we cannot of course ignore the possibility of its being

[1] Τὸ μέντοι Σικελικὸν τάλαντον ἐλάχιστον ἴσχυεν, τὸ μὲν ἀρχαῖον, ὥς Ἀριστοτέλης λέγει τέτταρας καὶ εἴκοσι τοὺς νούμμους τὸ δὲ ὕστερον δυοκαίδεκα, δύνασθαι δὲ τὸν νοῦμμον τρία ἡμιωβόλια. (Hultsch, Reliq. Metrol. Scrip. 300.)

[2] Cf. Hucher, L'Art Gaulois, p. 19 and Pl. I.

the form of the Phoenician gold standard, which we found above was employed by the Carthaginians both in Sicily and Africa; in other words it may be maintained that the Gauls followed the standard on which the Phocaeans of Massalia struck their *silver* coinage. As, however, the coins of Massalia were drachms of about 55 grains the probability is not very high that the Gauls had no gold standard of their own for gold until they got one from the *silver* of Marseilles.

The Teutonic tribes who likewise issued imitations of the Philippus also followed a standard of 120 grs. for coins, from which it is likely that they as well as the Gauls employed a unit of 120 grs. for gold before they ever began to strike money.

We have now taken a survey of the most ancient gold standards we can find throughout the wide regions through which the common system of weights of after years prevailed, extending in our range from the heart of Asia to the shores of the Atlantic.

Our results will best be seen in the following table:

	Grains.
Egyptian gold ring standard	127
Mycenaean	130—5
Homeric talent (or "Ox-unit")	130—5
Attic gold stater (the sole standard for gold)	135
Thasos	135
Rhodes	135
Cyzicus	130
Hebrew standard	130
Persian Daric	130
Macedonian stater	135
Bactrian stater	130—2
Indian standard (7th cent. A.D.)	140
Phoenician gold unit (double)	260
Carthaginian	120
Sicily and Lower Italy	130—5
Etruscan	130—5
Gaulish unit	120
German	120

A glance at the table will suffice to show the truth of the proposition which we laid down as the object of this chapter, viz., that over the whole of the area with which we are dealing, the same unit with but little variations and fluctuations was employed for the weighing of gold.

Having proved the universal employment of the ox as a chief unit of barter, the universal distribution of gold, the priority of that metal both in discovery and in being weighed, and finally, in the preceding pages, the remarkable fact that to all intents and purposes the same unit of weight during many centuries was employed in its appraising, we advance to our next proposition, that this uniformity of the gold unit is due to the fact that in all the various countries where we have found it, it originally represented the value in gold of the cow, the universal unit of barter in the same regions.

It will of course be hardly possible for us to find data for a direct proof that in all the countries given in our table as employing the gold unit, that unit really represented the value of the ox. In some cases we shall be able to produce a fair amount of evidence more or less direct, whilst in others owing to the necessity of the case the evidence will be almost wholly inferential. Finally we shall be able to bring forward a very cogent form of proof by demonstrating the absolute necessity felt by barbarous persons of equating a ready made weight standard, which is being taken over from their neighbours, to the older unit of barter, and likewise the necessity felt by semi-civilized peoples under certain circumstances, even when long accustomed to the use of coined money, of returning to the animal unit as a means of fixing the standard of their coinage.

Starting first with the Greeks, we have already seen at an early stage in this work that the talent of the Homeric Poems was the equivalent of the ox, the older barter name being as yet the only term used in expressing prices of commodities, and the term talent being confined to the small piece of gold.

Passing next to the Italian Peninsula and Sicily, although possessed of certain definite statements as regards the value in *copper* of an ox in the fifth century B.C., nevertheless, owing to the uncertainty which still exists as regards the relative value of gold, silver and copper at Rome, we shall encounter considerable obstacles in our attempt to find the value of an ox in *gold*.

As Dr Theodore Mommsen[1] has laid down certain propositions in reference to inter-relations in value of the metals at Rome, which were generally received until a very recent period, when Mr Soutzo[2], in a clever brochure, put forward views of a widely different character which have met with the approval of some competent critics, and as the matter is still *sub judice*, I think it best, after briefly giving the historical evidence for the value of cattle, to give the views of both these writers.

The Law known as Aternia Tarpeia (451 B.C.) dealt with questions of penalties; certain notices of it fortunately preserve for us some valuable material. Cicero[3] says, "Likewise popular was the measure brought forward at the Comitia Centuriata in the fifty-fourth year after the first consuls (451 B.C.) by the consuls Sp. Tarpeius and A. Aternius concerning the amount of the penalty." To the same law Dionysius of Halicarnassus refers[4]: "They ratified a law in the Centuriate Assembly in order that all the magistrates might have the power of inflicting punishment on those who were disorderly or acted illegally in reference to their own jurisdiction. For till then not all the magistrates had the power, but only the Consuls. But they did not leave the penalty in their own hands to fix as much as they pleased, but they themselves defined the amount, having appointed as a maximum limit of penalty two oxen and thirty sheep. And this law continued to be kept in force by the Romans for a long time." Festus (*s. v. Peculatus* p. 237 ed. Müller) says: "Peculation (*peculatus*), as a name for public theft, was derived from *pecus* 'cattle,' because that was the earliest kind of fraud, and before the coining of copper or silver the heaviest penalty for crimes was one of two sheep and thirty oxen. That law was enacted by the Consuls T. Menenius Lanatus and P. Sestius Capitolinus. As regards which cattle, after the Roman people began to use coined money, it was

[1] *Histoire de la Monnaie Romaine*, I. 236.
[2] *Étude des Monnaies de l'Italie antique.*
[3] *De Rep.* II. 35, 60.
[4] x. 50.

provided by the Tarpeian Law that an ox should be reckoned at 100 asses, a sheep at 10 asses."

Again Aulus Gellius[1] has a curious notice, too long to quote in full, which ends "on that account afterwards by the Aternian Law ten asses were appointed for each sheep, one hundred for each ox."

Cicero and Dionysius are probably right (as Niebuhr thinks) in saying that Tarpeius and Aternius fixed the number of animals. C. Julius and P. Papinius, who were Consuls in 429 B.C., to whose reckoning of fines (*aestimatio multarum*) Livy refers (IV. 30), probably changed the penalties in cattle into money equivalents. Festus and Gellius have evidently muddled their authorities, having interchanged the words *sheep* (*ovium*) and *cows* (*bovum*). But the important thing is that both are agreed in giving the value of the cow at 100 asses.

Now Dr Hultsch (*Metrologie*², 19. 3), following Mommsen, shows that gold being to silver as $12\frac{1}{2} : 1$, the small talent, called the Sicilian, of which we have just spoken, confined exclusively to gold, would be exactly equivalent to a Roman pound of silver ($135 \times 3 \times 12\frac{1}{2} = 5062$ grains of silver; whilst the Roman lb. = 5040 grs.). Since at Rome, previous to the reduction of the As in 268 B.C., a *Scripulum* of silver was equivalent to a pound of copper or *as libralis*, and there are 288 *Scripula* or *scruples* in the pound, it follows that the pound of silver or its equivalent the Sicilian gold talent was worth 288 *asses librales*. This gold talent = 3 Attic staters (or ox-units), therefore 1 Attic stater = 96 *asses librales*. But we learned from Festus and Gellius that the value of the cow fixed in 429 B.C. was 100 asses. From this it appears that the value of the ox on Italian soil at this period was almost exactly the same as the traditional value which it had in the Homeric Poems, and which it continued to have in the Delian sacrifices in later times. The mere difference between 96 and 100 asses calls for no elaborate comment. It is enough to remark after Hultsch, that the further we go back the cheaper copper appears to be in relation to silver. This fact will easily explain any dis-

---

[1] Aulus Gellius, XI. 1. 2. 3; Plutarch, *Poplic.* 11, says a cow = 100 ὀβολοί, a sheep 10 ὀβολοί.

crepancy. Thus Mommsen's view that silver was to copper as 288 : 1 gives us a most interesting result.

Let us now turn to Mr Soutzo's view on the same subject. He maintains that at no time was the relation between silver and copper greater than 120 : 1, basing his argument on the assumption (which we shall find to be against the statements of the ancient writers) that when the first silver *denarius* or 10-*as* piece was coined in 268 B.C., as the *as* at that time weighed only two *unciae*, or one-sixth of a pound, silver was to copper as 120 : 1. He also argues from the fact that in Egypt, under the Ptolemies, the same relations existed between silver and bronze. He likewise maintains that the relation between gold and silver in Italy and Sicily at this period was as 16 : 1, from which it follows that gold was to copper as 1920 : 1. This of course gives us as the value of a cow about 390 grains of gold, that is about three gold staters, or ox-units. We would certainly be able to prove that at no time or place in the ancient world was a cow of so great a value in gold.

I shall refrain from any discussion of the merits of either view for the present. I will only add one observation: Mr Soutzo (p. 17) regards the Italian weight standards as borrowed from the East, and starts with bronze as the earliest stage in the history of the weights. The only clearly defined unit of Roman growth according to him is the Centupondium, which he says is the same as the Assyrian talent. From this the Romans obtained their own libra or pound by dividing their talents into 100 parts instead of 60. We shall find hereafter that this is an untenable position, but meantime it is interesting to find the Centupondium, or sum of 100 *asses* taken by an unprejudiced writer as the basis of the Roman system in the light of the fact that the ancient Roman value of the cow is likewise 100 *asses*. If Mr Soutzo was right, our thesis finds complete support, as it would plainly appear in that case that, although the Italians received their weight-unit ready made, they found it nevertheless necessary to equate the new metallic unit so obtained to the cow, the older unit of barter.

In Sicily we have an opportunity not merely of finding the approximate value of a cow in gold without having to deal with the disturbing question of the relative value of copper and silver, but also of showing that Soutzo's relation of 120 : 1 as that between silver and copper in early Italy must certainly be wrong, and that Mommsen's view is in the main correct. The famous Sicilian poet Epicharmus has left us a line: "Buy me straightway a nice heifer calf for ten *nomoi*[1]." As regards the value of the *nomos*, or *nummus* (νόμος or νοῦμμος), Pollux supplies us with some definite information.

In passage (IX. 87) already quoted he says: "Yet the Sicilian talent was the least in amount, the ancient one, as Aristotle says, weighed four and twenty *nummi*, but the later one twelve; now the *nummus* is worth three half obols." These three half obols plainly mean the ordinary half obols of the Attic standard. As the Attic drachm is $67\frac{1}{2}$ grains (normal), 65 grains in actual coins, the $\frac{1}{6}$ or obol = 11 grains roughly speaking; three half obols therefore weigh $16\frac{1}{2}$ to 17 grains. Accordingly, if we take the weight of the *nummus* or *litra* at 16 to 17 grains of silver, we shall not be wide of the mark. The price then of a good heifer calf was 10 *nummi* or 160 to 170 grains of silver. The term *moschos* (calf) is used rather vaguely by various Greek writers, but fortunately by the aid of the Sicilian poet Theocritus, we are certain that it means a calf of the first year not yet weaned; for he speaks[2] of putting the *moschos* to the cows to suck. From what we have seen (p. 32) of the relative values of cattle of different ages, it is tolerably certain that no full-grown cow would be worth less than six or more than ten calves of the first year. Hence the Sicilian cow, at the end of the sixth century B.C., must have been worth from 960—1020 to 1600—1700 grains of silver. We cannot tell exactly what was the ratio between gold and silver in Sicily or Italy at this time, but as we find it was 14 to 1 in Attica in 440 B.C., the probability is that it was not very far from that in Sicily. It certainly must have been at some point between 15 : 1 and 12 : 1. Taking it at 12 : 1, the value of the cow would range from 80 to $141\frac{3}{4}$ grains of gold, whilst in the ratio of 15 : 1 the range is from 64 to 113 grains of

---

[1] Pollux, IX. 80, εὐθὺς πρίω μοι δέκα νόμων μόσχον καλάν.
[2] Theocr. IX. 3, μόσχως βουσὶν ὑφέντες.

gold. It is thus absolutely certain that the value of a cow in Sicily in the sixth century B.C. must lie within the limits of 64 to 141 grains, and if the calf of Epicharmus is a suckling, the range in the value of the cow must be from 113 to 140 grains. This is all we require for practical purposes, and it will be admitted that the value of a cow in Sicily comes very close to our Homeric ox-unit of 130—5 grains.

We are now in a position to test the truth of Mr Soutzo's hypothesis. It will be conceded that at the beginning of the fifth century B.C., the cow must have had about the same value both in Italy and Sicily. The cow in Italy was worth 100 Roman pounds of copper, in Sicily about 1650 grains of silver. If Soutzo is right in saying that silver was to copper as 120 : 1 on multiplying 1650 by 120 we ought to get a result in copper corresponding to 100 Roman pounds: $1650 \times 120 = 198000$. Taking the Roman pound before it was raised at about 5000 grs. the Sicilian cow was worth 39 pounds of copper $\left(\dfrac{198000}{5000} = 39\right)$.

It is absurd to suppose that even at any time the Italian cow could have been worth $2\frac{1}{2}$ times the Sicilian. Let us now apply the same test to Mommsen's doctrine, and multiply 1650 grs. of silver by 300. (I take this as being more likely than 288 to have been the relation between copper and silver in the fifth century B.C.). $1650 \times 300 = 495000 \div 5000 = 99$ pounds of copper. The result is too striking to admit of our coming to any other conclusion than that Mommsen is right.

Next let us examine his doctrine that in ancient Italy gold was to silver as 16 : 1. Mr Soutzo[1] supports this view by three arguments: (1) that when Rome in the course of the Second Punic War issued gold coins for the first time, gold was to silver as 16 : 1; (2) Mr Head[2] has shown that at Syracuse under

---

[1] Mr Head (*Coinage of Syracuse*), *Numismat. Chronicle*, New Series, Vol. XIV., thinks that under Dionysius the Elder (406—367 B.C.) and his successors gold was to silver as 15 : 1 at Syracuse, whilst in the time of Agathocles (317—289 B.C.) it was as 12 : 1. We can however hardly take the evidence of the coin weights as sufficient, when we consider the extraordinary devices to which Dionysius resorted to raise money, causing coins of tin to pass as silver, making the silver coins bear a double value etc. as is related by Aristotle, *Oeconomica*, II. 21.

[2] *Op. cit.* 26.

the despot Dionysius (405—345 B.C.) gold was to silver as 15 : 1; (3) that certain symbols on the gold coins of Etruria when interpreted as referring to silver *litrae* give the proportion between the metals as 16 : 1. The same answer can dispose of the first two arguments. The state of affairs both at Rome in B.C. 207, and at Syracuse under Dionysius, was quite exceptional. Rome was in a state of bankruptcy, her subjects largely in revolt, the Lex Oppia (215 B.C.) prevented women from wearing more than half an ounce of gold ornaments[1]. It is therefore irrational to treat as normal the relation found to exist between the metals at such a crisis.

Similarly at Syracuse the relations between the metals were completely upset by the wild conduct of Dionysius, who forced his subjects to take coins of tin at the same rate as though they were silver. Moreover any evidence to be drawn with reference to the ratio between silver and gold at Syracuse in the time of Dionysius is completely nullified by the fact that in the reign of Agathocles (B.C. 307) gold was to silver as 12 : 1[2]. It is evident therefore that if in 207 B.C. gold was to silver all over Italy as 16 : 1, there must have been a great appreciation of gold. Are we not then justified in regarding the ratio of 16 : 1 as exceptional, and that of 12 : 1 as the more regular? That great fluctuations in the relations of the metals did take place in Italy, we know from a statement of Polybius that in his own time in consequence of the great output of gold from a mine in Noricum gold went down one-third in value. Silver was scarce in Central Italy, for it was only after the conquest of Magna Graecia that Rome found herself in a position to issue a silver currency. On the other hand there must have been a large and constant supply of gold coming down from the gold-fields of the Alps in exchange for the bronze wares of Etruria. Now as at Athens, where silver was so plenty and gold in earlier days scarce, the ratio was never higher than 15 : 1, it is impossible to suppose that in Northern and Central Italy, where the conditions were contrariwise, the ratio can ever have been in ordinary times higher than 12 : 1.

[1] Livy xxxiv. 1. Valer. Max. 9. 1. 3.
[2] Head, *Op. cit.* 160.

It is quite possible that after the Gauls got possession of Northern Italy, the supply of gold which reached Etruria and Latium may have been considerably reduced, and this would perfectly explain the relation existing at a certain period between gold and silver coins in Etruria, supposing that Soutzo's interpretation of the symbols is correct. But as we have no literary evidence to check off any deductions drawn from the coins, it is impossible for us to say whether the symbols on the gold pieces refer to units of silver or bronze.

Returning to the Kelts, the close kinsfolk of the Italians, the reader will recollect that the Gauls struck their imitations of the stater of Philip of Macedon on a standard of 120 grs., 15 grains lower than the weight of the archetype. Now similar but still more barbarous imitations of Philip's gold stater are found in Germany. These Rainbow dishes (*Regenbogen-*

FIG. 20. "REGENBOGENSCHÜSSEL"
(ancient German imitation of the Stater of Philip of Macedon).

schüsseln), as they are popularly termed in allusion to the picturesque superstition that a treasure of gold lies at the foot of the rainbow, and also to their scyphate form, are found in especial abundance in Rhenish Bavaria and Bohemia. Like the Gaulish imitations of the Philippus from which they are copied, they follow a standard of 120 grs. (and like the Gauls the Germans struck quarters of this coin, a division wholly unknown to the Greeks)[1]. In the region just indicated dwelt the ancient Alamanni, and there can be no doubt that it was this people who issued the coins found there. Now the Alamanni were among the barbarians who after, having overrun the provinces of the Roman Empire, committed to barbarous Latin their immemorial laws and institutions. In the Laws of the Ala-

[1] Mommsen (Blacas), *Histoire de la Monnaie romaine*, III. 275.

manni the best ox is estimated at five *tremisses*[1], that is 1⅔ *solidi*, or in other words 120 grs. of gold, the medium ox = 4 *tremisses* = 96 grs. The coincidence that the value of the ox in gold is the actual weight of the coins of the Alamanni is too striking to admit of any other explanation than that the gold coins of this people were struck on the native standard, the ox-unit. The Keltic and Teutonic tribes were so intermixed that we may plausibly infer that the Gauls had reduced the weight of the Philippus to 120 grs. because owing to gold being less plentiful and cattle more abundant to the north of the Alps, from a very remote time the ox-unit throughout Gaul and Germany was slightly lower than along the Mediterranean.

In the Laws of the Burgundians the value of an ox is set at 2 *solidi* = 144 grs. of gold[2]. This of course is considerably more than that of the Alamannic ox, but when we consider the late period at which the laws of the Barbarians were compiled, and the various recensions which they underwent, the strange fact is that the ox should have varied so little in its relation to gold from the Homeric ox-unit of at least 1000 B.C.

Passing into Scandinavia we once more, even so late as the eighth century A.D., find the same strange agreement in value. In the ancient Norse documents (where the cow is the unit of value as we have already seen) it is reckoned at 2½ ores (ounces) of silver = 1078 grains. But we likewise know from the same sources that gold stood to silver as 8 : 1; accordingly the cow was worth 134 grs. of gold[3].

Besides the Hellenes and Italians there was another people who strove for the mastery of all the Western Mediterranean. The ancient city of Tyre had sent out many colonies into the far West, when the nascent power of Hellas had already begun to assert its superiority in the Aegean. Trade grew and flourished between the colonies and the mother city in Phoenicia; thus there was unbroken intercourse between remote Gades and her Eastern mother until after the destruction of the latter

---

[1] Pertz, *Monumenta Historica Germaniae*, Vol. III. Lex Alamannorum, *lib. sec.* LXXX. *summus bovis 5 tremisses valet cett.*

[2] Pertz, *Op. cit. Leges Burgundiorum*, p. 534: pro bove solidos 2 cett.

[3] Schive and Holmboe, *Norges Mynter* (Christiania, 1865), pp. i—iv.

(720 B.C.). Henceforward the headship of the Phoenician cities of the West falls into the hands of Carthage, the scene of the last great act and final catastrophe in the drama of Phoenician history. At the very time, nay some say on the very day, when the Greeks of the East were destroying the host of Xerxes in the Strait of Salamis, the Hellenes of the West led by brave Gelon of Syracus were repelling a great army of Carthaginians before the walls of Himera, and during the third and fourth centuries B.C. the Greeks of Sicily lived in constant danger from the Carthaginians, who held the western part of the island with their factories of Lilybaeum, Drepanum and Motyé, until at last they were finally expelled from the island by the resistless might of Rome (241 B.C.).

Could we but learn the estimate put upon the ox by the Phoenicians or Carthaginians, we would get a fair index to its value over a wide extended area. For as in earlier times the Phoenician influence extended from Tyre to Gades, linking both east and west, so in later days Carthage extended her power over all North Africa from the Pillars of Herakles to the confines of Egypt, and over Southern Spain.

Some forty years ago the longest Phoenician inscription yet known was found at Marseilles. The inscription seems to have belonged to a temple of Baal, and contains directions touching sacrifices and certain payments to be made to the officiating priest. Chemical analysis of the stone has demonstrated that it is of a kind not found in France, but known in North Africa. Hence M. Renan thought that it had been brought as ballast in some ship. The names of two Suffetes stand at the head of the inscription, which seems along with other evidence to point to its having been engraved at Carthage. On palaeographical grounds its date is placed in the fourth century B.C., but why it came to Massalia seems still inexplicable. It is possible that in the fourth century B.C. there was a considerable body of Carthaginians resident at Massalia, just as on the other hand we know that there was a large Greek community residing at Carthage. If that were so, the Carthaginians would naturally keep up the worship of Baal at Marseilles, and would regulate the temple worship in accordance with the practice of the mother city. The stone in that case may have been imported to serve as an official declaration of the rules to be observed in sacrifices. Movers and Kenrick regarded the sums of money named in connection with the victims as composition for the animals named, whilst the editors of the *Corpus Inscriptionum Semiticarum* (Vol. I. Pt. I. p. 217) regard them as fees to be paid to the priests for the performance of the sacrifices, saying that it is analogous to the directions for the burnt offerings, peace offerings and thank offerings contained in Leviticus i—vii. The few lines of the inscription with which we are concerned I shall translate from the Latin version given in the *Corpus*.

"Concerning an ox, whether it is a whole burnt offering, or deprecatory offering or a thank offering, there shall be to the priests ten shekels of silver, and if it is a whole burnt offering, in addition to the fees this weight of flesh, three hundred; and if it is a peace offering the first cuts and additions, the appurtenances thereof, and the skin and the entrails, carcase and the feet, and the rest of the flesh shall belong to the giver of the sacrifice.

"Concerning the calf without horns, concerning an animal which is not castrated, or a ram, whether it is a whole burnt offering, or a peace offering, or a thank offering, there shall be to the priests five shekels of silver, and if it be a whole burnt offering in addition to the fee this weight of flesh, one hundred and fifty.

"Concerning a he-goat or a she-goat, whether it is a whole burnt offering, etc. there shall be to the priest one shekel of silver two *zer*.

"Concerning a sheep or kid or goat, whether it is etc., there shall be etc. $\frac{3}{4}$ shekel one [*zer*] of silver.

"Concerning a tame bird, or wild bird, $\frac{3}{4}$ shekel and two *zer*."

Let me here remark that in Leviticus there is no mention whatsoever made of any fees to the priest, also that whilst according to the above version the giver of the victim gets the skin, in Leviticus (vii. 8) it is the priest who gets it as his perquisite, as seems also to have been the practice in Greece. For we know that the Spartan kings, who in their capacity of priests offered all sacrifices at Sparta, always got the skins as their payment[1]. That the sums mentioned are really the prices of the victims is made almost certain by the fact that at the famous Phoenician temple of Aphrodite at Eryx in Sicily the victims were kept ready by the priests to be sold to worshippers who wished to sacrifice, as we know from a curious story told by Aelian[2].

Whilst it would be of great importance for my purpose to have been able to regard the sums mentioned in the inscription as the actual value set upon the animals, even if we simply regard them as fees they still give us some aid. For as it is most unlikely that the fee for sacrificing would exceed the value of the victim to be sacrificed, we thus can obtain a minimum limit of value. We may then safely assume that the value of the ox was not less than 10 shekels of silver. On the other hand we shall find from Exodus what must have been the maximum value among the Hebrews at a comparatively late date. As the Punic ox cannot have been worth less than 1350 grs. of silver, and the Hebrew not more than 1760 grs., it is almost certain that the value of the ox at Carthage lay between these limits.

The pieces of silver mentioned in the inscription are probably ordinary silver didrachms of the Attic standard. The Carthaginians had coined silver in Sicily on the Attic standard from about 410 B.C., but issued no silver coins at Carthage itself until after the acquisition of the Spanish Silver Mines (241 B.C.), although gold, electrum, and bronze coins were minted. In Greece Proper in the 4th century B.C. gold was to silver as 10 : 1; we may therefore not be far wrong if we assume a

---

[1] Herod. vi. 57. See evidence of this collected by Stengel, Die griechische Sakralaltertümer, pp. 29 sq. 81 sq. (Iwan Müller's Handbuch, Vol. v. pt. iii.)

[2] Hist. Animal. x. 50, τά γε μὴν ἱερεῖα ἑκάστης ἀγέλης αὐτόματα φοιτᾷ καὶ τῷ βωμῷ παρέστηκεν, ἄγει δὲ ἄρα αὐτὰ πρώτη μὲν ἡ θεός, εἶτα ἡ δύναμίς τε καὶ ἡ τοῦ θύοντος βούλησις. εἰ γοῦν ἐθέλοις θῦσαι οἶν, ἰδού σοι τῷ βωμῷ παρέστηκεν οἶς, καὶ δεῖ χέρνιβα κατάρξασθαι· εἰ δὲ εἴης τῶν ἁδροτέρων καὶ ἐθέλοις θῦσαι βοῦν θήλειαν ἢ καὶ ἔτι πλείους, εἶτα ὑπὲρ τῆς τιμῆς οὔτε σὲ ὁ νομεὺς ἐπιτιμῶν ζημιώσει οὔτε σὺ λυπήσεις ἐκεῖνον· τὸ γὰρ δίκαιον τῆς πράσεως ἡ θεὸς ἐφορᾷ. καὶ εὖ καταθεὶς ἴλεων ἕξεις αὐτήν· εἰ δὲ ἐθέλοις τοῦ δέοντος πρίασθαι εὐτελέστερον, σὺ μὲν κατέθηκας τὸ ἀργύριον ἄλλωι, τὸ δὲ ζῷον ἀπέρχεται, καὶ θῦσαι οὐκ ἔχεις.

similar ratio between the metals to have held at Carthage about the same period. That silver was scarce is shown by the fact that they did not coin it, although issuing gold, electrum and bronze. Ten silver didrachms would therefore = 1 gold didrachm of 135 grs., which is of course our ox-unit. This is a remarkable result, and of itself would make one believe that the sum represents the real value of an ox, which the practice at Eryx puts beyond doubt. We know that at Athens the people who were bound to provide the public sacrifices supplied very wretched oxen, so we need not be surprised to find precautions taken by the priests of Baal to ensure that proper animals should be provided for the altar, especially as they themselves got a share of the flesh.

Next let us see if that most ancient of all known civilized lands, Egypt, can produce from her store of monumental records any evidence for our purpose. Professor Brugsch[1], in his *History of Egypt under the Pharaohs*, gives from inscriptions a list of the prices of various commodities about 1000 B.C.: a slave cost 3 *ten* 1 *ket* of silver; an ox 1 *ket* of silver (= 8 *ten* of copper); a goat cost 2 *ten* of copper; 1 pair of fowls (geese ?) cost $\frac{1}{3}$ *ten* of copper; 1 *hotep* of wheat cost 2 *ten* of copper; 1 *tena* of corn of Upper Egypt cost 5—7 *ten* of copper; 1 *hotep* of spelt 2 *ten* of copper; 1 *hin* of honey 8 *ket* of copper; 50 acres of arable land 5 *ten* of silver. Of course there must be more or less uncertainty about some of these statements owing to the imperfect knowledge which we as yet possess. At first sight the reader naturally wonders how it is possible to calculate the value of the ox as here given, which is only 1 ket of silver, that is, the Egyptian ox of 1000 B.C. was only worth 140 grains of *silver*, whilst an ox hitherto has been worth about the same amount in *gold*. At first sight this is enough to stagger us, but a moment's reflection makes the matter very intelligible. We have already noticed (p. 59) that at a certain stage in the history of the metals silver was far scarcer than gold, and that its rarity combined with its beauty no doubt made it to be eagerly sought and held in great esteem. We saw that the Arabs of the Soudan down to the present day prefer silver

[1] *Egypt under the Pharaohs* (2nd edit. Engl. transl.), Vol. II. p. 199.

to gold; whilst in the earlier part of the present century when Japan was opened to European commerce the Japanese eagerly exchanged gold for silver at the rate of one to three, and even less, as they possessed no native silver, and were charmed with the beauty of the little known metal[1]. Marco Polo also tells us that "in the province of Carajan (the modern Yunnan) gold is so plenty that they give a saggio of gold for only six of the same weight of silver;" and of the province of Zardandan, five days west of Carajan, he says, "I can tell you they give one weight of gold for only five of silver[2]."

It is almost certain that in all countries at one stage silver must have been of higher value than gold; afterwards as its production became greater, it became equal in value, and finally, little by little, much less valuable, until at last the relation between the metals is 1 : 22. Of course we must add that there must have been always certain fluctuations, according as a sudden increase of output of one or other of the metals altered temporarily their relations. We have evidence that silver in early times in Egypt was held in higher esteem than gold. Thus Erman[3] says that according to ancient Egyptian notions silver was the most costly of the precious metals; for they always in an enumeration mention it before gold, and in the tombs ornaments of silver are of far rarer occurrence than those of gold. This circumstance is simply and sufficiently explained (thinks Erman) by the fact that Egypt herself possesses no deposits of silver, but must have obtained the metal from Cilicia. Under the 18th dynasty (1400 B.C.), the Phoenicians supplied Egypt with silver and under the new empire the supply had so increased that it was now evidently cheaper than gold, for the later texts always name silver after gold, just as we do. We have previously noticed the paucity of silver articles in the tombs at Mycenae which are commonly dated 1400 B.C.

It is therefore reasonable to suppose that towards the end of the Second Millennium B.C. gold and silver were almost of equal value, not alone in Egypt, but in other parts likewise of the

---

[1] Sir Rutherford Alcock, *The Capital of the Tycoon*, I. 281.
[2] Marco Polo, Yule's Transl. II. pp. 62 and 70.
[3] *Aegypten und ägyptisches Leben in Alterthum*, p. 611.

ancient world. The great supply of silver had not yet been obtained which in the 10th century B.C. made silver at Jerusalem like stones. "As for silver," says the sacred writer, "it was nothing accounted of in the days of Solomon" (900 B.C.)[1], who had "made silver and gold at Jerusalem as plenteous as stones[2]." By this time silver had become very cheap in Egypt likewise. At least if we can at all rely on the author of the books of Chronicles. For the king's merchants "fetched up and brought forth out of Egypt a chariot for six hundred shekels of silver, and an horse for one hundred and fifty: and so brought they out horses for all the kings of the Hittites and the kings of Syria[3]."

The shekel here meant is probably that of 130—135 grains, while the price of the ox in Brugsch's list is 1 ket or 140 grains. At a moderate computation this would make a horse worth 150 oxen, if our documents were contemporary. But from lists of relative prices in ancient and modern times it is preposterous to suppose that at any time or in any place such a remarkable difference in value existed between the horse and the cow. From this it follows that if Brugsch is right in his translation of his Egyptian text, the latter must date from several centuries before 1000 B.C., when as yet silver was of the same or almost the same value as gold. Finally, we have no means of knowing the age of the ox, but as it is equal in value to only four goats, it is possible that it was not a full-grown animal. I have dealt with this point at some length, and have little positive gain to show, but it is necessary to put before the reader all data which may aid in our search, and still more necessary to do so in the case of evidence which seems to present serious difficulties.

Unfortunately for us the Old Testament gives very scanty information on the question of the cost of various commodities, and in no place do we get any information regarding the price of cattle. For in the account of the purchase of the threshing-floor and oxen of Ornan the Jebusite by king David, there is a discrepancy in price between the Second Book of Samuel (xxiv. 24) and First Chronicles (xxi. 25), the former making

---

[1] 1 Kings x. 21. [2] 2 Chron. i. 15.
[3] 2 Chron. i. 17.

the sum 50 shekels of silver, the latter "six hundred shekels of gold by weight," and in any case, as we do not know the number of oxen used in threshing or the value of the floor and threshing instruments, it is impossible for us to draw any inference. In the Book of Exodus, however, we obtain the value of a slave, from which we may at least get an approximate idea of the value of an ox: "If the (wicked) ox shall push a manservant or a maidservant; he (the owner of the ox) shall give unto their master thirty shekels of silver, and the ox shall be stoned" (xxi. 32). Here, as in the ancient laws of Wales and elsewhere, the value of the male and female slave is the same, and thirty shekels or pieces of silver seems to have been the conventional price of a slave among the Hebrews. To this Zechariah (xi. 12) seems to allude, "So they weighed for my price thirty pieces of silver," in reference to which the Evangelist writes: "Then was fulfilled that which was spoken by Jeremy the prophet, saying, And they took the thirty pieces of silver, the price of him that was valued, whom they of the children of Israel did value" (Matt. xxvii. 9). The average slave among the Homeric Greeks (as we saw above) was worth about three oxen, amongst the Irish three, among the modern Zulus about 10, and among the wild tribes of Annam seven (pp. 24—5). Allowing three oxen as the value of a slave among the Hebrews, the ox is worth 10 shekels (ancient) = 1300 grains of silver = 130 grains of gold, taking gold to silver as 10 : 1, which at an early period was probably the regular ratio in parts of Asia Minor. The result thus reached gives us once more the Homeric ox-unit as the value of the Hebrew ox. It is certain that it cannot have been higher, although we cannot show that it may not have been less.

The cow is estimated in the Commentary on Vendîdâd, Fargard, IV. 1—2 at 12 *stirs* or *istirs*.

Our task must be now to find out the weight of this *istir*. *Istir* or *stir* is identified with Greek στατήρ (as *dirham* is with Greek δραχμή).

The Pahlavi Texts, translated by Dr West, naturally afford us the readiest means of discovering our object[1].

[1] *Sacred Books of the East*, Vols. v., xviii., and xxiv.

THE VALUE OF A COW. 149

Sins or equivalent good works	I Shayast l. 1	II XL 1	III XVI. 1–3	IV XVI. 5	V Spiegel Rivaya	VI Spiegel Rivaya
Srôshô-Karanam	—	1 dirham 2 mads	—	3 coins and a half	—	—
Farmân	weight of 4 stirs and each stir has 4 dirhams	3 dirhams of 4 mads	3 coins of 5 annas some say, 3 coins	a Farmant is a Srô-shô-Karanâm	7 stirs	8 stirs
Agerept	1 dirham	33 stirs	53 dirhams	16 stirs	12 stirs	
Avôirist	1 dirham	the weight of 33 dir-hams	78 dirhams	25 stirs	15 stirs	
Aredûs	30 stirs	30 stirs	30 stirs	30 stirs		
Khôr	60 stirs	60 stirs		60 stirs		
Bâzâî	90 stirs	90 stirs		90 stirs		
Yât	180 stirs	180 stirs		180 stirs		
Tanâpûhar	300 stirs	300 stirs		300 stirs		

There are in the Shayast-la-Shayast various lists of sins and good works. These sins or good works are put in the golden balance and weighed, in which case the *stir* is a weight, whilst in other cases we have a money evaluation. As much confusion arises from variations in the lists, it will be best to tabulate the different lists, and thus get a synoptic view of the whole.

On looking at the table, we find that all our authorities are in complete harmony as to the amounts of the last five; Aredûs is 30 *stirs*, Khôr = 60, Bâzâî = 90, Yât = 180, and Tanâpûhar = 300 *stirs*. Let us first consider these. We must remember that on the third night after death the soul is judged by having its sins and good works weighed, and according as the one or other predominates, is the ultimate destiny of the soul foul or fair. It is thus essentially a scale of *weights*, not of *coins*. The arrangement of the numbers at once speaks for itself. 30 *stirs* = ½ *mina* on the Babylonian system, as will be seen on p. 251. 60 *stirs* (Khôr) = 1 *mina*, 90 *stirs* (Bâzâî) = 1½ *minae*, 180 (Yât) = 3 *minae*, and finally we get 300 *stirs* (Tanâpûhar) = 5 *minae*. What then is the weight of the *stir*? It is none other than the light Babylonian shekel (130 grains Troy).

Now let us approach the bewildering tangle of the first four degrees. It is evident that there are mistakes of numerals in some cases, e.g. in Column I., where the Agerept and Avoîrist are made equal, both being only $\frac{1}{18}$ of the first degree or Farmân, and also in Col. II. we have the Agerept greater than the Avoîrist and Aredûs. But in Columns III. IV. and V. we get some elements of regularity. Two of them at least introduce coined money, thus giving us an indication that it is owing to the constant effort to make the lower weight conform to the monetary units of the various periods at which the Commentaries were written that the confusion has in great part arisen. We find the Farmân = 3 *dirhams* of 4 *mads*, to 3 coins of 5 annas, and to 3½ coins. Dr West, calculating the anna on the basis of the old rupee of Guzarat (Pt. III., p. 180), makes the coin of Col. IV. = 50 grains Troy, the old rupee being less than its present weight (180 grains). The Farmân in this case is 150 grains. The 3 *dirhams* of 4 *mads* each probably are the same in amount. So too are the three coins and a half of Col. IV. In which case each coin must weigh 43 grains (150 ÷ 3½ = 42$\frac{6}{7}$), that is the regular weight of the *dirhams* struck by the Arab conquerors of Persia. Comparing Cols. III. and IV., we shall find the Agerept worth respectively 53 *dirhams* and 16 *stirs*, the Avoîrist set at 73 *dirhams* and 25 *stirs*. We find then a very close approximation in comparative values. The same proportion for all practical purposes exists between the coin of 5 annas (50 grains) and the coin of 43 grains, as between the 53 *dirhams*, and 16 *stirs* and 73 *dirhams* and 25 *stirs*. But it is evident that in Col. III. the coin of 5 annas is a thing quite distinct from the *dirhams* mentioned in the same table, or else why is there a difference in nomenclature? The *dirham* is probably the usual *dirham* of 43—40 grains. But as we find 53 of these *dirhams* = 16 *stirs* of Col. IV. accordingly the *stir* of Col. IV. = 132 grains Troy, which is plainly the Babylonian shekel, and 73 *dirhams* = 25 *stirs*. This gives an average for the *stir* of 126 grains Troy, which again points directly to the light shekel of 130 grains Troy, or in other words to the weight of the Daric. Another piece of evidence in the same direction is the fact that the Sassanide kings struck their silver coins on the so-called Attic standard, which of course was identical with that in use from the earliest times in Asia, as the standard of the Daric. The founder of the Sassanide Dynasty, Ardeshir, struck his first gold coins on this standard (staters of 135·0), whilst all the silver coins of this dynasty are half-staters (65 grains) of the same standard. The statement in Col. I. that each *stir* has four *dirhams* probably refers to a later period, when 4 *dirhams* of the ordinary Muhammedan standard (43 grains Troy) were equivalent to a rupee (180—170 grains).

If it should be objected that the *istir* of the Avesta is the old Persic silver standard of 172 grains, my reply is that as it is evident from what we have seen above that in this *weight* system there were *sixty* staters in the *mina*, this must be the *weight*, not the silver *coin*, as there were only *fifty* staters in the *money* mina.

The ox of the Zend-Avesta according to tradition is therefore rated at 12 *stirs* or staters of 130 grains of silver each. From the time of Alexander right down to the third century after Christ it is probable that all through the Eastern Mediterranean and Asia Minor gold was to silver as 12 : 1. If this were so, the ox of the Avesta was worth 130 grs. of gold, that is the weight of a Daric, and of the Homeric ox-unit.

Such then are the approximate results that we have been able to obtain regarding the value in gold of an ox in various parts of the ancient world. Of course I do not pretend that they have the same force as if they represented the value of the ox everywhere in one particular epoch, or as if we had found the ox directly equated to gold in every case. But on the other hand the persistency of prices in semi-civilized countries is a fact well known: for example, prices have changed but very slightly in India[1] during a long course of years, for although the silver rupee has sunk to about two-thirds of its nominal value in exchanges for gold, it purchases as much as ever in India. It is likely therefore that the conventional value of the ox would have remained unchanged for a long period of time, and the fact that our approximate values taken from various countries and from various centuries so closely coincide is a strong indication that such was the case.

Savages are still more conservative in their ideas of the relative value of certain articles; and when once a standard price has been fixed for certain commodities, it is almost impossible to get them to change.

Thus I am told by Mr W. H. Caldwell that, when he gave half-a-crown to a Queensland black for the first specimen of a certain kind of animal brought into camp, henceforth he had to pay the same amount for every specimen, even when they came in considerable numbers. So with the early men of Asia and Europe who first possessed cattle, and later on gold. Once a certain amount of gold was taken as the recognized value of a cow of certain age, the idea would become strongly rooted that so much gold was the proper equivalent of a cow. And it would only be in the lapse of centuries and with the develop-

[1] *Report of the Royal Commission appointed to enquire into the recent changes in the relative values of the precious metals.* 1st Report, p. 60 (1866).

ment of cities and general commerce that the price of cattle would begin to fluctuate.

But even when such variation in price arose, it made no difference as regards the weight standard. The unit had already long been fixed and it remained unaltered, just as the beaver skin of account still means only two shillings, although a real beaver skin is now worth many times that amount.

Another reason why the price of cattle would remain stationary would be that in early times as all the cows were kept under more or less similar conditions of food, and there was no attempt at the development of superior breeds, there would be little difference in the value of animals of the same age.

The connection between the cow and the gold unit is rendered all the more probable not merely by the fact so often noticed that the words for *money* in different languages originally meant *cattle*, but by the remarkable fact that the earliest known weights are in the form of cattle. The relation between *weight* and money must always be close, but it comes still more prominently into view, when as yet there is no coinage, but gold and silver pass by weight alone. If then the value of a cow formed the first gold unit, we can at once understand why the first weights took the form of oxen and sheep.

It was not for mere artistic reasons, for whilst such animal weights appear on Egyptian paintings, the numerous known Egyptian weights are of a very conventional form, as we shall find below. Doubtless the horns and ears made a cow's head exceedingly ill-suited for a weight, and in course of time utility prevailed over the traditional idea that the weight unit ought to take the shape of the animal, whose value in gold it was meant to represent.

The following table sums up briefly the results of this chapter:

Homeric ox-unit	= 130—135 grains of gold.
Roman ox (5th cent. B.C.)	= 135 ,, ,,
Sicilian (5th cent. B.C.)	= 135 ,, ,,
Ancient German	= 120 ,, ,,

Ancient Gaulish	= 120 grains of gold.
Phoenician? (4th cent. B.C.)	= 135 ,, ,,
Egyptian (1500 B.C.?)	= 140 grains of silver = 140 grains of gold(?).
Hebrew	= 130 grains of gold.
Zend-Avesta	= 130 ,, ,,
Burgundian	= 140 ,, ,,
Alamannic	= 120 ,, ,,
Scandinavian[1] (8th cent. A.D.)	= 128 ,, ,,

As has been remarked before, I do not include the values of the ox or cow in the ancient Laws of Wales or Ireland, since from the insular position of Britain and Ireland the principle that we must have unbroken touch between the various peoples in order to have a constant unit does not apply. There could be no free flow of trade in cattle between Britain and the continent until the development of steam navigation.

It is worth noting that the value of a buffalo at the present day among the Bahnars of Annam is almost the same as that of the ancient ox. The buffalo is reckoned at 280 hoes[2], that is 28 francs = £1. 2s. 4d. Taking gold at the rate of twopence per grain, the value of the buffalo in gold is 134 grs. Troy.

[1] This is almost exactly the weight of the *ortug*, into 3 of which the *ora* (ounce) of 410 grs. was divided. The *ortug* of gold being 136·7 grs., and the value of a cow being 128 grs. of gold, it is hard not to believe that there was a connection between them. (See App. C.)

[2] See above, p. 24.

CHAPTER VII.

THE WEIGHT SYSTEMS OF CHINA AND FURTHER ASIA.

> Subiectos Orientis orae
> Seras et Indos.
> HOR. *Carm.* I, 12, 56.

WE have now found that within the area where our weight standards arose the ox was universally diffused, and regarded as the chief and most general form of property and medium of exchange; that over the same area gold was found to be more or less equally distributed in antiquity; that the metallic unit is found in all cases adapted to the chief unit of barter, whether that be ox or reindeer, beaver skin, or squirrel, as soon as peoples have learned the use of metal; and finally that over our special area from the Atlantic to Central Asia the cow at various times and places retained a value which fluctuated only from 120 to 140 grains of gold. When therefore we recall the fact, also pointed out above, that the gold unit employed from Gaul to Central Asia was one that only fluctuated from 120 to 140 grains, and when we recollect further that this unit in the ancient Greek Epic is called not a talent but an *ox*, when prices, and not merely the actual ingots of gold are mentioned, the conclusion follows that not merely in Greece but in all the other countries the gold unit represented originally simply the conventional value of the cow as the immemorial unit of barter.

Next follows an important question, How was the primitive weight standard fixed? In other words, how did mankind arrive at the general opinion that a weight of gold of about

130 English grains was the equivalent to the conventional value of the animal?

If we could but discover a region in which the weight and monetary systems still in use are essentially independent of our Graeco-Asiatic standards, and where it could be proved that the monetary system is an independent native development, and where this development is of such recent date that the record has been preserved in a written document, not merely reaching us in the dim form of a tradition, blurred and broken in the long and misty space of years that lie between us and those who first shaped our system, we would undoubtedly discern more clearly the stages of its evolution.

The Chinese empire with the neighbouring peoples who have participated in its civilization afford us just the case which we desire. It will be seen from what follows that not merely the monetary system of China, but her weight system is of an origin almost wholly unaffected by Western influences.

We saw above that the earliest form of money in Greece took the form of *spits* or small rods of copper, no doubt of a specified size; we found in Annam that iron hoes, in mediaeval India iron formed into large-sized needles, in modern times in Central Africa pieces of iron of given dimensions, bars of iron among the Hottentots and among the peoples of the West Coast of Africa, brass rods of fixed length in the region of the Congo, and pieces of a precious wood likewise of fixed dimensions, have served or do still serve as media of exchange, and as units by which the values of other commodities are measured. In all these cases mere *measure* not *weight*, is the method of appraisement. As the archaic Greek "spit" or *obolus* of bronze eventually became a round bronze coin, familiar to us as Charon's fee, and in still later times under the abbreviation *ob.* as the accountant's symbol for a half-penny, as *d.* (*denarius*) denotes the penny, so we shall find that the common Chinese copper coins pierced with a square hole in the centre have had an almost identical history.

At the time when the Chinese made their great invasion into South-eastern Asia (214 B.C.) they still were employing a bronze currency under the form of knives, which were 135 millimetres (5⅔ in.) in length, bearing on the blade the character *minh*, and furnished with a ring at the end of the handle for stringing them. Under the ninth dynasty (479—501 A.D.) they used knives of the same form and metal, but

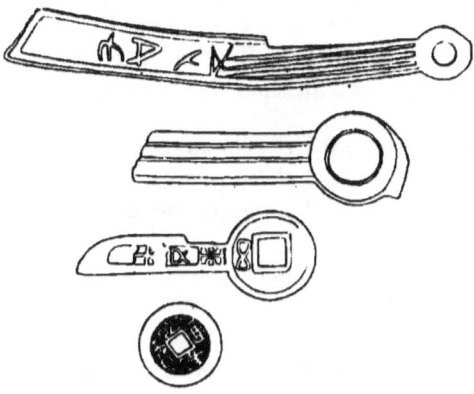

Fig. 21. CHINESE KNIFE MONEY
(showing the evolution of the modern Chinese coins).

180 millim. (7⅛ in.) in length, furnished with a large ring at the end of the handle and inscribed with the characters *Tsy Kú-u Hoa*. Next the form of the knife was modified, the handle disappeared, and the ring was attached directly to the blade, but now as weight was regarded of importance, its thickness was increased to preserve the full amount of metal, and the ring became a flat round plate pierced with a hole for the string[1]. Later on these knives became really a conventional currency, and for convenience the blade was got rid of, and all that was now left of the original knife was the ring in the shape of a round plate pierced with a square hole. This is a brief history of the *sapec* (more commonly known to us as *cash*) the only native coin of China, and which is found everywhere from Malaysia to Japan[2].

[1] J. Silvestre, "Notes pour servir à la recherche et au classement des monnaies et des médailles de Annam et de la Cochin-Chine Française." *Excursions et Reconnaissances*, No. 15 (1883), p. 395.

[2] H. C. Millies, *Recherches sur les monnaies des Indigènes de l'Archipel Indien et de la péninsule Malaie* (La Haye, 1871).

Except where foreign coins such as American silver dollars are employed, all payments in silver and gold are made by weight, the only money being the copper *cash*. The Chinese metric system, like our own, is based on natural seeds or grains of plants. Thus ten of a kind of seed called *fên* (the Candarin) probably placed sideways make 1 *ts'un* (the Chinese inch[1]), just as our forefathers based the English inch on 3 barleycorns placed lengthwise. So with their monetary system,

10 *li*[2] (copper cash) = 1 *fên* (*Candarin*) of silver.
10 *fên* = 1 *chi'en* (*mace*).
10 *chi'en* = 1 *liung* (or *tael* or Chinese ounce).

This *liung* or, as it is more commonly called, *tael* is the maximum monetary weight. Hence we hear always of payments in silver as being 1000 or 2000 ounces and so on, but never in the higher commercial units of the *catty* or pound, and *pical* or hundredweight, to which we shall come immediately. But though the Chinese never employed any coinage of gold or silver, beyond all doubt they have possessed and employed both metals for almost an incalculable time in the form of ingots of rectangular shape, and of very accurately fixed dimensions. The maximum unit employed in commercial relations between China, Cochin-China, Annam and Cambodia is the *nên* or *bar*. It is of course among her less advanced neighbours that we can best see how the system developed and worked. For whilst China herself now reckons exclusively by the *tael* or ounce, Annam and Cambodia still employ ingots of fixed weights and dimensions as metal units almost to the present time. Thus when Msg. Taberdier in 1838 published his account of the money of Annam, they had no coins except the ordinary cash or *sapec* with a square hole in its centre, and which is there made of zinc and called *dong*[2], they had no coinage in the proper sense of the term. However they employed ingots of gold and silver of a parallelopiped shape. Five sizes of ingots were employed for both gold and silver alike.

[1] Sir Thomas Wade's *Colloquial Chinese Course*, I. p. 213 (2nd ed.).
[2] J. Silvestre, *Op. cit.* p. 308 seqq.

Gold.

1. *Nén-Vang, loaf of gold* = 10 *lu'ong* or *taels* (ounces).
2. *Thoi-Vang* or *Nua Nén-Vang* = 5 *lu'ong*.
3. *Lu'ong-Vang, nail of gold* = 1 *lu'ong* (39·05 grammes).
4. *Nua-Vang, half nail of gold* = ½ *lu'ong*.
5. The quarter *lu'ong* = ¼ *tael* (9·762 gram.).

Silver.

1. *Nén-bac, loaf of silver* = 10 *lu'ong* or *taels*.
2. *Nua Nén-bac, half loaf of silver* = 5 *lu'ong*.
3. *Lu'ong* or *Dinh-bac, nail of silver* = 1 *tael*.
4. Half *Lu'ong*, half nail = ½ *tael*.
5. Quarter *Lu'ong* = ¼ *tael* (9·762 gram.).

The lowest unit then was the quarter *nail* of 152½ grains troy, whilst the largest was the *nén* of 6500 grains. These ingots did not circulate freely but were generally kept in wealthy families as reserve treasure.

In very similar manner in Greece and Italy gold and silver, fashioned into talents and bars or wedges, were employed side by side with the bronze *oboli* or *spits* which served as the ordinary currency of every-day life.

We have now seen that the highest unit employed for silver and gold is the *Nén* or bar of ten *taels* or ounces. Before going further it will be convenient to describe briefly what we may term the Chinese system of *avoirdupois* weight. Then we shall give the system borrowed from the Chinese and used in Cambodia and Cochin-China.

Chinese.

10 *fên* = 1 *ch'en*² (mace).
10 *ch'en*² = 1 *liang, tael* or ounce.
16 *tael* = 1 *chin*, commonly known as catty, = 1⅓ lbs. English.
100 catties = 1 *tan* or *shih*⁴, commonly known to us as the *picul* (= 133⅓ lbs. English).

Cambodia. Money system.

60 cash or sapecs of zinc = 1 *tien*.
10 *tien* = 1 string.
10 strings = 1 *nên* or bar of silver (90 francs).

The *nên* is an ingot of silver of parallelopiped form, which is invariably worth 100 strings of zinc cash[1]. This *nên* is subdivided for money of account as follows:

1 *nên* (375 grammes) = 10 *denh*.
1 *denh* = 10 *chi*.
1 *chi* = 10 *hun*.
1 *hun* = 10 *li*.

They employ a coin of silver called a *prac-bat* or *preasat*, worth 4 strings or $\frac{1}{25}$ *nên*[2].

The Mexican piastre, which circulates also, is worth on the average about 6 strings of cash.

1 gold ingot = 16 *nêns* of silver.

The half ingot of gold is also used = 8 ingots of silver.

The unit of commercial or *avoirdupois* weight is the *catty* (called by the Cambodians the *neal*) or pound.

1 *néal* (catty) (600 grammes) = 16 *tomlongs* or *taels* (ounces).
1 *tomlong* (37·5 grammes) = 10 *chi* (of 3·75 grammes).
1 *chi* = 10 *hun*.

The preceding weights are plainly borrowed from the Chinese, whilst the following are regarded as native in origin.

1 *pey* = 0·292 grammes.
4 *pey* = 1 *fuong* (1·174 grammes).
2 *fuong* = 1 *slong* (2·344 grammes).
4 *slong* = 1 *bat* (9·375 grammes).
4 *bat* = 1 *tomlong* (37·5 grammes).

For heavy merchandise they employ the *hap* or *picul*.

There are three varieties of *picul*: (1) that of the weight of 40 strings of cash (= 100 catties), (2) that of 42 strings, (3) that of 45 strings.

[1] J. Moura, *Le Royaume du Cambodge*, 1. p. 323 (Paris, 1883).
[2] This coin bears on one side the sacred bird Hangsa, on the other a picture of an ancient palace of the kings.

It will be noticed that the first-mentioned is simply the standard of the Chinese *picul* of 133⅓ lbs. English, whilst the others are native.

In Annam we found that the ingots of gold and silver, consisting of ten *luongs* or *nails*, were called *nén*. The *luong* was equal in weight to the Chinese *liung*, and Cambodian *tomlong*, and was also called *dinh* (*dinh-bac*, nail of silver), thus being identical with the ten *denh* into which the Cambodian *nén* or bar is divided.

In Laos[1] we again find the Chinese *picul* as the highest weight unit. It is divided into 100 catties (here called *Chang*) of 600 grammes each (1⅓ lb. Eng.).

1 *picul*   = 100 *catties*.
1 *catty (chang)* = 10 *damling* (60 grammes).
1 *damling*   = 4 *bat* (15 grammes).
1 *bat*   = 4 *chi* (3·75 grammes).
1 *chi*   = 10 *hun*.

All these or their equivalents are used as money of account. "If there is but little coin in Laos," says M. Aymonier, "there are monies of account in abundance." In the south-west of the country, Bassak and Attopoeu, Cambodian currency is employed, and they count by the *nén* or bar of silver.

1 *nén* = 10 *denhs* (money of account).
1 *denh* = 10 strings of *cash*.

The *string* is also money of account and is worth the same as the string of Annam, which is equal to the *sling* or Siamese franc (which is worth 75 or 80 centimes). The *nén* is also divided into 100 *chi*, and as there are 100 strings in the *nén*, the string of cash is equivalent to a *chi* of silver (3·75 gram.). The Siamese coins known also to Cambodia were the weight and money units of the ancient Cambodians, who probably weighed their precious metals. In Laos all of them except the *tical* are only monies of account. The *tical* or *bat* which under the ancient round form[2] was called *clom* in Cambodia is

[1] E. Aymonier, *Notes sur le Laos*. Saigon, 1885.
[2] For an account of the various kinds of Siamese coins of the bullet shape cf. Msg. Pallegoix, *Description du royaume Thai ou Siam*, I. 256 (Paris, 1854).

actually struck as a small piastre in Cambodia and Siam in imitation of European money. This *tical* is worth 4 Siamese *slings*, but the only monetary division of it known in Laos is the local *lat* or small ingot of copper.

4 copper *lats* = 1 silver *tical* (= 4 *sling* = 3 francs).
4 *tical* = 1 *damling*.
20 *damling* = 1 *catty* (*chang*).
50 *catties* = 1 *picul*.

The *chang* or *catty* of silver is a double one, hence 50 *catties* of silver are equal to 100 *catties* of ordinary commercial weight.

The *cutty* of silver thus weighs 1200 grammes instead of 600 grammes.

They likewise use the *moeun* of silver = 10 *changs* = ⅕ *picul*, but more generally the *moeun* is used as a measure of capacity which contains 20 *catties* of shelled rice, but as a measure of capacity it varies and is sometimes equal to 20 *catties*, sometimes to 25 *catties* of rice. That it really is a measure of capacity incorporated at a later date into the weight system like our own *bushels*, *barrels* and *quarters*, is made probable by the fact that in the provinces of Tonlé, Ropon, and Melou Préy they employ a *tramem* or *bag* containing 10 Cambodian *catties*, and in the province of Siphoum the *moeun* is sometimes the name given to a bag or pannier of a cubit in depth, and a cubit in width at the mouth. It is usually called *kanchoen* (*pannier*), and contains 25 *catties* of rice, and 36 *kanchoen* make a *cartload*.

We learn from another part of Laos an interesting fact which also throws some light on the development of the larger weight units from measures of capacity. For since in some parts of that country the cocoanut is used as the measure of capacity, and as *neal*, the native Cambodian name for the *catty*, means simply a cocoanut, it looks as though this was the real origin of the catty universally employed over all Further Asia. This likewise gives us the reason why the catty of silver is twice the weight of a catty of rice. If a weight unit is derived from a measure of capacity, according to the nature of the substance or liquid with which the measure is filled, the weight unit derived will be heavier or lighter, just as the Irish barrel of wheat is 6 stones heavier than the barrel of oats. A cocoanut, or bamboo-joint filled with silver will give a far heavier weight unit than if it is weighed when filled with rice.

We have now had a survey of the monetary and weight systems of China, Annam, Cambodia and Laos, and everywhere found that the *nén* or bar of 10 *taels* is the highest known metallic unit, and that except in Laos the counting of money even by the catty or pound is unknown, the Chinese themselves only employing the *tael* as their highest monetary unit, the catty being kept as in Annam and Cambodia itself for ordinary goods. This is borne out by the practices in the weighing of gold. In Attopoeu, the region where gold is found, 8 *chi* ($=2$ *ticals* or *bats* $= 4$ *slings* $= 30$ grammes) are exchanged for a bar of silver ($= 100$ *chi* $= 375$ grammes). M. Aymonier thinks that the gold *bat*, that is to say the weight in gold of a *tical* (15 grammes, 234 grains Troy), must have been the unit for weighing gold, as formerly it was necessary to give a gold *bat* in order to marry a girl of the blood royal. This gets considerable support from the fact that in Sieng-Khan the gold *bat* has only the weight of a *sling* or *chi* ($58\frac{1}{2}$ grains Troy), that is the quarter of a *tical*, and the weight of the *tical* or *bat* is called a *damling*. In fact they hardly reckon gold in any other way than by this small *damling* which is only the weight of a *tical* (234 grains Troy). In reference to my argument that as gold is the first of all things to be weighed, the primitive weight unit is certain to be small, as no man has, as a rule, any need to weigh his gold by the hundredweight or large mercantile talent, this fact that the highest unit for weighing gold in Attopoeu is so small, not even reaching the weight of the Graeco-Phoenician heavy gold shekel or double ox-unit of 260 grains, is of considerable importance.

This region supplies us with yet another point which can help to clear up the history of early metallic currency. The iron ingots which come from the Cambodian provinces of Kompong Soai form a special kind of money. These ingots are not weighed, but they have the length of the space between the base of the thumb and the tip of the forefinger, they are in breadth two fingers, and one finger in thickness in the middle, thinning off to either end. Three of these ingots = 1 *chi* = 1 *sling* = 1 string of cash; thus 12 ingots = 1 *tical* of silver. These ingots are also counted by bags of 20; thus 1 *nén* or bar of silver = 15 bags = 300 ingots of iron.

At Bassak the iron ingot is replaced by the *lat*, the copper ingot of Laos, which varies in value in the different moeungs (provinces) according to its size. Here is a remarkable confirmation of my contention that it was only at a period considerably later than the weighing of gold that the scales were employed for copper and iron, the catty being kept as in Annam and Cambodia for ordinary goods.

We can now make a further advance in our quest of the first beginnings of money and weights in this interesting region. There are many wild tribes in Annam and Laos, who still employ no method save that of barter, when dealing one with another, although when they touch on the more civilized regions they have to conform their native systems in some degree to the more developed currency of their neighbours, from whom they have to procure the few luxuries of their simple life. We saw above that among the wild tribesmen all articles have a well-defined relationship to each other, some particular article being usually taken as the common measure of all the rest, or rather two or three so that they may have units for estimating their more common as well as their more valuable possessions. So in Annam the buffalo often serves as the general unit of value for the more valuable articles. Thus a large chaldron is worth three buffalos, a handsome gong two buffalos, a small gong one buffalo, six copper dishes one buffalo, two lances one buffalo, a rhinoceros horn eight buffalos, a large pair of elephant's tusks six buffalos, a small pair three buffalos[1]. Thus the buffalo which takes the place of the ox in China and South-Eastern Asia, is used as the commercial unit in like fashion as we found the ox employed among the Homeric Greeks, the ancient Italians, the ancient

[1] E. Aymonier, *Cochin-Chine Française. Excursions et Reconnaissances*, Vol. x. No. 24 (1885), p. 317.

Irish, and the modern Ossetes. But the Annamites themselves employ as currency the silver bar and string of cash as we saw above: accordingly when the hill tribes have dealings with the people of the plain the full grown buffalo is reckoned at a bar of silver, or, its equivalent, 100 strings of cash[1], while the small buffalo is set at fifty strings.

Thus the Orang Glaï have often to buy a pair of elephant's tusks at the cost of eight buffalos or eight bars of silver. Taxes are paid in buffalos; thus the Tjrons of Karang pay a buffalo for each house, or compound for the whole village by a payment of ten buffalos whose horns are at least as long as their ears[2]. Here then we find that exactly as the ancient Irish when they borrowed the Roman system of *unciae* and *scripula* (*unga* and *screapall*) equated the ounce of silver to their own unit, the cow, so we find these wild tribes of Annam forced to adapt their primitive unit to the metallic unit of their more cultured neighbours. Again, the Bahnars of Annam, who dwell on the borders of Laos, have much the same system. With them the highest unit is the *head*, *i.e.* a male slave, who is estimated, according to his strength, age and skill, at 5, 6, or 7 buffalos, or the same number of kettles, as the buffalo and the kettle have the same value, which naturally varies with the size and age of the animal and the quality of the kettle. A full grown buffalo, or a large kettle, is worth seven glazed jars of Chinese shape with a capacity of 10 to 15 litres each. One jar is worth 4 *muks*. The *muk* was originally the name of some special article, but now is simply used as a unit of account. Each *muk* is worth 10 *mats*, or iron hoes, which are manufactured by the Cédans, and which form the sole agricultural implement of the wild tribes of all these regions. This hoe is the smallest monetary unit used by the Bahnars, and is worth about one penny in European goods. This *mat* or hoe serves them as small currency and all petty transactions are carried on by it. Thus a

---

[1] Aymonier, *ibid.*

[2] This mode of estimating the age of the buffalo by the length of its horns may throw some light on the young ox *suis cornibus instructus* of the Marseilles inscription (p. 143).

large bamboo hat costs 2 hoes, a Bahnar knife 2 hoes, ordinary arrows are sold at 30 for 1 hoe and so on. A large elephant is worth from 10 to 15 "*heads*" or slaves, whilst a horse costs 3 or 4 kettles or buffalos. When we read of such a state of human society we seem to be transported back into that far away Homeric time, and as we hear of slaves, and kine, chaldrons and kettles we think of the old Epics with their tale of slaves valued in beeves, and "crumple-horned shambling kine, and tripods" and "shining chaldrons." In the light of such analogies we at last can understand the significance of the 10 axes and 10 'half-axes" which formed the first and second prizes in the *Iliad*[1] when Achilles "set out for the archers the dark-hued iron, and put down 10 axes and 10 half-axes." Who can doubt that these axes and half-axes played much the same part in the Homeric system of currency as the hoes do at this present moment in that of the Bahnars of Annam? Probably such too were the 12 axes which Penelope[2] brought out from the treasure chamber to serve as a target for the suitors in their contests with the bow of Ulysses. The hoe is thus the lowest unit of currency among the Bahnars. From the known interrelations of all the articles of daily life it is easy to estimate how many hoes any even of their more costly possessions is worth. Thus the full-grown buffalo = 7 jars = 28 *muks* = 280 hoes, or about £1. 3s. 4d. of our money. All these transactions require no use of weights, being reckoned by bulk or tale. But now comes the most interesting feature for us, a people in the complete stage of barter, but who actually possess, work and traffic in gold.

In all the streams on the side next Laos the wild people wash gold, men, women and children all alike joining in this laborious industry, and employ as 'cradles' little baskets made of bamboo. The gold is sold in dust at the *rate of the weight in gold of one grain of maize for one hoe*. Here then we have finally run to ground one of the principal objects of our quest. We have a primitive people, who carry on all their trade by means of barter, who have no currency in the precious metals, but who employ as their most general unit of small value the

[1] xxiii. 850 *sq*. [2] *Od*. xxi. 76.

iron hoe. They are found to weigh one thing and one only, namely gold, and for that purpose they do not employ any weight standard borrowed from China or Annam, but equate a certain amount of gold to the unit of barter, and then fix as a constant that amount of gold by balancing it against a grain of the corn that forms one of the chief staples of their subsistence. Nature herself has supplied man with weights of admirable exactitude ready to his hand in the natural seeds of plants, and as soon as he finds out the need of determining with great care the precious substance which he has to win with toil and hardship from the stream, he takes the proffered means and fashions for himself a balance and weights.

We saw that a buffalo was worth 280 hoes; it is therefore an easy task for a Bahnar to tell its worth in gold. It was equally simple for the first Aryan or Semite who framed the gold shekel standard to compute the exact amount of gold which would represent the value of an ox. But perhaps we have not reached the earliest stage of all in the development of a standard for the sale of gold. I ventured to put forward in 1887 the suggestion that the way in which the amount of gold which represented the value of a cow was first fixed approximately was by *measuring* it in some way, as for instance by taking the amount which would fit in the palm of the hand, somewhat in the fashion that rustics measure gunpowder or shot for a gun. What was then but a mere guess may be now regarded as fairly certain. That excellent observer, M. Aymonier, notes that the Tapak tribe, who live at a distance of six days' journey from Attopoeu, wash gold. The women wade into the streams (after having first carefully placed five flowers or five leaves at the foot of a tree close by the stream to ensure good luck). Each dips a water-tight bag into the sand at the bottom of the stream, and after a long series of rewashings and cleansings at last gets the gold dust in a state of purity[1]. The savages carry it to Attopoeu, and sell it at the rate of 9 *chi* of gold for a *nén* or bar of silver (= 100 *chi*). The relative value in Attopoeu is 8 *chi* or two *bats* of gold to one bar (= 100 *chi*) of silver, or as they express it one *tical* of gold is changed for 12 *ticals* of silver.

[1] E. Aymonier, *Notes sur le Laos*, p. 33.

"The *tical* of gold is," it is said, "equivalent to the weight of 32 grains of a peculiar kind of rice of the country, with large grains and of a red colour, which is called ivory rice." Here we have the weighing by natural grains as before, but Aymonier adds (p. 35) that "the natives relate that gold was formerly so abundant that without *weighing it people were content to measure* it. A little stick of gold an inch broad and a span long *was exchanged against a buffalo.*"

We found the Bahnars equating a small quantity of gold to their smallest unit of barter, the hoe; now we find that in the wild parts of Laos the unit of gold, before weights of natural grains were employed, was based by measurement upon the buffalo, the chief unit of barter. Thus we have found among the remote peoples of Further Asia the very method of fixing a metallic unit, which I have endeavoured to prove was that followed by the Aryan and Semitic races in arriving at that shekel of gold, which was the common standard of all the civilized peoples of the ancient world, and which was the parent of all our mediaeval and modern systems.

CHAPTER VIII.

How were Primitive Weight Units fixed?

Ordiar ex minimis.
Carm. de ponderibus.

WE have seen that the Chinese system of weights is based upon natural seeds of plants, and we have actually found the wild hillsmen of Annam and Laos weighing their gold dust by grains of maize and rice. But it may be urged by the advocates of a Babylonian scientific origin based on the one-fifth of the cube of the royal ell, which in turn is based upon the sun's apparent diameter, that the Chinese names of weights are merely conventional terms taken from the name of certain seeds, and on the other hand that the mere fact that a very barbarous people like the Bahnars of Annam weigh their gold dust by grains of rice is no evidence that people in a higher stage of culture were content with such rude metric standards. I propose to show in this chapter that it has been the actual practice of peoples as far advanced in civilization as the ancient Greeks or Italians, to employ seeds as weights down to the present day in Asia, that it was the general practice in the middle ages, that it was likewise the practice of the Romans of the empire, of the Greeks, and finally that such too was the practice of the Assyrians themselves at a period long before the bronze Lion weights were ever cast, or the stone Duck weights were carved. If I succeed in proving this proposition, the doctrine that the art of weighing was scientific must give place to the contention that it was purely empirical.

As we have found among the barbarians of Asia the first beginnings of the art of weighing by the employment of grains of rice and maize, it is best for us to take first in order some other Asiatic countries lying towards the same region.

The great islands of the Indian Archipelago, singularly rich in all endowments of nature, have for ages enjoyed a high degree of culture. Conveniently placed, they have received all the advantages of contact with the civilization of China, India, and even that of the Arabs from the distant west of Asia. Never were people more favourably situated for obtaining foreign systems of weights and measures, if they felt so disposed, than the Malays of Java and Sumatra and the other islands of the Indian Archipelago. That admirable observer, John Crawfurd, writing in 1820 says[1]: "In the native measures everything is estimated by bulk and not by weight. Among a rude people corn would necessarily be the first commodity that would render it a matter of necessity and convenience to fix some means for its exchange or barter. The manner in which this is effected among the Javanese will point out the imperfection of their methods. Rice, the principal grain, is in reaping nipped off the stalk with a few inches of the straw, tied up in sheaves or parcels and then housed or sold, or otherwise disposed of. The quantity of rice in the straw which can be clenched between the thumb and the middle finger is called a *gagam* or handful, and forms the lowest denomination. Three *gagams* or handfuls make one *pochong*, the quantity which can be clenched between both hands joined. This is properly a sheaf. Two sheaves or *pochongs* joined together, as is always the case, for the convenience of being thrown across a stick for transportation, make a double sheaf or *gedeng*. Five *gedengs* make a *songga*, the highest measure in some provinces, or twenty-four make an *hamat*, the more general measure. From their very nature these measures are indefinite and hardly amount to more accuracy than we employ ourselves when we speak of sheaves of corn. In the same district they are tolerably regular in the quantity of grain or straw they

---

[1] *History of the Indian Archipelago* by John Crawfurd, F.R.S. Vol. I., p. 271.

contain, but such is the wide difference between different districts or provinces, that the same nominal measures are often twice, nay three times as large in one as in another. For the *hamat* or larger measure perhaps about eight hundred pounds avoirdupois might be considered a fair average for the different provinces of Java. This may convey some loose notion of the quantities intended to be represented. For dry and liquid measures they may naturally have recourse to the shell of the cocoanut and the joint of the bamboo which are constantly at hand. The first called by the Malays *chupa* is estimated to be two and a half pounds avoirdupois. The second is called by some tribes *kulch* and is equal to a gallon, but the most common bamboo measure is the *gantung*, which is twice this amount. To those exact and business-like dealers, the Chinese, and in a less degree to the Arabs and people of the east coast of the Indian Peninsula, the Indian islanders are chiefly indebted for any precision we find in their weights. In all the traffic carried on between the commercial tribes and foreigners, the Chinese weights, though occasionally under native names, are constantly referred to. The lowest of these, called sometimes by the native name of Bungkal, but more frequently by the Chinese name of Tahil [*tael*], varies from twenty-four pennyweights nine grains to thirty pennyweights and twenty grains. Ten of these make a *kati* [*catty*] or about twenty ounces avoirdupois; one hundred *katis* make a *pikul* or $133\frac{1}{3}$ lbs. avoirdupois, and thirty *piculs* make one *koyan*. Of these the *kati* and the *pikul*, because they are constantly referred to in considerable mercantile dealings, are the only well-defined weights. The *koyan* by some is reckoned at twenty *pikuls*, by others at twenty-seven, twenty-eight and even at forty. The Dutch are fond of equalizing it with their own standards and consider it as equal to a *last* or two tons.

"The *Bahara*, an Arabic weight, is occasionally used in the weighing of pepper, but its amount is very indefinite, for in some of the countries of the Archipelago it amounts to 396 lbs., and in others to 560 lbs."

Elsewhere he says[1], "The *picul* is strictly a Chinese weight

[1] P. 275.

as its amount shows, though the term happens in this case to be native. Its meaning in the vernacular languages is a natural load or burthen, and when used in this primitive sense it, without reference to the Chinese weight, is not found to exceed eighty pounds avoirdupois." This is a fact of great importance as we shall see when we come to the development of the *mina* and *talent* of Graeco-Asiatic commerce.

Finally Crawfurd says, "The nice question of weighing gold, the only native commodity which could not be estimated by tale or bulk, has given rise to the use of weights among the natives themselves. Grains of rice are still occasionally used in the weighing of gold in the neighbourhood of the gold mines in Sumatra" (p. 274).

I have quoted at full length these passages in order that the reader may accept with fuller confidence statements so instructive as regards the origin of weight, the first object to be weighed, and the origin of the *picul*, or as we may call it the *talent* of Eastern Asia. Nine years before Crawfurd wrote there had appeared William Marsden's admirable *History of Sumatra*[1]. He gives us far fuller information on the subject of gold than Crawfurd has done. Thus he writes: "In those parts of the country where traffic in this article (gold dust) is considerable, it is employed as currency instead of coin; every man carries small scales about him, and purchases are made with it so low as to the weight of a grain or two of *padi*. Various seeds are used as gold weights, but more especially these two: the one called *rakat* or *saga-timbañgan* (*Glycine abrus* L or *abrus maculatus* of the Batavian trans.), being the well-known scarlet pea with a black spot, twenty-four of which constitute a *mas*, and sixteen *mas* (mace) a *tail* (tael): the other called *saga puku* and *kondori batang* (*Aden anthera pavonia* L), a scarlet or rather coral bean much larger than the former, and without the black spot. It is the candarin weight of the Chinese, of which one hundred make a tail and equal, according to the tables published by Stevens, to 5·7984 gr. Troy, but the average weight of those in my posses-

---

[1] *History of Sumatra* by William Marsden, F.R.S. (London, 1811), p. 171.

sion is 10·50 Troy grains. The tāil differs however in the northern and southern parts of the island, being at Natal, Padang, Bencoolen and elsewhere twenty-six pennyweights six grains. At Achin the *bangkal* of thirty pennyweights twenty-one grains is the standard. Spanish dollars are everywhere current and accounts are kept in dollars, *sukus* (imaginary quarter dollars) and *kepping* or copper cash, of which four hundred go to the dollar. Besides these there are silver *fanams*, single, double and treble (the latter, called *tali*), coined at Madras, twenty-four *fanams* or eight *talis* being equal to the Spanish dollar, which is always valued in the English settlements at five shillings."

He adds that copper is sold by weight (*picul*), and that tin, which was accidentally discovered in 1710 by the burning of a house, is exported for the most part in small pieces or cakes called *tampangs*, sometimes in slabs (p. 172), and furthermore they purchase bar iron by measurement instead of by weight (p. 176).

Several points of great importance are to be noticed in the foregoing statements. Firstly, that whilst for foreign trade with the Chinese they employ the Chinese weight, which we know always by its Malay name of *picul*, a well-defined weight standard of 133⅓ lbs. avoirdupois, they had evidently a native unit of weight, their own *picul*, which simply means and actually was as much as a man can carry on his back, and which, as we saw, rarely exceeds 80 lbs. avoirdupois. This seems to give us an insight into the manner in which the most primitive highest weight unit is arrived at. A man's load is one of those natural standards which will vary according to race and climate, and the conditions under which the load has to be borne. Thus, the average weight of the load borne by a dock porter who has to endure the strain for only some few yards, will of course be far higher than that carried by the porters of travellers in Central Africa, where the load has to be borne day after day on a march of several hundred, or a thousand miles. Thus in the case of the Madis, a pure negro tribe, the average load seems to be about 50 pounds, which they can carry "20 miles a day for eight or ten consecutive days without shewing any signs of distress[1]." The Chinese, the superiors in science of all Eastern Asia, have carefully adjusted this "*load*," and it makes, as we have seen above, their highest weight unit. Its particular amount is probably due to the fact that, having carefully fixed the weight of the smaller units, the candarin, the mace, the *liung* or *tael*, and the *catty*, their pound, they simply took the hundredfold of the *chang* or *catty* as the standard for their highest unit, and thus that which at an earlier stage was just as vague and fluctuating as the *picul*, or back-loads in use still among the less-advanced peoples of the Indian Archipelago, became a fixed scientific unit. Secondly, we must notice that the Malays have not followed the Chinese in the subdivisions of the *catty*. For whilst in China 16 *taels* or ounces go to the catty, the Malays follow more strictly the decimal system, and make their catty simply the tenfold of the *tael* or ounce. This same method of division we found already in Annam, and not only in Annam but also in Cambodia and Laos we found the silver *nên* or bar, invariably consisting of ten such parts, corresponding in weight to the Chinese *tael*, sixteen of which go to the catty.

It would appear, then, that here we have a combination of units of weight and units of capacity. The higher gold and silver unit, the *nên*, is simply the tenfold of the lower unit, the *tael* or ounce, while the *catty*, which is never employed in China in estimating gold or silver, but is a genuine commercial unit, was probably originally some natural unit of capacity. We saw strong evidence of this in Cambodia, where the name for this weight is *neal* or cocoanut, and we have just found the cocoanut as the chief unit of dry measure amongst the Malays of the Indian Seas. It was probably found that 16 times the *tael* or ounce came nearer to the weight of the contents of a cocoanut or bamboo joint (whatever kind of matter they may have weighed in it for this purpose, whether rice, or water), than the original 10 ounces, which formed the *bar*, the highest genuine weight unit. Sixteen was likewise a convenient number, its factors being numerous, and it could be

[1] R. W. Felkin, 'Notes on the Madi or Moon tribe of Central Africa.' *Proceedings of Royal Society of Edinburgh*, Vol. XII. pp. 303, *seqq.*

divided in four portions, each of which contained four other units. It will presently be a question as to whether similar influences have not produced our pound avoirdupois, with its 16 sub-multiples.

M. Moura found a difficulty regarding the Cambodian *neal* or cocoanut *catty*; because a *neal* of rice only weighs half the weight, at which the *neal* is rated as a weight. But we saw in Java that the *chapa* or cocoanut measure is estimated at 2½ pounds avoirdupois. It is then not improbable that some liquid or substance far heavier than rice was used to fill the cocoanut, when the value of its contents was being ascertained by weighing so as to serve as a general unit. The same variation in weight, owing to the different nature of its contents, has, as mentioned before, given rise in Ireland to *barrels* of various weights. Thus a *barrel* of wheat contains 20 stone avoirdupois, a *barrel* of potatoes 24 stone, a *barrel* of barley 16 stone, and a *barrel* of oats 14 stone. This diversity simply arose from comparative lightness or heaviness of the different commodities which were measured by one and the same unit of capacity: the barrel itself, having been fixed by a process of measurement, similar to that by which the milk-pan was regulated among the Welsh, and the pannier among the natives of Laos. The principle by which higher units of capacity or weight are formed is likewise well illustrated by the instance given above of the *cartload* of rice, which is simply regarded as the multiple of the pannier or bag, which forms the smaller unit for rice. The size of the *cartload* would be conditioned by the size of the cart usually employed, which in turn would depend on a variety of other things, such as the nature of the country, or its roads, or the kind of animals employed for draught. The vagueness in amount of the *koyan* or multiple of the *picul* noticed by Crawfurd, may thus meet with a reasonable explanation.

We may now return to the mainland of Asia, where we shall find in the weight system of the Hindus at least one remarkable point of affinity with that of Sumatra. Marsden has told us that the *rakat* or scarlet pea with a black spot is one of the chief weights employed for gold in Sumatra. This *rakat* is none other than the *ratti*, which is usually taken as the basis of the modern Hindu weight system. "This weight," says that eminent scholar Colebrooke[1], "is the lowest denomination in general use, commonly known by the name *ratti*, the same with *rattika*, which, as well as *raktika*, denotes the red seed as *krishnala* indicates the black seed of the *gunjá*-creeper." Mr Thomas has shown the true weight of the *ratti* is 1·75 grains[2].

Many different standards have been used in India for various purposes, one for the weighing of gold, another for the weighing of silver, another used by jewellers, and yet another by the medical tribe, but all alike start from the *ratti*.

"The determination of the true weight of the *ratti* has done much both to facilitate and give authority to the comparison of the ultimately divergent standards of the ethnic kingdoms of India. Having discovered the guiding unit, all other calculations become simple, and present singularly convincing results, notwithstanding that the bases of all these estimates rest upon so erratic a test as the growth of the seed of the *gunja*-creeper (*Abrius precatorius*) under the varied influences of soil and climate. Nevertheless the small compact grain, checked in early times by other products of nature, is seen to have the remarkable faculty of securing a uniform average throughout the entire continent of India, which only came to be disturbed when monarchs like Shîr Shâh and Akbar in their vanity raised the weight of the coinage without any reference to the numbers of *rattis*, inherited from Hindu sources, and officially recognized in the old, but entirely disregarded and left undefined in the reformed Muhammadan mintages[3]." We shall learn shortly that in its uniformity the *ratti* does not differ from other seeds such as wheat and barley. Probably, however, the fact that the *gunja*-creeper was found everywhere in India gave it its position of a universal standard.

---

[1] H. T. Colebrooke, *On Indian Weights and Measures* (Miscellaneous Essays edited by Prof. E. B. Cowell, 1873), Vol. I. 528—543.

[2] *Numismatic Chronicle*, IV. 131 (N. S.).

[3] Thomas, *Initial Coinage of Bengal*, II. p. 6 (*Royal Asiatic Journal*, Vol. VI.).

Those who wish to study the elaborate systems of later times employed in India can consult the works of Colebrooke and Thomas already referred to.

The legislators Manu, Yájnavalkya, and Nárada trace all weights from the least visible quantity which they concur in naming *trasareṇu* and describing as the very small mote, "which may be discovered in a sunbeam passing through a lattice." Writers on medicine proceed a step further, and affirm that a *trasareṇu* contains 30 *paramáṇu* or atoms. The legislators above-named proceed from the *trasareṇu* as follows:

8 *trasareṇus* = 1 *likshá*, or minute poppy-seed.
3 *likshás* = 1 *raja-sarshapa*, or black mustard-seed.
3 *raja-sarshapas* = 1 *gaura-sarshapa*, or white mustard-seed.
6 *gaura-sarshapas* = 1 *yava*, or middle-sized barley-corn.
3 *yavas* = 1 *kṛishnala*, or seed of the *gunjá*.

But as we want to learn what was the actual usage of the Hindus, instead of dealing with the mere theoretic statements of late authors, I shall at once quote in full the tables given in the *Lilavati* of Brahmegupta, who wrote his Algebra and Arithmetic about 600 A.D.[1]

MONEY (*by tale*). Twice ten cowries[2] are a *cácíní*; four of these are a *pána*, sixteen of which must here be considered as a *dramma*, and in like manner a *nishká*, as consisting of sixteen of these.

WEIGHT. A *gunjá* (or seed of *Abrus*), is reckoned equal to two barley-corns (*yavas*). A *valla* is two *gunjas*, and eight of these are a *dharana*, two of which make a *yadyanaca*. In like manner one *dhataca* is composed of fourteen *vallas*.

Half ten *gunjas* are called a *masha* by such as are conversant with the use of the balance; a *karsha* contains sixteen of what

---

[1] Algebra with Arithmetic and Mensuration translated from the Sanskrit of Brahmegupta and Bhascara by H. T. Colebrooke (London, 1817).

[2] Down almost to the present day a system of currency, similar to that shown in the *Lilavati* prevailed in Assam. "Gold continues to pass current in small uncoined round balls, usually weighing one *Tola*," there was a silver coinage also, and cowries passed as money. W. Robinson, *Descriptive Account of Assam*, pp. 249 and 267 (London, 1841).

are called *mashas*, a *pala* four *karshas*. A *karsha* of gold is named *suvarṇa*.

This is quite in harmony with the *weight* of *gold* as given by the legislators:

5 *krishnalas* or *raktikas* = 1 *másha*.
16 *máshas* = 1 *karsha, aksha, tolaka*, or *suvarṇa*.
4 *karshas* or *suvarṇas* = 1 *pala* or *nishka*.
10 *palas* = 1 *dharana* of gold.

Yájnavalkya adds that according to some 5 *suvarṇas* = 1 *pala*.

All the authorities seem agreed in regarding the term *suvarṇa* as peculiar to gold, for which metal it is also a name.

We learn thus that the Hindu standards were fixed by means of natural seeds, and at no period do they, clever mathematicians as they were, seem to have made any effort at obtaining a mathematical basis for their metric systems.

We also observe that the weight known as the *suvarṇa* or *gold* weight *par excellence* is the weight of a *karsha* or 80 *gunjás*, which, if we take the *gunjá* = 1·75 grains Troy, gives the weight of the *suvarṇa* as 140 grains. I have already (p. 127) taken the original Hindu gold unit as not far from this amount. From the *Lílávati* we may now with little misgiving assume it to have been such.

Lastly, let us observe that the barley-corn appears as the basis of the system in the tables of Brahmegupta and Bhascara, although the *raktika* evidently overmasters it in the course of time. This is very interesting, for it indicates that the Hindus had learned the art of weighing in a comparatively northern region, where barley was the chief cereal under cultivation. If the system had been invented in the more southern parts of India, the grain of rice, the staple of life in the southern regions, would certainly have appeared as the sub-multiple of the *raktika*, instead of the barley. As a matter of fact, rice-grains seem to have been occasionally used locally, for Colebrooke remarks that "it is also said that the *raktika* is equal in weight to four grains of rice in the husk." This supposition is completely in accord with what we found in Persia, where the modern weight system for gold, silver and medicine runs thus:

3 *gendum dsho* (barley-corn) = 1 *nashod*.
4 *nashod* (a kind of pea, lupin?) = 1 *dung*.
6 *dung* = 1 *miscal*[1].

Although the *miscal* and *habba* denote Arabic influence, we may, without straining probabilities, conjecture that the use of the *barley-corn* here as well as in India, where we found it at a period anterior to Muhammadan conquest, indicates that in Persia it existed likewise from the earliest times. The close relationship between the ancient Hindus and ancient Persians makes it all the more likely. It is also pointed out that formerly the *nashod* was divided into *three* instead of four grains. As the Arabs divide their *karat* into four *habbas*, it is all the more likely that the 3 barley-corns = 1 *nashod* belong to the ancient system.

The Arab weight system is based on the grain of wheat, four of which make a *karat* (the seed of the carob or St John's Bread)[2]. Occasionally in the Arab writers mention is made of a karat divided into 3 *habbas*[2]. The weight of the karat remains unchanged, but the grains in this case are barley grains, since, as we shall see presently, 3 grains of barley are equal to 4 grains of wheat ($.063 \times 3 = 4.047 \times 4$).

It will now be most convenient for us to begin in the extreme west, and once more from that work back towards the coast of the Aegean Sea, in which our chief interest must always be centred.

Whether the Kelts of Ireland had any indigenous weight system or not, we have no direct evidence, although we do know as a fact that when Caesar landed in Kent he found the Britons employing coins of gold and bronze, and bars (or according to some MSS. *rings*) of iron adjusted to a fixed weight. However the earliest Irish documents reveal that people using

---

[1] Martini, *Metrologia*, p. 770. Formerly the *nashod* = 3 *habbi* of ·063 gram which is just the weight of the barley grain, whereas ·047 the weight assigned to the *gendum* is that of a grain of wheat.

[2] Queipo, *Essai sur les Systèmes Métriques et Monétaires des anciens peuples* I. 360 (Paris, 1859).

a system of weights for silver directly borrowed from the older Roman system (although it is likely that they had a native standard for gold). As the *solidus* and *denarius* became the chief units of Europe from the time of Constantine the Great (336 A.D.), the Irish probably received their system at an earlier date.

1 *unga* (*uncia*) = 24 *screapalls* (*scripula*).
1 *screapall* = 3 *pingiuns*.
1 *pingiun* = 8 grains of wheat[1].

When we pass to England, the very word *grain* which we employ to express our lowest weight unit, would of itself suggest that originally some kind of *grain* or *seed* was employed by our forefathers in weighing, but as the grain in use among us is the *grain Troy*, and as we have not yet learned its origin, it will not do to argue vaguely from etymology. But a little enquiry soon brings us to a time when the grain Troy did not as yet form the basis of English weights, and when a far simpler method of fixing the weight of the king's coinage was in vogue. It was ordained by 12 Henry VII. ch. v. "that the bushel is to contain eight gallons of wheat, and every gallon eight pounds of wheat, and every pound twelve ounces of Troy weight, and every ounce twenty sterlings, and every sterling to be of the weight of thirty-two grains of wheat that grew in the midst of the ear of wheat according to the old laws of this land[2]." Going backwards we find that in 1280 (8 Edward I.) the penny was to weigh 24 grains, which by weight then appointed were as much as the former 32 grains of wheat. By the Statute *De Ponderibus*, of uncertain date but put by some in 1265, it was ordained that the penny sterling should weigh 32 grains of wheat, round and dry, and taken from the midst of the ear. Going back a step still further we find that by the Laws of Ethelred, every penny weighed 32 grains of wheat[3], and as the pennies struck by King Alfred weigh 24 grains Troy, we may assume without hesitation that they were struck on the same

---

[1] *Ancient Laws of Ireland*, Vol. IV. 335, (Book of Aicill), O'Donovan's Supplement, s.v. *pingiun*.

[2] Ruding, *Annals of the Coinage of Great Britain*, II. 58.

[3] Ruding, *op. cit.* I. 369.

standard of 32 grains of wheat. Thus from Alfred (871—901) down to Henry VII. (1485—1509), we find the penny fixed by this primitive method, and the actual weight of the coins, as tested by the balance at the present day, affords proof positive of the method.

But all the standards of mediaeval Europe (with the exception of the Irish) were based on the gold *solidus* of Constantine the Great[1]. The *solidus* (itself weighing 72 grains Troy or $\frac{1}{72}$ of the Roman pound) was divided into 24 *siliquae*. The *siliqua*, or as the Greeks called it *keration* (κεράτιον, from which comes our word *carat*), was the seed of the *carob*, or as it is often called, *St John's Bread* (*Ceratonia siliqua* L.). Thus the lowest unit in the Roman system, as it is usually given, is found to be the seed of a plant. The same holds of the Greek system, for the *drachma* is described as containing 18 *kerata* or *keratia*, whilst according to others "it contains three *grammata*, but the *gramma* contains two *obols* and the *obol* contains three *kerata*, and the *keras* contains four *wheat grains*[2]." From this we see that the *keration* or *siliqua* was further reduced to 4 *sitaria*, or grains of wheat, whilst from another ancient table of weights[3] we learn that the *siliqua* likewise equals 3 barley-corns (*siliqua grana ordei* iii). Hence it appears that 3 barley-corns = 4 wheat grains. Thus both Greek and Roman systems just like the English and Irish take as their smallest unit a grain of corn. This also throws important light on the origin of that mysterious thing, the Troy grain. We saw above (8 Edward I.) that at the time of its introduction into England that 24 grains Troy = 32 grains of wheat, that is the Troy grain stands to wheat grain as 3 : 4. But as we have just seen that the *siliqua* = 3 barley-corns, and also = 4 wheat-corns, it follows that 3 barley-corns = 4 wheat-corns. And as 3 Troy grains = 4 wheat-corns, it likewise follows that 3 Troy grains = 3 barley-corns, or in other words, the barley-corn and

---

[1] Marquardt, *Röm. Staatsverwaltung*, II. p. 30.

[2] *Fragm.* ap. Hultsch, *Metrol. Script.* I. 248, ἡ δὲ δραχμὴ κέρατα ιη'. ἄλλοι δὲ λέγουσιν· ἔχει γραμμὰς τρεῖς...τὸ γράμμα ὀβολοὺς β'. ὁ δὲ ὀβολὸς κέρατα γ'. τὸ δὲ κεράτιον ἔχει σιτάρια δ'.

[3] Hultsch, *Op. cit.* II. 128.

Troy grain are the same thing. It thus appears that the Troy grain is nothing more than the barley-corn, which was used as the weight unit in preference to the grain of wheat in some parts of the Roman empire. Furthermore this relation between barley-corns and wheat-corns can be proved to be a fact of Nature. In September, 1887, I placed in the opposite scales of a balance 32 grains of wheat "dry and taken from the midst of the ear," and 24 grains of barley taken from ricks of corn grown in the same field at Fen Ditton, near Cambridge, and I thrice repeated the experiment; each time they balanced so evenly that a half grain weight turned the scale. The grain of Scotch wheat weighs ·047 gram, the Troy grain = ·064, ·047 × 4 = 188, ·064 × 3 = 192. Practically 4 wheat grains = 3 Troy grains.

Before passing from the Greek and Roman standards I may add that even higher denominations than the *siliqua* were expressed by the seeds of plants. The Romans made the lupin (*lupinus*) = 2 *siliquae* and under its Greek name of *thermos* (θερμός), it was assigned a like value (*Metrol. Script.* I. 81). In the *Carmen de Ponderibus* (*Metrol. Script.* II. 16), 6 grains of pulse (*grana lentis*) are made equal to 6 *siliquae*, and a like number of grains of spelt are given a similar value.

We next advance towards the East and take up the Semitic systems. We have already had occasion to touch upon that of the Arabs when dealing with the modern Persians. "There can be little doubt," says Queipo (I. 360), "that the Arab system of weight was based on the grain of wheat." The *habba* was their smallest unit. Four *habbas* are equal to 1 *karat*, the latter of course representing the *keration* or *siliqua*, and the former the 4 *sitaria* or *wheat-grains*, which we saw were its equivalent. This is the most ordinary value given to the karat in Makrizi and the other Arabic writers on Metrology, but occasionally we find the karat made equal to only 3 grains, which of course are barley-corns. We saw above that in the Persian system the *nashod* was formerly divided into 4 *habbi* of ·048 gram (which is plainly the weight of the wheat-grain), whilst now it is divided into 3 grains each of ·063 which represents the barley-corn, or in other words the Troy grain of ·064 gram. Of course the objection might be raised that as the Arabs had borrowed their higher denominations such as the *dirhem* (δραχμή) and *dinar* (*denarius*, δηνάριον), from the Greeks and Romans, and as their standard weight the *mithkal* is nothing more than the *sextula* or $\frac{1}{6}$ of the Roman ounce, employed in the eastern Empire under the name of *exagion* (ἐξάγιον, whence comes the *saggio* of Marco Polo), so too their wheat-corns and barley-corns were not of their own devising, but likewise adventitious. After what we have seen above (p. 166) to be the practice of primitive people in the selling of gold, a traffic in which the Arabs had been engaged for many ages, it would seem hardly necessary to reply to such an argument, but as a more complete answer can be given in the course of the last portion of this enquiry, we shall deal with it in that place.

We now come to the Assyrians themselves, from the discovery of whose weights in the shape of lions and ducks, the whole modern theory of a scientific origin for all the weight standards of the Greeks as well as Asiatics and Egyptians has had its origin. But even within this sacred precinct of *à priori* metrology the irrepressible grain of corn springs up vigorously, although almost choked by the abundant crop of tares which have been sown around it. If we find that a Semitic people, who were the ancients of the earth before Pelops passed from Asia into Greece, or Romulus had founded his Asylum, employed the wheat grain as their lowest weight unit, we may then well argue that ages before the birth of the Prophet and the Arab conquest of Egypt and Syria, the Semitic folks employed grains of corn to form their lowest weight unit.

M. Aurès[1], a well-known Assyrian metrologist, has recently set forth the Assyrian system in its latest and most advanced stage. Following the veteran Assyriologist, M. Oppert, he finds that the Assyrians used a denomination lower than the obol. In the Museum of the Louvre there is a small Assyrian weight of the "duck" kind, which bears on its base the Assyrian character of 22 *grains* $\frac{1}{2}$. The ideogram translated *grain* is evidently meant to represent some kind of corn with a rounded end. The weight of this object is ·95 gram (14$\frac{2}{3}$ grains

[1] *Recueil de travaux relatifs à la Philologie et l'Archéologie Egyptienne et Assyrienne*, Vol. x. fasc. 4, p. 157.

Troy). The weight is a ¾ obol, and therefore 30 grains went to the obol. This is the obol of the heavy Assyrian system, of which we shall presently speak. For the sake of clearness, I take M. Aurès' table.

> 30 grains = 1 obol.
> 6 obols = 1 drachm.
> 2 drachms = 1 shekel.
> 10 drachms = 1 "stone."
> 60 „ = 1 *light* mina.

For our present purpose it is quite sufficient to call attention to the fact that this grain which forms the lowest unit of the Assyrian scale weighs ·042 gram (·95 ÷ 22·5) which is a very close approximation to the weight of the *wheat-grain* (·047). Making allowance for some loss which the weight may have sustained, it seems impossible to doubt that we have here the wheat-grain being used to form the smallest unit as it is in the modern Arabic system. The double obol of the Assyrians weighs 30 grains; we shall also find that the Hebrew *gêráh* or obol (twenty of which made a shekel), weighed exactly 15 *grains of wheat*, that is the Hebrew *gêráh* is the light obol which stood side by side with the heavy obol of 30 grains in the Assyrian system. Let us treat the matter from a slightly different point of view: As the *light* Assyrian obol contained 15 *Assyrian* grains, the *light* shekel contained 180 *Assyrian* grs. But as we know that this light Assyrian shekel weighed 8·4 grams, or 131 grains *Troy*, and as we know that the *Troy* grain is really the barley-corn and likewise that 3 barley-corns = 4 *wheat* grains, it is obvious that 131 grains Troy = 175 *wheat* grs. nearly, a very close approximation to the 180 *Assyrian* grs. Again as 180 *Assyrian* grs. = 8·4 grams, the *Assyrian* grain weighed ·046 gram, that is almost exactly the weight of a *wheat* grain (·047 gram).

But let us see for a moment in what fashion M. Aurès accounts for the presence of corn-grains in a system so elaborately scientific as he and his school maintain.

Starting as usual with the old assumption that all weight standards come from the measures of capacity and all measures of capacity in their turn are derived from the linear measures, he proceeds thus: The Assyrian ideogram which represents *tribute*, likewise represents *talent*. Tribute being paid in corn, no doubt the idea of weight first arose as the people carried their quota of corn on their backs to the receipt of custom. They accordingly weighed the measure (*bar*), which contained the proper amount of corn and took it as their weight unit, and then proceeded to make subdivisions of it. When their weight system was thus fixed, for convenience instead of going to the trouble of adjusting weights they took 30 grains of corn which would be just equivalent to the weight of an obol. After the many historical instances quoted in the preceding pages in which the methods of appraising the value of corn and other dry commodities have been set out, and also the manner in which corn grains have been employed for fixing the higher standard, as for instance in the adjustment of the English bushel in the reign of Henry VII., the reader will feel that M. Aurès has simply inverted the true order of events, and that as we found the natives of Annam and the Malays of the Indian Archipelago making their first essay in weighing by means of a grain of maize, or rice, or *padi*, so the ancient inhabitants of Mesopotamia made their first beginning, and as we have found everywhere that gold, the most precious of objects, was the first thing to be weighed, and as it only existed in small quantities, thus requiring but a very small unit of weight, so the Assyrians likewise began to weigh gold first of all, employing the natural seeds of corn, and only in process of time arrived at higher units by multiplying the smaller.

To all the evidence collected from Asia and Europe we can likewise add a fact of great importance from Africa. We saw that it was highly probable that the Carthaginians traded for gold to the West Coast of Africa, and beyond all reasonable doubt the natives of the Gold Coast have for ages been acquainted with that metal. Now it can be proved that these peoples, whilst employing no weights for any other mercantile transaction, used the seeds of certain plants for weighing their gold; thus Bosman writing two centuries ago says, "Having treated of gold at large, I am now obliged to say something concerning the gold weights, which are either pounds, marks, ounces or angels........... We use here another kind of weights which are a sort of beans, the least of which are red spotted with black and called Dambas; twenty-four of them amount to an angel, and each of them is reckoned two stiver weights; the white beans with black spots or those entirely black are heavier and accounted four stiver weights: these they usually call Tacoes, but there are some which weigh half or a whole gilder, but are not esteemed certain weights, but used at pleasure and often become instruments of fraud. Several have believed that the negroes only used wooden weights, but that is a mistake; all of them have cast weights either of copper or tin, which though divided or adjusted in a manner quite different to ours; yet upon reduction agree exactly with them[1]".

I am informed by Mr Quayle Jones, Chief Justice of Sierra Leone, that at the present day, a seed called the *Taku* (with a black spot) is employed by the natives of the Gold Coast for weighing gold. He also tells me that small quantities of gold are measured by a quill in ordinary dealings in the market[2]. I learn from another private source that 6 Takus = 1 ackie (20 ackies = 1 ounce). From Bosman's equating the bean with the red spot to 2 stiver-weights, we can deduce its weight as 2 grs. troy; this result combined with the colour of the bean would make us *à priori* conclude that the Damba was the *Abrus precatorius*, so familiar to us already under its Hindu name of *ratti*.

Here we have a primitive people with a weight system of their own based on the Damba and Taku, just as the Hindu is based on the *ratti*, and here too we have another proof that the first of all articles to be weighed is gold. From Bosman we also learn that gold in small quantities was not always weighed, for he says of the inferior gold which was mixed with

---

[1] Bosman, *Guinea, Letter VI.* (*Pinkerton's Voyages*, Vol. xvi. p. 374).

[2] Although I have made many enquiries and Dr Thiselton Dyer of Kew has taken much trouble in the matter, I am unable to give the reader the botanical names of the Taku and Damba. Dr Dyer thinks the Damba is our old friend the *Abrus precatorius*, the Indian *ratti*, confirming the opinion I had previously formed from its weight. These seeds are commonly known as crabs' eyes.

silver or copper, that it is cast into fetiches (small grotesque figures). " These fetiches are cut into small bits by the negroes of one, two, or three farthings. The negroes know the exact value of these bits so well at sight, that they never are mistaken, and accordingly they sell them to each other without weighing as we do coined money[1]." This recalls the practice as regards silver among the Tibetans at the present day.

Crossing to the eastern side of Africa we find the natives of Madagascar employing a system, the basis of which is a grain of rice. "The Malagasy have no circulating medium of their own. Dollars are known more or less throughout the island: but in many of the provinces trade is carried on principally by an exchange of commodities. The Spanish dollar, stamped with the two pillars, bears the highest value. For sums below a dollar the inconvenient method is resorted to in the interior, of weighing the money in every case. Dollars are cut up into small pieces, and four iron weights are used for the half, quarter, eighth, and twelfth of a dollar. Below that amount, divisions are effected by combinations of the four weights, and also by means of grains of rice, even down so low as one single grain— "Vary vray venty," one plump grain, valued at the seven hundred and twentieth part of a dollar[2]. The grain of rice therefore weighs $\frac{4}{7}$ gr. troy ('036 gram). As gold is not found in Madagascar[3] the natives could not weigh it first of all things; but they have carried out the principle of taking silver, the most precious article they possessed, as the first object to be weighed.

In this chapter, therefore, we have sought the method by which weight standards are fixed among primitive and semi-civilized peoples; we have studied the system or systems of China, Cochin-China, Cambodia, Laos and the great Islands of the Indian Ocean. Everywhere we have received the self-same answer, everywhere the lowest unit is nothing more than a natural seed or grain. We found in two places in the area

[1] *Op. cit.* 373. "The fetiches they cast in moulds made of a black and heavy earth into what form they please." (p. 367.)
[2] Ellis, *History of Madagascar*, I. p. 335.
[3] *Op. cit.* I. p. 6.

studied, amongst the Tapaks of Annam and the Malays of Sumatra, the art of weighing in its earliest infancy; only one product, gold, as yet being weighed, and the weight unit employed for it being a grain of rice or maize. We found that this smallest natural unit of gold was amongst the Bahnars equated to the smallest unit of barter in use among them, the hoe, whilst their highest unit was the buffalo; and that by a simple process based on the known relation existing in value between the hoe, the *muk*, the jar, and the buffalo, there was no difficulty in arriving empirically at the exact value in gold of a buffalo. We found also that the two higher units of weight the *picul*, and the *catty*, which in almost every case were found to be confined to the ordinary merchandise, were beyond reasonable doubt not originally multiples of the lower the *tael*, but were really natural units obtained by a totally different process; the *picul* being the amount which an average man can conveniently carry on his back, the *catty*, as seen especially in the case of the *neal* of Cambodia, being nothing more than the cocoa-nut shell used as the ordinary measure of capacity, as a gourd of a certain kind is employed at Zanzibar, as the hen's egg was employed by the Hebrews and also by the ancient Irish, as the cochlea or mussel shell was taken by the Romans as the basis of their measures of capacity, and as possibly the gourd itself under its name of *Kyathos* formed the lowest unit of capacity among the Greeks. We saw clearly that the catty has never become a weight-unit for precious metals among the Chinese, Annamites or Cambodians; the first named never having used any higher unit for such purpose than a bar of ten *taels*, and at the present day for the most part contenting themselves with the *tael* or ounce, whilst the two latter still use the *nên* or bar with its subdivisions into 10 *denhs*, or in other words, use as their highest monetary unit the tenfold of the *tael* or ounce. We likewise found that in Annam among the less advanced peoples there was considerable evidence to show that the *bat* or tical was originally the highest unit used for gold, and that this name *bat* was applied to weights of different amount; thus the *chi* which in commercial weight is only the quarter of a *bat*, is itself called the gold *bat*. The *bat* itself was the third of the *tael*. We also found the bar of silver, the common monetary unit at the present moment, equated to the buffalo, the common unit of barter among the Bahnars, and finally we had a distinct tradition that not so long ago the wild tribesmen who win the gold dust from the sands of their native brooks did not as yet even weigh the metal by means of the grains of maize which are now employed, but that they measured off a small rod of gold an inch long as the equivalent of a buffalo.

From all these facts it seems easy to trace the history of the development of weight standards in Further Asia; the first stage in trafficking in gold seems to be one purely by measure, then comes that of weighing by means of grains of corn, the weight in gold of one or more grains of corn being taken in the ordinary way of barter like other articles in the common scale of exchange. A multiple of the higher unit the *bat* was formed, possibly based on the slave as the multiple of the buffalo. This multiple is threefold of the *bat*, in that respect offering a strange analogy to the gold talent of Sicily, Magna Graecia, and Macedonia, which is the threefold of the Homeric ox-unit, and which, as I have conjectured, may have represented the value of a slave, as we certainly know as a fact that the highest unit in the Irish system, the *cumhal*, which represented the value of three cows or three ounces of silver, was neither more nor less than an *ancilla* (or ordinary *slave-woman*): the tenfold of this *tael* was the highest unit employed for either gold or silver by the most advanced peoples in this region, and is very well known as the *nên* or bar. All other goods were long appraised by measurement, the lowest unit of capacity being the cocoa-nut or the joint of the bamboo, the former known certainly to the Cambodians, the latter to the Chinese, whilst both are equally familiar to the Malays. The weight of the contents of the bamboo or cocoa-nut was presently taken, the standard employed being the *tael*, or highest unit yet employed for the precious metals. The weight of the contents would depend on the nature of the substance or liquid employed, for instance rice or some other kind of grain, or water. Thus the Chinese equate their catty to 16 taels; no doubt too convention came in at a later stage, and even though the contents might not actually weigh 16 taels, it was found convenient for practical purposes to regard some suitable multiple of the tael, such as 16, as the legal weight of the catty. A similar process was carried out in the case of the *picul* in the more advanced communities; a *load* was equated to the most convenient multiple of the catty, and as it was found that 100 catties gave a sufficiently near approximation to the ordinary load which a man could carry on his back, 100 catties were made the legal contents of the *picul* of trade.

We also learned how currency in baser metals such as copper or iron takes its origin. The history of the ordinary copper *cash* of the Chinese, which can be clearly traced step by step, brings us back to a time when a bronze knife, one of the most requisite articles of daily life, formed the ordinary small currency of the Chinese, just as the Greek *obolos* originally was an actual *spike* made of copper or iron, and just as the Bahnars of Annam still use the hoe as their lowest monetary denomination, an implement likewise similarly employed by the Chinese at an early period, as miniature hoes at one time used as true currency put beyond doubt. We also saw the negroes of Central Africa employing iron made into pieces ready to be cut into two hoes, and we also found those on the West Coast of Africa and the Hottentots employing bars of iron in a raw state, as a kind of currency. We also saw one most important feature possessed by all those in common, viz. the fact that in the determination of the value of the bar, the ingot, the piece of iron made in the shape of two hoes, and the bronze knife, not weight but linear measurement based on the parts of the human body, was the method invariably employed.

We then advanced to Western Asia and Europe and found everywhere alike the weight standards fixed by means of the seeds of plants. The process likewise was made perfectly plain. We did not find the highest denomination taken as the unit and the lowest reached by a long process of subdivisions, and finally for convenience sake described as consisting of so many grains of corn, as the brilliant French *savant* assumes in the case of the Assyrians: on the contrary we found that the bushel of Henry VII. was reached by first fixing the weight of the penny sterling by means of 32 grains of wheat, round and dry and "taken from the midst of the ear of wheat after the old laws of the land." Again the Irish Kelts did not say that the *unga* or ounce must contain so many *screapalls*, and each *screapall* so many *pingiuns*, but they proceeded in quite the reverse way first fixing the weight of the *pingiun* by eight grains of wheat. We may then well assume that such too was the process among Greeks, Romans, Arabs and Hindus. Brahmegupta, and the legislators quoted above support this view by starting always with the smallest unit. It is only when we come to the system of Babylon we are asked to reverse the process, to admit that the idea of weights began with corn, the very commodity of all others which, according to all the instances previously quoted, was the last to be valued by weight, and which even amongst ourselves at this present moment can hardly be said to be regarded as an article appraised by weight. But furthermore if the Assyrians regarded the Talent as their unit, and their lesser denominations as its subdivisions, why did not the maker of the weight mentioned above inscribe it as $\frac{3}{4}$ obol, or by some other term to indicate that it was essentially regarded as a fraction of a higher denomination, and not as a multiple of a lower? But the ancient Assyrian who made the weight must plainly have regarded it in the latter light, for otherwise he would not have engraved on it 22 *grains* $\frac{1}{2}$, actually resorting to the fraction of a grain. The only reasonable explanation of his conduct is that he was as firmly impressed with the idea that the basis of his system was the grain of corn (wheat) as were Brahmagupta, or Henry VII.'s parliament with the idea that the barley-corn and wheat-corn were the bases of their respective systems. If the objection be raised that the grains of corn were only devised in days long after the scientific fixing of weight standards, my answer is that if it was necessary to employ natural seeds as a means of determining the accuracy of scientifically obtained units, *à fortiori* it was necessary for mankind to have employed such seeds as their first step in the establishing of a system of weights.

No simpler idea connected with weight could have struck

the primitive mind. The difficulty experienced by savages in counting beyond 3 or 4 is met by them by the use of counters. We are all familiar with the use of *pebbles* or small stones among the Greeks and Romans. Our own word *calculate* is simply an adaptation of the Latin *calculare* to count by pebbles (*calculi*). Some nations, probably all, have been unable to form abstract names for their numerals, and the name of the concrete object which they habitually employed as a counter has become firmly embedded as a suffix in the names of their numerals. Thus the Aztec numerals end in *tetl*, a *pebble*, because they employed small stones as counters. Similarly the Malays whom we found weighing gold by means of grains of *padi* employ that word as a numeral suffix, because they employed grains of rice for their *calculations* or, to speak more accurately, *seminations*. In the case of this people we find coincident the most primitive forms of numeration and of weighing, both processes being carried on by means of the same simple instrument, which Nature put ready to hand in the corn which formed their daily sustenance.

If any one still maintains that the Indian Islander or Tapak of Annam learned the art of weighing by grains from the Chinese, and would maintain that the latter either invented for themselves or borrowed from Babylonia a scientifically devised weight system, I will go a step further and try to produce some evidence of the process by which weight standards are arrived at, by seeking instances in a region so isolated as to be beyond the reach of all suspicion of having borrowed from Babylon.

From what I have said above, we cannot expect to find any such community in the Old World. The New World on the other hand supplies us with what we desire. When the Spaniards under Cortes, conquered the Aztecs of Mexico, that people, although in a high state of civilization, had as yet no system of weights. In consequence of this want the Spaniards experienced some difficulty in the division of the treasure, until they supplied the deficiency with weights and scales of their own manufacture. There was a vast treasure of gold, which metal, found on the surface or gleaned from the beds of rivers, was cast into bars, or in the shape of dust made part of the regular tribute of the southern provinces of the empire. The traffic was carried on partly by barter, and partly by means of a regulated currency of different values. This consisted of transparent quills of gold dust, bits of tin cut in the form of T, and bags full of cacao containing a specified number of grains[1].

From this we get an insight into the first beginnings of weights. Some natural unit (and by natural I mean some product of nature of which all specimens are of uniform dimension) is taken, such as the quill used by the Aztecs. The average-sized quill of any particular kind of bird presents a natural receptacle of very uniform capacity. These quills of gold-dust were estimated at so many bags containing a certain number of grains. The step is not a long one to the day when some one will balance in a simple fashion quills of gold dust against seeds of cacao, and find how much gold is equal to a nut. Nature herself supplies in the seeds of plants weight-units of marvellous uniformity. If any one objects to my assumption that the Aztecs were on the very verge of the invention of a weight system, my answer is that another race of America, whose political existence ceased under the same cruel conditions as that of their Northern contemporaries, I mean the Incas of Peru, who were in a stage of civilization almost the same as that of the Aztecs, had already found out the art of weighing before the coming of the Spaniards, although they were inferior to the Mexicans in so far as they had not a well-defined system of hieroglyphic writing, nor of currency such as the latter possessed. Scales made of silver have been discovered in Inca graves[2]. The metal of which they are made shows that they were only employed for weighing precious commodities of small bulk.

Unfortunately I can find no record of weights having been found along with the silver scales in the Inca graves. If the weights were simply natural seeds, they would easily perish, or even if perfect when the tombs were opened, would be simply

[1] Prescott, *Conquest of Mexico*, p. 44.
[2] Prescott, *Peru*, p. 56.

regarded as part of the ordinary supply of food placed with the dead in the grave. But I forbear from laying the slightest stress on negative evidence of such a kind.

But beyond doubt we have on the American continent, far removed from connection with Asia, a series of facts closely harmonising with what we have found in Further Asia, and also among the peoples of Hither Asia, Europe and Africa. The Aztecs are still measuring gold, but the Incas have invented the balance. The Incas have no alphabet, the *quipus* as yet being their greatest advance towards a means of keeping a record of the past. It follows that it is possible for the human race to invent a system of weighing before it has made any advance in letters or science. Hence it is logical to infer that the civilized races of Asia and Europe could have discovered a means of weighing gold long before the Chaldean sages made a single step in their astronomical discoveries, or a single symbol of the cuneiform syllabary had as yet been impressed on brick or tablet.

Weights of various grains.

	grammes
Troy Grain	·064
Barley	·064
Wheat	·048
Rice	·036
Carob	·192 = 3 barley = 4 wheat
Lupin	·384 = 2 carobs
Maize (ordinary)	·128 = 2 barley
Ratti	·128 = 2 barley
Rye	·032 = $\frac{1}{2}$ barley

CHAPTER IX.

STATEMENT AND CRITICISM OF THE OLD DOCTRINES.

Nec Babylonios
Tentaris numeros.
Hor. *Carm.* I. 11. 2.

WE now proceed to the statement and criticism of the old doctrines of the origin of metallic currency and weight standards. To enter into an elaborate account of the various shades of doctrine held by the followers of Boeckh would be useless and wearisome, for as they all alike are agreed in starting from an arbitrary scientifically obtained unit, it matters not as far as my object is concerned. Certain metrologists lay down that Egypt borrowed her system from Babylon, whilst others[1] again declare that Egypt is the true mother of weight standards, and this battle is raging hotly at the present moment. Thus but recently Professor Brugsch has written a vigorous article (in the *Zeitschrift für Ethnologie*[2]) to prove that the Chaldeans borrowed their system from Egypt. But the Assyriologists were not prepared to assent to a doctrine which placed the Babylonians in an inferior position. Accordingly Dr C. F. Lehmann (*Zeitschrift für Ethnologie*, 1889, p. 245 *seqq.*) has made an elaborate defence of the original doctrine first propounded by Boeckh and developed and expounded by Dr Brandis and Dr Hultsch. This Assyrio-Egyptian struggle for pre-eminence has at present no importance for our enquiry, as it is based almost entirely on *à*

---

[1] Nissen, "Griechische und römische Metrologie" (Iwan Müller's *Handbuch der classischen Alterthumswissenschaft* I. 663 *seq.* or separately, Nordlingen, 1886).

[2] "Das älteste Gewicht," 1889, pp. 1—9, 34—43.

priori assumptions, although when we come eventually to deal with the question of efforts at systematization which arose at a later stage in the evolution of weight and measure standards, it will be necessary for us to examine the respective claims. At present we are engaged in searching for an historical basis, and as both the Assyriologists and Egyptologists alike unite in deriving all weights from a deliberate scientific attempt on the part of a highly civilized people, they are perfectly agreed in the principle, the soundness of which it is the object of the present investigation to test. The ablest exponent in this country of the German theory is Dr B. V. Head, who has given an admirable summary of the position of that school in his Introduction to his great work, *Historia Numorum* (p. xxviii.). To ensure a fair statement of the doctrine for the reader, it will be better for me to give here Mr Head's exposition in preference to any summary of my own, as any statement by the critic of the doctrine to be criticized is always liable to the suspicion of being *ex parte* and consequently inadequate. Such a suspicion is avoided by letting as far as possible our opponents state their position in their own words.

"For many centuries before the invention of coined money there can be no doubt whatever that goods were bought and sold by barter pure and simple, and that values were estimated among pastoral people by the produce of the land, and more particularly in oxen and sheep.

"The next step in advance upon this primitive method of exchange was a rude attempt at simplifying commercial transactions by substituting for the ox and the sheep some more portable substitute, either possessed of real or invested with an arbitrary value.

"This transitional stage in the development of commerce cannot be more accurately described than in the words of Aristotle, 'As the benefits of commerce were more widely extended by importing commodities of which there was a deficiency, and exporting those of which there was an excess, the use of a currency was an indispensable device. As the necessaries of Nature were not all easily portable, people agreed for purposes of barter mutually to give and receive some article which, while it was itself a commodity, was practically easy to handle in the business of life; some such article as iron or silver, which was at first defined simply by size and weight, although finally they went further and set a stamp upon every coin to relieve them from the trouble of weighing it, as the stamp impressed upon the coin was an indication of quantity.' (*Polit.* I. 6. 14—16, Trans. Welldon.)

"In Italy and Sicily copper or bronze in very early times took the place of cattle as a generally recognized measure of value, and in Peloponnesus the Spartans are said to have retained the use of iron as a standard of value long after the other Greeks had advanced beyond this point of commercial civilization.

"In the East, on the other hand, from the earliest times gold and silver appear to have been used for the settlement of the transactions of daily life, either metal having its value more or less accurately defined in relation to the other. Thus Abraham is said to have been 'very rich in cattle, in silver and in gold' (Gen. xiii. 2, xxiv. 35), and in the account of his purchase of the cave of Machpelah (Gen. xxiii. 16), it is stated that 'Abraham weighed to Ephron the silver which he had named in the audience of the sons of Heth, four hundred shekels of silver, current with the merchants.'

"As there are no auriferous rocks or streams in Chaldaea, we must infer that the old Chaldaean traders must have imported their gold from India by way of the Persian Gulf, in the ships of Ur frequently mentioned in cuneiform inscriptions.

"But though gold and silver were from the earliest times used as measures of value in the East, not a single piece of coined money has come down to us of these remote ages, nor is there any mention of coined money in the Old Testament before Persian times. The gold and silver 'current with the merchant' were always weighed in the balance; thus we read that David gave to Ornan for his threshing-floor [including oxen and threshing instruments] 600 shekels of gold by weight (1 Chron. xxi. 25).

"It is nevertheless probable that the balance was not called into operation for every small transaction, but that little bars of silver and of gold of fixed weight, but without any official mark (and therefore not coins) were often counted out by tale, larger amounts being always weighed. Such small bars or wedges of gold and silver served the purposes of a currency, and were regulated by the weight of the shekel or the mina.

"This leads us briefly to examine the standards of weight used for the precious metals in the East before the invention of money.

"*The metric systems of the Egyptians, Babylonians, and Assyrians.*

"The evidence afforded by ancient writers on the subject of weights and coinage is in great part untrustworthy, and would often be unintelligible were it not for the light which has been shed upon it by the gold and silver coins, and bronze, leaden and stone weights which have been fortunately preserved down to our own times. It will be safer, therefore, to confine ourselves to the direct evidence afforded by the monuments.

"Egypt, the oldest civilized country of the ancient world, first claims our attention, but as the weight system which prevailed in the Nile valley does not appear to have exercised any traceable influence upon the early coinage of the Greeks, the metrology of Egypt need not detain us long....

"The Chaldaeans and Babylonians, as is well known, excelled especially in the cognate sciences of arithmetic and astronomy. On the broad and monotonous plains of Lower Mesopotamia, says Professor Rawlinson, where the earth has little to suggest thought or please by variety the 'variegated heaven,' ever changing with the times and the seasons, would early attract attention, while the clear sky, dry atmosphere, and level horizon, would afford facilities for observations so soon as the idea of them suggested itself to the minds of the inhabitants. The records of these astronomical observations were inscribed in cuneiform character on soft clay tablets, afterwards baked hard and preserved in the royal or public libraries in the chief cities of Babylonia. Large numbers of these tablets are now in the British Museum. When Alexander the Great took Babylon, it is recorded that there were found and sent to Aristotle a series of astronomical observations extending back as far as the year B.C. 2234. Recent investigations into the nature of these records render it probable that upon them rests the entire structure of the metric system of the Babylonians. The day and night were divided by the Babylonians into 24 hours, each of 60 minutes, and each minute into 60 seconds—a method of measuring time which has never been superseded, and which we have inherited from Babylon, together with the first principles of the science of astronomy. The Babylonian measures of capacity and their system of weights were based, it is thought, upon one and the same unit as their measures of time and space, and as they are believed to have determined the length of an hour of equinoctial time by means of the dropping of water, so too it is conceivable that they may have fixed the weight of their *talents*, their *mina*, and their *shekel*, as well as the size of their measures of capacity, by weighing or measuring the amounts of water, which had passed from one vessel into another during a given space of time. Thus, just as an hour consisted of 60 minutes and the minute of 60 seconds, so the talent contained 60 minae, and the mina 60 shekels. The division by sixties or sexagesimal system, is quite as characteristic of the Babylonian arithmetic and system of weights and measures, as the decimal system is of the Egyptian and the modern French. And indeed it possesses one great advantage over the decimal system, inasmuch as the number 60, upon which it is based, is more divisible than 10.

"About 1300 years before our era the Assyrian empire came to surpass in importance that of the Babylonians, but the learning and science of Chaldaea were not lost, but rather transmitted through Nineveh by means of the Assyrian conquests and commerce to the north and west as far as the shores of the Mediterranean Sea. Let us now turn to the actual monuments. Some thirty years ago Mr Layard discovered and brought home from the ruins of ancient Nineveh a number of bronze lions of various sizes which may now be seen in the British Museum. With them were also a number of stone objects in the form of ducks[1]."

[1] The whole series of these ancient weights was some years ago subject to

From this double series of weights Mr Head infers that there were two distinct minae simultaneously in use during the long period of time which elapsed between about B.C. 2000, and B.C. 625. "The heavier of these two minae appears to have been just the double of the lighter. Brandis is probably not far from the mark in fixing the weight of the heavy mina at 1010 grammes, and that of the light at 505 grammes.

"It has been suggested that the lighter of these two minae may have been peculiar to the Babylonian, and the heavier to the Assyrian empire; but this cannot be proved. But nevertheless it would seem that the use of the heavy mina was more extended in Syria than that of the lighter, if we may judge from the fact that most of the weights belonging to the system of the heavy mina have in addition to the cuneiform inscription an Aramaic one.

"The purpose which this Aramaic inscription served must clearly have been to render the weight acceptable to the Syrian and Phoenician merchants who traded backwards and forwards between Assyria and Mesopotamia on the one hand, and the Phoenician emporia on the other.

"*The Phoenician traders.*

"The Phoenician commerce was chiefly a carrying trade. The richly embroidered stuffs of Babylonia and other products of the East were brought down to the coasts, and then carefully packed in chests of cedarwood in the markets of Tyre and Sidon, whence they were shipped by the enterprising Phoenician mariners to Cyprus, to the coasts of the Aegean, or even to the extreme West.

"Hence the Phoenician city of Tyre was called by Ezekiel (xxvii.) 'a merchant of the people for many isles.'

---

a careful process of weighing in a balance of precision by an officer of the Standard Department and the result was published by Mr W. H. Chisholme in the *Ninth Annual Report of the Warden of the Standards* 1874—5, where a complete list of all of them may be found.

All the more important pieces had however been weighed many years before, and it need only be stated that the results of the process of re-weighing under more favourable conditions are in the main identical with those formerly arrived at by Queipo and the late Dr Brandis.

"But the Phoenicians in common with the Egyptians, the Greeks and the Hebrews etc. with whom they dealt were at no time without their own peculiar weights and measures upon which they appear to have grafted the Assyrio-Babylonian principal unit of account or the weight in which it was customary to estimate values. This weight was the 60th part of the *manah* or mina.

"The Babylonian sexagesimal system was foreign to Phoenician habits. While therefore these people had no difficulty in adopting the Assyrio-Babylonian 60th as their own unit of weight or shekel, they did not at the same time adopt the sexagesimal system in its entirety but constituted a new mina for themselves consisting of 50 shekels instead of 60. In estimating the largest weight of all, the *Talent*, the multiplication by 60 was nevertheless retained. Thus in the Phoenician system as in that of the Greeks 50 shekels (Gk. *staters*) = 1 Mina, and 60 Minae or 3000 shekels or staters = 1 Talent.

"The particular form of shekel which appears to have been received by the Phoenicians and Hebrews from the East was the 60th part of the heavier of the two Assyrio-Babylonian minae above referred to. The 60th of the lighter for some reason which has not been satisfactorily accounted for seems to have been transmitted westwards by a different route, viz. across Asia Minor, and so into the kingdom of Lydia.

"*The Lydians.*

"'The Lydians,' says E. Curtius (*Hist. Gr.* I. 76), 'became on land what the Phoenicians were by sea, the mediators between Hellas and Asia.' It is related that about the time of the Trojan Wars and for some centuries afterwards, the country of the Lydians was in a state of vassalage to the kings of Assyria. But an Assyrian inscription informs us that Asia Minor, west of the Halys, was unknown to the Assyrian kings before the time of Assur-banî-apli, or Assurbanipal (circ. B.C. 666), who it is stated received an embassy from Gyges, king of Lydia 'a remote' country, of which Assurbanipal's predecessors had never heard the name. Nevertheless that there had been some sort of connection between Lydia and Assyria in ancient times is probable, though it cannot be proved.

"Professor Sayce is of opinion that the mediators between Lydia in the west, and Assyria in the east, were the people called Kheta or Hittites. According to this theory the northern Hittite capital Carchemish (later Hierapolis) on the Euphrates, was the spot where the arts and civilization of Assyria took the form which especially characterises the early monuments of Central Asia Minor.

"The year B.C. 1400 or thereabouts was the time of greatest power of the nation of the Hittites, and if they were in reality the chief connecting link between Lydia and Assyria it may be inferred that it was through them that the Lydians received the Assyrian weight, which afterwards in Lydia took the form of a stamped ingot or coin.

"But why it was that the light mina rather than the heavy one had become domesticated in Lydia must remain unexplained. We know however that one of the Assyrian weights is spoken of in cuneiform inscriptions as the '*weight of Carchemish.*' If then the modern hypothesis of a Hittite dominion in Asia Minor turn out to be well founded, the *weight of Carchemish* might by means of the Hittites have found its way to Phrygia and Lydia, and as the earliest Lydian coins are regulated according to the divisions of the Light Assyrian mina this would probably be the one alluded to.

"From these two points then, *Phoenicia* on the one hand and *Lydia* (through Carchemish), on the other, the two Babylonian units of weight appear to have started westwards to the shores of the Aegean sea, the heavy shekel by way of Phoenicia, the lighter shekel by way of Lydia."

So far I have thought it but right to give Mr Head's exposition *in extenso*, that the enquirer may be enabled to fully grasp the principles of the orthodox school, before we enter on any criticism of them. I shall now treat more summarily all that remains to be said.

Let us briefly state the peculiar doctrines of two leading continental metrologists. The veteran Dr Hultsch derives all standards of weight thus: The royal Babylonian cubit was based on the sun's apparent diameter; the cube of this measure gave the *maris*, the weight in water of one-fifth of which was the royal Babylonian talent, which was divided into 60 *manehs* (*minae*) and each mina in turn into 60 shekels. For silver and gold however they formed their standard by taking *fifty* shekels to form a mina[1]: thus after elaborating with such care a scientific system, they abandoned it as soon as they came to deal with the precious metals.

M. Soutzo[2] in a clever essay has maintained that all the weight systems both monetary and commercial of Asia, Egypt, Greece, come from one primordial weight the Egyptian *uten* (96 grammes), or from its tenth, the *kat* (9·60 grammes). He ascribes the origin of these weights to an extremely remote epoch not far perhaps from the time of the discovery of bronze in Asia, and the invention of the first instruments for weighing: he considers also that bronze *by weight* was the first money employed in Asia, Egypt, and Italy, and that everywhere the decimal system of numeration has preceded the sexagesimal.

The evidence which we have produced in the earlier part of this work has I trust convinced the reader that gold, not copper, was the first object to be weighed; M. Soutzo's assumption that the *uten* is the primordial unit is upset even for the Egyptians themselves by the passage already cited from Horapollo (p. 129).

The invention of coinage.

The evidence of both history and numismatics coincides in making the Lydians the inventors of the art of coining money. At first sight it may seem surprising that none of the great peoples of the East, whose civilization had its first beginning long ages before the periods at which our very oldest records begin, should have developed coined money, acquainted as they indubitably were with the precious metals, both for ornament and exchange. But a little reflection shews us that it has been quite possible for peoples to attain a high

---

[1] *Metrologie*[2], p. 393.
[2] *Étalons pondéraux primitifs et lingots monétaires* (Bucharest, 1884), p. 49.

degree of civilization without feeling any need of what are properly termed coins. Transactions by means of the scales are comparatively simple, and as a matter of fact we shall find hereafter that even after a coinage had been for centuries established, men constantly had recourse to the balance in monetary transactions, just as down to the present moment the Chinese, who have enjoyed a high degree of culture for several thousand years, still have no native currency but their copper cash, foreign silver dollars being the only medium in the precious metals, whilst all important monetary transactions are carried on by the scales and weights. I may here likewise point out incidentally that where the supply of the precious metals is only sufficient to meet the demand for personal adornment, the establishment of a coinage in those metals will naturally be slow, whilst on the other hand where there is so abundant a supply of the metals, that there is more than sufficient for purposes of personal use, the tendency to produce a coinage will be much greater. If we enquire what were the metalliferous regions of Asia Minor, we at once find that Lydia above all other countries was especially rich in gold, or rather a natural alloy of gold and silver. The wealth of two Lydian kings, Gyges and Croesus, which has been through the ages a proverb consisted of vast quantities of this metal, which the Greeks called *electron* (ἤλεκτρον) or *white gold* (λευκὸς χρυσός, Herodotus, I. 50). The ancients regarded it as almost a distinct metal, doubtless because from their imperfect methods they experienced the greatest difficulty in extracting the pure metal. The pure gold in circulation in Asia Minor must have come from the valley of the Oxus, or the Ural mountains. Thus Sophocles speaks of "the electron of Sardis and the gold of Ind[1]." Even in the time of Strabo (A.D. 21), the process was regarded as so difficult that the great geographer thinks it worth while to quote from Posidonius (flor. 90 B.C.), the description of how the separation of the metals was effected (III. 146). It is there-

---

[1] Soph. *Antig.* 1038 *seqq.*

κερδαίνετ', ἐμπολᾶτε τὸν πρὸς Σάρδεων
ἤλεκτρον, εἰ βούλεσθε, καὶ τὸν Ἰνδικὸν
χρυσόν.

fore natural to find in Lydia, the land of gold, the first attempts at coined money.

"So far as we have knowledge," says Herodotus[1], "the Lydians were the first nation to introduce the use of gold and silver coin."

This statement is fully borne out by the evidence of Xenophanes[2], and also by the coins themselves, although some writers, e.g. Th. Mommsen[3], have held that it was in the great cities of Ionia, Phocaea and Miletus that money was first coined. "From the little we know of the character of this people (the Lydians) we gather that their commercial instinct must have been greatly developed by their geographical position and surroundings, both conducive to frequent intercourse with the peoples of Asia Minor, Orientals as well as Greeks."

About the time when the mighty Assyrian empire was falling into decay, Lydia, under a new dynasty called the Mermnadae, was entering upon a new phase of national life.

"The policy of these new rulers of the country was to extend the power of Lydia towards the West, and to obtain possession of towns on the coast. With this object Gyges (who, according to the story told by Plato, was a shepherd who owed his good fortune to the finding of a magic ring in an ancient tomb, and who was the founder of the dynasty of the Mermnadae, circ. B.C. 700) established a firm footing on the Hellespont, and endeavoured to extend his dominions along the whole Ionian coast. This brought the Lydians into direct contact with the Asiatic Greeks.

"These Ionian Greeks had been from very early times in constant intercourse, not always friendly, with the Phoenicians, with whom they had long before come to an understanding about numbers, weights, measures, the alphabet, and such like matters, and from whom, there is reason to think, they had received the 60th part of the *heavy* Assyrio-Babylonian mina as their unit of weight or *stater*. The Lydians on the other hand had received, probably from Carchemish, the 60th of the *light* mina.

---

[1] I. 94.   [2] Pollux, IX. 83.
[3] *Histoire de la Monnaie Romaine*, I. 15.

"Thus then, when the Lydians in the reign of Gyges came into contact and conflict with the Greeks, the two units of weight, after travelling by different routes, met again in the coast towns and river valleys of Western Asia Minor, in the borderland between the East and the West.

"To the reign of Gyges, the founder of the new Lydian empire as distinct from the Lydia of more remote antiquity, may perhaps be ascribed the earliest essays in the art of coining. The wealth of this monarch in the precious metals may be inferred from the munificence of his gifts to the Delphic shrine, consisting of golden mixing cups and silver urns, amounting to a mass of gold and silver such as the Greeks had never before seen collected together." This treasure was called the Gygadas, and is described by Herodotus[1].

"It is in conformity with the whole spirit of a monarch such as Gyges, whose life's work it was to extend his empire towards the West, and at the same time to hold in his hands the lines of communication with the East, that from his capital Sardes, situated on the slopes of Tmolus and on the banks of the Pactolus, both rich in gold, he should send forth along the caravan routes of the East and into the heart of Mesopotamia, and down the river valleys of the West to the sea, his native Lydian ore gathered from the washings of Pactolus and from the diggings on the sides of Tmolus and Sipylus.

"This precious merchandize (if the earliest Lydian coins are indeed his) he issued in the form of oval-shaped bullets or ingots, officially sealed or stamped on one side as a guarantee of their weight and value. For the eastern or land-trade the *light* mina was the standard by which this coinage was regulated, while for the western trade with the Greeks of the coast the *heavy* mina was made use of, which from its mode of transmission we may call the *Phoenician*, retaining the name *Babylonian* only for the weight which was derived from the banks of the Euphrates."

To prevent misapprehension, it may be advisable to mention that the standards here termed *Phoenician* and *Babylonian* are not to be confounded with the *heavy* and *light* shekels already

---

[1] Herod. 1. 14.

mentioned, but are the standards derived from the latter specially for silver, in the ways shown a little lower down.

Modern analysis of electrum from Tmolus shows that it consists of 27 per cent. of silver and 73 per cent. of gold[1]. It consequently stood to silver in a different relation from that of pure gold. Thus while gold stood to silver as 13·3 : 1, electrum would stand at 10 : 1 or thereabouts. Mr Head considers that "this natural compound of gold and silver possessed some advantages for coining over gold. In the first place it was more durable, harder, and less liable to injury and waste from wear. In the second place it was more easily obtainable, being a natural product; and in the third place, standing as it did in the proportion of about 10 : 1 to silver, it rendered needless the use of a different standard of weight for the two metals, enabling the authorities of the mints to make use of a single set of weights, and a decimal system easy of comprehension and simple in practice" (p. xxxiv.). The second of these reasons is probably the true one, the first being a good example of the tendency of even the most able modern writers to ascribe to early times ideas which are only the outcome of a far later period. The idea of getting a metal which will be more durable in circulation is purely modern, and not even received by Orientals in modern times. Thus the gold mohurs of India down to their latest issue were of pure gold, free from alloy (in consequence of which they are still sought after by the native Hindu goldsmiths in preference to the English sovereign, as the addition of alloy makes the latter less easy to work up into jewellery).

I allude to this here because we shall find in the course of our enquiry that most of the errors into which metrologists have fallen, are the consequence of their failing to recognize the great gulf which is fixed between the habits and ideas of a primitive community, slowly evolving principles which are now part and parcel of the common heritage of civilization, and an era like our own, when all progress is effected by the development and application of scientific principles long since discovered.

[1] Hultsch, *Metrol.*² 579.

Electrum was thus coined on the same standard as silver, one *talent*, one *mina* and one *stater* of electrum being consequently equal to ten *talents*, ten *minae*, or ten *staters* of silver. The weight of the electrum stater in each district would depend therefore on the standard which happened to be in use there for silver bullion, or silver in the shape of bars or oblong bricks, the practice of the new invention of stamping or sealing metal for circulation being in the first place only applied to the more precious of the two metals, electrum representing in a small compass a weight of uncoined silver ten times as bulky and ten times as difficult of transport.

The invention was soon extended to pure gold and silver, and there is good reason to believe that by the time of Croesus (568—554 B.C.) both these metals were used for purposes of coinage in Lydia.

The Greeks begin to coin money.

The clever Greeks of Asia Minor, who formed the portal through which so many of the arts of the East reached the Western lands, were not slow to adopt, and by reason of their superior artistic taste to improve, the great Lydian invention. To the Ionic cities such as Phocaea and Miletus we must probably ascribe the credit of substituting artistically engraved dies for the rude Lydian punch-marks, and at a somewhat later period of inscribing them with the name or rather the initial of the people or potentate by whom they were issued.

The official stamps by which the earliest electrum staters were distinguished from mere ingots consisted at first only of the impress of rude unengraved punches, between which the lump or oval-shaped bullet of metal was placed to receive the blow of the hammer. Subsequently the art of the engraver was called in to adorn the lower of the two dies, which was always that of the face or *obverse* of the coin, with the symbol of the local divinity under whose auspices the currency was issued.

As our object is to deal with coins from the point of view of metrology, the short summary here given of the genesis of the art of coining will suffice for our purposes.

Weight standards.

"Silver was very rarely at this early period weighed by the same talent and mina as gold, but, according to a standard derived from the gold weight, somewhat as follows:—

Gold was to silver as 13·3 : 1. This proportion made it difficult to weigh both metals on the same standard. That a round number of silver shekels or staters might equal a gold shekel or stater, the weight of the silver shekel was either raised above or lowered below that of the gold. The *heavy* gold shekel weighed 260 grains Troy, being the double of the *light* gold shekel, which weighed 130 grains Troy (8·4 grammes).

THE SILVER STANDARDS DERIVED FROM THE GOLD SHEKEL[1].

I. From the *heavy* gold shekel of 260 grains:

$260 \times 13·3 = 3458$ grains of silver.
3458 grains of silver = 15 shekels of 230 grains each.

On the silver shekel of 230 grains the *Phoenician* or Graeco-Asiatic *silver* standard may be constructed:

Talent = 690,000 grains = 3000 staters (or shekels).
Mina = 11,500 grains = 50 staters.
Stater = 230 grains.

II. From the *light* gold shekel of 130 grains we get the so-called Babylonian or Persian standard:

$130 \times 13·3 = 1729$ grains of silver.
1729 grains of silver = 10 shekels of 172·9 grains each.

On the silver shekel or stater of 172·9 grains the *Babylonic, Lydian*, and Persian *silver* standard may be thus constructed:—

Talent = 518,700 grains = 3000 staters = 6000 sigli.
Mina = 8645 grains = 50 „ = 100 „
Stater = 172·9 grains = 1 „ = 2 „
Siglos = 86·45 grains."

[1] Head, *op. cit.* XXXVI.

210 STATEMENT AND CRITICISM OF THE OLD DOCTRINES.

It is desirable "to take note of the fact that in Asia Minor and in the earliest periods of the art of coining, (a) the heavy gold stater (260 grains) occurs at various places, from Teos northwards as far as the shores of the Propontis; (β) the light gold stater (130 grains) in Lydia (Κροίσειος στατήρ) and in Samos (?); (γ) the electrum stater of the Phoenician *silver* standard, chiefly at Miletus, but also at other towns along the west coast of Asia Minor, as well as in Lydia, but never however in full weight; (δ) the electrum and silver stater of the Babylonic standard, chiefly if not solely in Lydia; (ε) the silver stater of the Phoenician standard (230 grains) on the west coast of Asia Minor[1]."

Here we may call attention to the fact that whilst Miletus struck her electrum staters on the Phoenician *silver* standard (their normal weight being 217 grains), the Phocaeans always from the infancy of coining employed for their electrum the *gold* standard of the *heavy* shekel (260 grains). But the proper time for discussing why the Lydians, Milesians and Phocaeans all struck their electrum coins of various standards, will come further on in our enquiry.

The coin-standards of Greece Proper.

Before we attempt to examine into the connection of the Homeric talent or ox unit, and the ancient systems of the East, it will be advisable to get a clear view of the coin-standards found in actual use in historical times, and to understand the common doctrine of the derivation of the same. As gold was not coined in Greece Proper until a comparatively late period, owing doubtless to the fact that there was no great supply of it to be had, and that all of it was required to meet the demand for personal adornment, the entire early coinage of Greece (with some few exceptions to be presently noted) consisted of silver. These silver issues were all struck on either of two systems; (1) the Aeginean, or Aeginetic, and (2) the Euboic, the stater of the former weighing about 195 grains, that of the latter about 135—130 grains. But it is a fact of paramount importance that gold, whenever and

[1] Head, *op. cit.* xxxvi.

wherever coined in Greece, was always on the Euboic standard, and there is likewise every reason to believe that gold bullion in the days before gold was coined was computed according to the same standard. Such at least was undoubtedly the case at Athens, as we learn from Thucydides[1], where he describes the resources of Athens both in coined and uncoined metal, and in the gold plates which overlaid the famous chryselephantine statue of Pallas Athene, the masterpiece of Pheidias, and the glory of the Acropolis; and such also, as we shall see, was the case, in the days of Solon.

All ancient accounts are agreed in the statement that Aegina was the first place in Hellas Proper which saw the minting of money. That island was famous from old time as the meeting-place of merchants, and as such under its ancient name of Oenone was glorified by Pindar[2]. Its position rendered it a most convenient emporium, where the merchantmen of Tyre met in traffic the traders from both Peloponnesus and northern Greece. Tradition makes its population a very mixed one: "It was called Oenone," says Strabo, "in ancient times, and it was settled by Argives, Kretans, Epidaurians, and Dorians[3]." According to a fragment of Ephorus, to be referred to presently, it was owing to the barren nature of the soil that the natives turned to trade.

All Greek tradition is unanimous in representing Pheidon of Argos as the first to coin money in Hellas Proper, and to have done so at Aegina. Much obscurity enshrouds the history and the date of Pheidon, owing to the conflicting accounts of the historians. For our immediate purpose it would be quite sufficient to state simply that he cannot have lived later than 600 B.C., but in consequence of some prevailing doctrines with regard to the history of Greek weights being based on inferences (probably quite unwarrantable) which have been drawn from the statements given about this despot, we must take a more elaborate survey of the sources.

[1] Thuc. II. 13.
[2] Ol. I. 75: Nem. IV. 46.
[3] VIII. 375, ὠνομάζετο δ' Οἰνώνη πάλαι, ἐπῴκησαν δὲ αὐτὴν Ἀργεῖοι καὶ Κρῆτες καὶ Ἐπιδαύριοι καὶ Δωριεῖς.

Pausanias[1], writing about 174 A.D. says that the Pisaeans in the eight Olympiad (747 B.C.) brought to their aid Pheidon of Argos, who of all despots in Hellas waxed most insolent, and that along with him they celebrated the festival. But now comes the testimony of Herodotus[2], who was writing circ. 440 B.C., and who tells us (VI. 127) that when Cleisthenes the despot of Sicyon held the *svayamvara* for his daughter Agariste; amongst the suitors who came from all parts of Hellas, was "Leocedes, son of Pheidon, the despot of the Argives, Pheidon, who had made their measures for the Peloponnesians, and had of all Greeks waxed to the greatest pitch of violence, he who expelled the Elean presidents of the games and himself held the festival." There cannot be the slightest doubt that both Pausanias and Herodotus refer to the same tyrant, but the dates are irreconcileable. As Cleisthenes, the Athenian law-giver, was the son of Agariste, her wooing cannot have been much earlier than 560 B.C., and consequently Pheidon must have reigned at Argos shortly before 600 B.C.

Weissenborn (followed by Ernst Curtius) has sought to cut the Gordian knot by emending the text of Pausanias, thus reading 28th instead of 8th Olympiad, which would make Pheidon help the Pisaeans in the year 668 B.C. But even this drastic remedy is hardly sufficient to meet the requirements of the statement of Herodotus.

Our earliest authority for the tradition that Pheidon coined at Aegina is a passage of Ephorus preserved by Strabo (VIII. 376)[3]: "Ephorus says that in Aegina silver was first struck by Pheidon; for it had become an emporium, inasmuch as its population, owing to the barrenness of the land, engaged in maritime trade; whence trumpery goods are called Aeginean ware." According to another passage of Strabo, which may be likewise from Ephorus, as it comes at the end of a long statement, the first part of which Strabo expressly declares is taken from

---

[1] VI. 22. 2, 'Ολυμπιάδι μὲν τῇ ὀγδόῃ τὸν 'Αργεῖον ἐπήγαγον Φείδωνα τυράννων τῶν ἐν Ἕλλησι μάλιστα ὑβρίσαντα κ.τ.λ.

[2] Φείδωνος δὲ τοῦ τὰ μέτρα ποιήσαντος τοῖς Πελοποννησίοισι καὶ ὑβρίσαντος κ.τ.λ.

[3] Ἔφορος δ' ἐν Αἰγίνῃ ἄργυρον πρῶτον κοπῆναί φησι ὑπὸ Φείδωνος, ἐμπόριον γὰρ γενέσθαι, διὰ τὴν λυπρότητα τῆς χώρας τῶν ἀνθρώπων θαλαττουργούντων ἐμπορικῶς, ἀφ' οὗ τὸν ῥῶπον Αἰγιναίαν ἐμπολὴν λέγεσθαι.

that writer: ("They say) that Pheidon of Argos, who was tenth in descent from Temenus, and who surpassed his contemporaries in his power, whence he recovered the whole of the inheritance of Temenus, which had been rent into several parts, and that he invented the measures which are called Pheidonian and weights and stamped currency, both the other kind and that of silver." It must be carefully observed that this is the only ancient passage which says a word about the invention of *weights* by Pheidon. If this statement can be taken as trustworthy we might very well conclude that Pheidon was the person who introduced the decimal principle and made 10 silver pieces instead of 15 equivalent to the gold stater. If however this is an addition of Strabo[1], who wrote about A.D. 1—21, and whose account of Greece Proper is the most defective portion of his great work, we cannot let this passage weigh against that already given from Herodotus, who is perfectly silent as regards the invention of *weights*. Furthermore there is the fact that Strabo does not venture to describe the *weights* as called *Pheidonian*, but carefully limits that appellation to the measures as we find also to be the case with Pollux, when he is describing various kinds of vessels: " and likewise a Pheidon would be a kind of vessel for holding oil, deriving its name from the Pheidonian measures respecting which Aristotle speaks in his Polity of the Argives[2]." Here again we find a clear mention of the Pheidonian measures, coupled with the high authority of Aristotle's treatise on the Constitution of Argos in his great "Collection of Polities," formed to serve as the material from which to build his great philosophic work on Politics.

There is again no mention of Pheidonian *weights* in the newly found Polity of the Athenians (which seems beyond doubt the same as that known to the ancients under the name of Aristotle), where it is stated that "in his (Solon's) time the

---

[1] Strabo VIII. 358, Φείδωνα δὲ τὸν Ἀργεῖον, δέκατον μὲν ὄντα ἀπὸ Τημένου, δυνάμει δὲ ὑπερβεβλημένον τοὺς κατ' αὐτόν, ἀφ' ἧς τήν τε λῆξιν ὅλην ἀνέλαβε τὴν Τημένου διεσπασμένην εἰς πλείω μέρη, καὶ μέτρα ἐξεῦρε τὰ Φειδώνια καλούμενα καὶ σταθμοὺς καὶ νόμισμα κεχαραγμένον τό τε ἄλλο καὶ τὸ ἀργυροῦν.

[2] Pollux Onom. X. 179, εἴη δ' ἂν καὶ Φείδων τι ἀγγεῖον ἐλαιηρόν, ἀπὸ τῶν Φειδωνίων μέτρων ὠνομασμένον, ὑπὲρ ὧν ἐν Ἀργείων πολιτείᾳ Ἀριστοτέλης λέγει.

measures (at Athens) were made larger than those of Pheidon" (c. 10)[1]. Although the writer refers to the Aeginetic coin-weights in the next clause, he does not refer to them as the Pheidonian.

Now let us pass on to a remarkable passage in the *Etymologicum Magnum* (s. v. ’Οβελίσκος).

"First of all men Pheidon of Argos struck money in Aegina; and having given them (his subjects) coin and abolished the spits, he dedicated them to Hera in Argos. But since at that time the spits used to fill the hand, that is the grasp, we, although we do not fill our hand with the six obols (spits) call it a *grasp full* (δραχμή) owing to the *grasping* of them. Whence even still to this day we call the usurer the spit-*weigher*, since by weights the men of old used to hand (money) over[2]." The writer of this passage evidently regards Pheidon as the first inventor of the art of coining but not of *weight* standards.

Finally the Parian Marble recounts that, "Pheidon the Argive confiscated the measures...and remade them and made silver coin in Aegina[3]." Such then is the body of evidence which we possess, all pointing to Aegina as the first place in Greece which saw a mint set up, and to Pheidon of Argos as the first to establish that mint. As we have pointed out above we have nothing but a very dubious statement of Strabo (which is coupled with another most certainly wrong, *i.e.*, that Pheidon was the inventor of every other kind of money as well as silver) as regards the invention of weights by Pheidon, although from the passage in Herodotus already quoted, metrologists one after another have assumed that the measures (μέτρα) meant a *metric system* in the modern sense, and have not hesitated to

---

[1] This enables us to understand why it was that in the truce at Pylus it was stipulated (probably by the Spartans) that they should be allowed to send in 2 *Attic* (not Peleponnesian) *choenikes* of barley meal for each of their men daily. By this arrangement the beleaguered men got a larger ration.

[2] πάντων δὲ πρῶτος Φείδων ’Αργεῖος νόμισμα ἔκοψεν ἐν Αἰγίνῃ· καὶ δοὺς τὸ νόμισμα καὶ ἀναλαβὼν τοὺς ὀβελίσκους, ἀνέθηκε τῇ ἐν ῎Αργει ῞Ηρᾳ, ἐπειδὴ δὲ τότε οἱ ὀβελίσκοι τὴν χεῖρα ἐπλήρουν, τουτέστι, τὴν δράκα, ἡμεῖς, καίπερ μὴ πληροῦντες τὴν δράκα τοῖς ἐξ ὀβόλους δραχμὴν αὐτὴν λέγομεν παρὰ τὸ δράξασθαι.

[3] Φείδων ὁ ’Αργεῖος ἐδήμευσε τὰ μέτρα...καὶ ἀνεσκεύασε καὶ νόμισμα ἀργυροῦν ἐν Αἰγίνῃ ἐποίησεν (l. 30).

build on this somewhat crazy foundation an elaborate Aeginetic system of weights and measures intimately related to each other.

We are then probably justified in assuming that Pheidon coined silver at Aegina. The numismatic evidence coincides with the literary authorities. The coins of Aegina are well known, for from first to last the symbol of the sea tortoise (χελώνη, from which they are called in vulgar parlance *tortoises*) is found on them. Why Pheidon set up his mint in Aegina instead of in his own city of Argos is not very difficult to understand. Argos was an inland town remote from the highways of commerce, and little in contact with the merchants of the Levant. On the other hand Aegina stood at the portal of central Greece, intercepting the trade of Athens and Corinth; in later days Pericles called it the "eyesore of the Piraeus." It would be probably here that the Greeks first saw the new invention of the East in the hands of the foreign traders, and it would be here, in a great emporium, that the need of a currency would be most felt. In an inland city like Argos or Sparta bars of bronze or iron would serve well for the small commercial transactions of a very primitive society, as we know that the iron currency actually did at Sparta in historical times. E. Curtius suggested (*Numism. Chron.*, 1870) that the tortoise on the Aeginetan coins, which is the symbol of Ashtaroth who was the Phoenician goddess both of the sea and of trade, may be an indication that the mint was set up in the temple of Aphrodite, which overlooked the great harbour of Aegina. Whilst his hypothesis as regards the origin of the tortoise type on the coin is probably wrong, it is quite possible that the coins were first struck in some temple, as we know that the great shrines of the ancient world served as banks and treasuries, as for example the temple of Athena at Athens, that of Apollo at Delphi, and that of Juno Moneta at Rome. The temple priests of Delphi and other rich shrines had at their command large stores of the precious metals, which in the earliest times doubtless were in the shape of small ingots or bullets, such as the gold talents mentioned in the Homeric Poems.

The temple shrines of Delphi and Olympia, Delos and Dodona were centres not merely of religious cult, but likewise of trade and commerce, just as the great fairs of the Middle Ages grew primarily out of the feast day of the local saint, merchants and traders taking advantage of the assembling together of large bodies of worshippers from various quarters to ply their calling and to tempt them with their wares. The temple authorities encouraged trade in every way; they constructed sacred roads, which gave facility for travelling at a time when roads as a general rule were almost unknown, and what was just as important, they placed these roads and consequently the persons who travelled on them under the protection of the god to whose temple they led in each case, thus affording a safe conduct to the trader as well as the pilgrim; again at the time of the sacred festivals all strife had to cease, the voice of war was hushed, and thus even amidst the noise of intestine struggles and international strife, peace offered a breathing space for trade and commerce. Hence the probability is considerable that the art of minting money, that is, of stamping with a symbol the ingots or *talents* of gold or silver which had circulated in this simple form for centuries, first had its birth in the sanctuary of some god.

On the whole then we may assume that the bullet-shaped coins of Aegina, which are undoubtedly the earliest coins of Greece Proper, are the Pheidonian currency mentioned in the ancient authors and on the Parian Marble. As silver was probably not at all plenty at Argos, but was brought to Aegina by the traders, Pheidon had every motive for minting at Aegina instead of at his own capital. The fact that the Romans struck silver coins in Campania before they issued any at Rome affords a curious parallel. A local supply of the metal offers the explanation in each case. "It may be also positively asserted that none of the Aeginetan coins are older than the earliest Lydian electrum money, and that consequently the date of the introduction of coined money into Peloponnesus must be subsequent to circ. 700 B.C. It follows that Pheidon was not the inventor of money, for already before his time all the coasts and islands of the Aegean must have been acquainted with the pale yellow electrum coins of Lydia and Ionia[1]."

[1] Head *op. cit.* xxxviii.

What then was the standard on which these early coins of Aegina were struck?

The heaviest specimens of these Aeginetan staters or didrachms weigh over 200 grains Troy, but these seem somewhat exceptional. The best numismatic authorities are agreed in setting the normal weight at 196 grains Troy; the drachm consequently weighs 98 grains, and the obol about 16 grains. The origin of this standard has caused much difficulty to metrologists. For it is not the standard of the Babylonian gold shekel of 130 grains, nor of the Babylonian silver shekel of 172 grains, nor again that of the Phoenician silver shekel of 230 grains. Various solutions have been proposed. Brandis[1] regards it as a raised Babylonian silver standard, 172·9 to 196 grains. Mr Head regards it as the reduced Phoenician standard; "The weight standard which the Peloponnesians had received in old times from the Phoenician traders had suffered in the course of about two centuries a very considerable degradation[2]." Others, like Mr Flinders Petrie (Encyclop. Britannica, *Weights and Measures*), regard it as Egyptian in origin. According to Herodotus (II. 178) the Aeginetans were on terms of friendly intercourse with Egypt; furthermore weights of this standard have been found in Egypt.

Again, Dr Hultsch (*Metrol.*[2] p. 188) regards it as an independent standard midway between the Babylonian silver standard (172·9 grs.) on the one hand, and the Phoenician silver standard (230 grs.) on the other, the old Aeginetan silver mina being equivalent in value to six light Babylonian shekels of gold ($130 \times 6 = 780$ grs. $= 10300$ grs. of silver), assuming that in Greece as in Asia Minor gold was to silver as $13\cdot3 : 1$.

All these theories labour under serious difficulties. Brandis' theory was overthrown easily as soon as attention was called to the well-defined heavy series of Aeginetic coins, he having been led to his opinion by a comparison of the heaviest specimen of the Babylonian standard with the lightest of the Aeginetic. Here incidentally we may call the readers' attention to the fact that in numismatics the weight of the heaviest specimens of any series must be regarded as the true index of

---

[1] *Op. cit.* 153. [2] *Op. cit.* XXXVIII.

the normal weight, for whatever may have been the inclination to mint coins of a weight lighter than the proper standard, we may rest assured that the ancient mint-master was no more inclined than his modern representative to put into coins of gold or silver a single grain more than the legal amount. Hence it is a most faulty and fallacious method when dealing with coin weights to take the average of a certain number of specimens as the true standard. Out of 30 specimens 29 may have lost more or less in weight by wear, whilst one may be a *fleur de coin*, perfect as at the moment when it left the die. No one can doubt that the evidence of that single coin as regards the standard is worth far more than that of all the remaining 29 examples. I have thought it well to call attention to this question of method as the vicious principle of arriving at standards by taking the average is still found in works of men of great eminence.

Next let us consider the probability of the derivation of the Aeginetic standard from Egypt. The fact that weights of like standard have been found in that country, although superficially plausible, in reality is of little force as evidence of borrowing. For unless we find that the Egyptians used those weights for weighing *silver*, even the *prima facie* case breaks down at once. As a matter of fact there is no evidence up to the present that these weights were so employed, although there is some evidence of their being employed for gold (Flinders Petrie, *op. cit.*). But even granting that the Egyptians used the same standard as the Aeginetans for silver, it does not at all follow that there has been borrowing on either side. On the principle laid down below it will be seen that it is quite possible for two peoples to evolve a like *silver* standard perfectly independently of each other. But the real difficulty which besets the theory of an Egyptian origin is that if the Aeginetans were to borrow their standard from abroad, the people from whom they would in all probability have obtained it were not the Egyptians, with whom they had but slight relations directly, but rather the Phoenicians, with whom they were in constant intercourse.

It cannot be proved that at any time the Egyptians were a

maritime people trading round the coasts of Greece. There was undoubtedly intercourse between Greece and Egypt, but that intercourse was through the medium of the shipmen of Tyre. Why should then the Aeginetans adopt a standard from abroad which differed from that of the Phoenicians with whom they were in constant commercial relations? Again, if there is any connection between the importation of weight standards and the commencement of coinage, it may be urged that whilst it was from the Phoenicians the Aeginetans learned the art which had been originated in Asia Minor, or at all events from the Greeks of the coast of Asia Minor who coined electrum money on the Phoenician standard, we ought naturally to find the Greeks of Aegina using this standard for their earliest coinage rather than a standard borrowed from Egypt, which most certainly was very backward in developing the art of coining, seeing that it was not until after the conquest of that country by Alexander the Great (B.C. 330) that money was there struck for the first time[1].

Passing by for the moment Mr Head's view, let us next deal with that of Dr Hultsch. This theory has the great merit of granting that the Greeks were capable of evolving a *silver* standard for themselves from a knowledge of the relative value of gold and silver, whilst the other theories assume that they borrowed blindly ready-made standards, which they for some unknown reason either raised according to Brandis, or degraded according to Head. But Dr Hultsch is met by two crucial difficulties. (1) Why should the Aeginetans have taken six light Babylonian shekels of gold and arbitrarily made them the basis of their new silver standard? (2) But the fatal objection is that whereas Hultsch's theory depends on gold being to silver in the same relation (13·3 : 1) in Greece Proper as it was in Asia Minor, as a matter of fact it can be proved that the precious metals there stood in a very different relation to each other. In the *Journal of Hellenic Studies*, 1887, I gave some reasons for believing that in early times gold was to silver in Greece in the relation of 15 : 1. For whilst gold was plentiful in Asia,

[1] Of course it is quite possible that the Persians issued coins in Egypt after their conquest, but these coins cannot be regarded as really Egyptian.

at no place in Greece Proper were there auriferous deposits. Hence it is probable that gold had to silver a higher relative value in Greece than it had in Asia. Certain archaeological discoveries recently made at Athens add great strength to the view which I then put forward. At a meeting of the Berlin Academy of Science in 1889 Dr Ulrich Köhler discussed certain fragments of inscriptions which refer to the famous statue of Athena, wrought in gold and ivory by Pheidias for the Parthenon. By combining with a fragment published by M. Foucart (*Bullet. de Corresp. Hell.* 1889, p. 171), another fragment previously copied by himself, Dr Köhler arrived at the result that the fragments relate to the purchase of materials for the construction of the statue, that is of gold and ivory. The gold purchased is described both according to its weight and according to the price ($\tau\iota\mu\acute{\eta}$) paid for it in Attic silver currency (whilst the ivory is only described by the value or price). The sum paid for gold amounted to 526·652 drachms, 5 obols, the weight of the gold being 37·618 drachms: from this we learn that the relative value of gold to silver at that time was as 14 : 1. According to Thucydides (II. 13), forty talents of gold were used in the making of the statue, whilst according to the more explicit statement of Philochorus the amount was forty-four. The image was dedicated at the great Panathenaic festival of the year 438 B.C. As not more than 10 to 11 talents of gold were used in the three years to which the fragments refer, Köhler draws the inference that the construction of the statue commenced in the same year as that of the Parthenon (447 B.C.), and that Pheidias was engaged on his great work for fully nine years.

We thus know now the relative value of silver and gold in Attica about 450 B.C. But we must not regard this as the relation which existed at earlier times. It was only after the Persian wars that Athens had got possession of the island of Thasos with its rich gold mines, and the equally rich districts on the Thracian coast. The fact of her coming into the possession of such wealthy gold-producing regions must have materially lowered the price of gold in Athens. We know how the development of the mines of Pangaeum by Philip of Macedon in the following century lowered the value of gold throughout Greece, for by the time of Alexander the relative value of the two precious metals was as 10 : 1. In the sixth century B.C. gold was so scarce in Greece that when the Spartans wanted to make a dedication in gold they had to send to Asia to obtain a sufficient supply of the metal[1]. Hence if we conclude that in earlier times the relative value of gold to silver in Greece proper was as 15 : 1, we shall not be far from the truth. At all events it is put beyond doubt that the relation was higher than that of 13·3 : 1, and accordingly Dr Hultsch's theory of the origin of the Aeginetic silver standard, which is based on that relation falls at once to the ground, unless he can shew that such a standard, based on six light gold Babylonian shekels had been previously fixed in Asia or Egypt, and thence adopted by the Greeks without any regard to the relative value existing in Greece itself between the precious metals. But as a matter of fact Dr Hultsch does not make any such attempt. Thus this essay at a solution breaks down.

On the other hand if we make the very slight and very probable assumption that the early Greeks had formed a definite idea of the relative value of gold and silver, which they would have determined exactly on the same principle as they would arrive at a notion of the relative value of any other two commodities, which they were in the habit of giving and taking in exchange, that is by the simple principle of supply and demand, we shall find a ready solution without having to resort to either Egypt or Babylon. If gold was to silver as 15 : 1 in Greece, it follows that the Homeric talent, the earliest Greek standard, being about 135 grains, ten silver pieces of 202 grains each would be equivalent to *one* gold unit

$$135 \times 15 = 2025 \text{ grs. of silver.}$$
$$\frac{2025}{10} = 202\text{·}5 \text{ grs. of silver.}$$

This gives a singularly close approximation to the weight of the existing coins of the Aeginetic standard of the earliest and heaviest kind. Taking the Homeric talent at 130 grains of

[1] Herod. I. 62.

gold, by the same process we obtain 10 silver pieces each of the weight of 195 grains ($130 \times 15 = 1950$; $1950 \div 10 = 195$ grs.

The second standard which we find in Greece at the beginning of the historical epoch was the Euboic. This standard was used for both *silver* and *gold*. The ordinary account of its origin is as follows: "From Ionia possibly through Samos the Euboeans imported the standard by which they weighed their silver. This standard was the light Assyrio-Babylonian gold mina with its shekel or stater of about 130 grains. The Euboeans having little or no gold transferred the weight used in Asia for gold to their own silver, raising it slightly at the same time to a maximum of 135 grains, and from Euboea it soon spread over a large part of the Greek world by means of the widely extended commercial relations of the enterprising Euboean cities. This may have taken place towards the close of the eighth century and before the war which broke out at the end of that century between Chalcis and Eretria, nominally for the possession of the fields of Lelantum, which lay between the two rival cities[1].

This Euboic standard of 135—130 grains is seen at once to be identical in weight with the Homeric talent.

Several difficulties (irrespective of the fact that there was no need for the Greeks to borrow from Asia a standard which they themselves already possessed from very early times) meet this theory.

(1) If the Euboeans derived their standard from Ionia why did they not rather adopt the Phoenician standards, on which we have already seen the great Ionian cities based their coinages of gold, silver, and electrum? Some very early electrum coins found at Samos (Head, *op. cit.* XLI.), have suggested that that island formed the link. "The theory," says Mr Head, "that Samos was the port whence the Euboeans derived the gold standard subsequently used by them for silver,

---

[1] Head, *op. cit.* p. XL. Professor Percy Gardner (*Types of Greek Coins*, p. 2), regards the Euboic standard as 130, which he thinks was raised to 135 grs. by Solon when the latter introduced (as he supposes) the Euboic system at Athens.

rests upon the weight of some very early electrum coins (about 44 grs.) which have been found in the island of Samos, and of the earliest Euboean coins, Euboea and Samos having been two of the greatest colonizing and maritime powers of the Aegean Sea. Thus I think we may account for the fact that the towns of Euboea, when they began to strike silver money of their own, naturally made use of the standard which had become from of old habitual in the island, precisely in the same way as Pheidon in Peloponnesus struck his first silver money on the reduced Phoenician standard which was prevalent at the time in his dominions." But as a matter of fact the recognized Samian coins are of the Phoenician standard (220 grs.) in its slightly reduced state as found at Miletus (Head, *op. cit.* 515). This being so it would indeed be strange if the Euboeans from occasionally coming in contact with Lydian coins at Samos would have adopted that standard in preference to that in use in the great cities of Ionia with which their commerce directly lay.

(2) Why did the Euboeans take the Lydian *gold* standard of 130 grs. for their own electrum and silver instead of the Lydian *silver* standard of 172·9 grs.? According to Mr Head's view, as we have seen above, the early Lydian electrum was struck on the standard of 172 grs. (the so-called Babylonian silver) when meant for circulation in the interior of Asia Minor, but on the Phoenician standard for circulation in trade with the Greeks of the coast of Ionia.

(3) We may ask the question, why did the Euboeans if they were taking over a ready-made standard which had no relation to any standard which they themselves already possessed, adopt the *gold* standard of 130 grs. instead of the electrum and silver standard which was in use among all the Greek cities with which they traded?

We can now conveniently revert to the theory that the Aeginetan *silver* standard was a reduced Phoenician. Much has been written about *degradation of coin weights* and *reduced standards*. It may be therefore well to clear our notions on the subject by asking ourselves what do we mean by such terms. Both the terms and the process are equally familiar to those at all acquainted with the history of mediaeval coinage.

The king then controlled, as for instance in England, the mintage. If the sovereign thought fit to reduce the amount of silver in the groat from 80 to 72 grains his subjects had no alternative but to take the new and lighter pieces as equivalent to four pennies sterling. The sovereign thus was able to relieve an exhausted treasury, making a considerable profit off every groat and penny put into circulation. Again, the impecunious monarch might resort to another method of making a profit, by debasing the coinage, and might issue one such as the fourth of Henry VIII., of exceeding base silver, and again his subjects could simply grumble and take the new money. These groats and pennies passed as such within the realm, but when the question of foreign exchange came, the matter assumed an entirely new complexion. Would a shrewd Flemish merchant from Antwerp accept a base or a reduced English groat at the same rate for which it passed current in England? Of course he did no such thing, and the scales were at once called into use, and the silver changed hands not by tale, but by *weight*. Now the condition under which such a degradation or debasing of the coinage as we have described can take place is that a state or country shall be of such considerable magnitude that it has room within its own borders to employ a large amount of coin in internal trade without much necessity of external commerce. Did such conditions exist among the Greek states of antiquity? There is another condition, namely, sovereign power vested in the hands of a monarch possessed of unlimited authority, who has a direct personal interest in the profit to be made from the degradation of the coinage, and who has power sufficient to enable him to force his debased coinage on a reluctant people. Did such conditions exist in any of the Greek states of antiquity? Nowhere in Greece Proper do we find them fulfilled, but if we turn to Sicily we get a good example of the practice so often followed in after centuries by the mediaeval monarchs. The tyrant Dionysius there put an arbitrary value on gold in relation to silver: for although this relation was probably not more than 12 : 1, this despot raised it perforce to 15 : 1[1]. He

---

[1] Head, *Coinage of Syracuse*, p. 71.

also issued a coinage of tin, according to Aristotle[1], which he perhaps forced his subjects to take as equivalent to silver coins of like size. In later years again when Timoleon liberated Syracuse and the democracy was once more restored, the state issued a coinage of electrum instead of that of pure gold, which had previously been in currency, by this means making a profit of 20 per cent.[2] It is hardly necessary to point out that whilst this coinage of Dionysius might pass for an artificial value within the dominions of Syracuse, the moment a Syracusan came to make payment to a foreign merchant, its factitious value vanished and the transaction took place according to the current value of the metals. So as long as the English penny remained of good weight and quality it found ready currency on the continent, and the potentates of Flanders issued numerous imitations of them known as *esterlings*, but when the English silver penny became debased all foreign imitations ceased[3]. Now the Greek states of Greece Proper were very small in extent, and seldom had a very strong central authority. The area being limited it was absolutely necessary for them to have constant dealings with their neighbours. It would have been difficult for any government in republican times to have forced on its citizens a debased silver currency, and even had this been possible, any benefit derived therefrom would have been counterbalanced by the great drawback arising to trade. If Athens had reduced her famous "Owls" or as they were otherwise called "Maidens" (from the head of Pallas Athena), by five grains, her credit would have suffered and her merchants have gained nothing by it, as the balance would have been at once resorted to, and allowance would have had to be made on each coin of the new debased standard. We who live in modern times are too apt to forget the readiness with which men in older days had resort to the scales, although at this moment large transactions in gold between bankers and financiers are carried out by weight. Only so late as the beginning of this century, when the gold coinage of the country was in a wretched state, every farmer and

[1] Arist. *Oeconomica*, II. 21.   [2] Head, *op. cit.* p. 26.
[3] Chautard, *Imitations des monnaies au type esterling* (Nancy, 1871).

trader went to fairs in Ireland equipped with a pocket balance (which was adjusted for the guinea, half-guinea, sovereign, half-sovereign, and gold seven-shilling-piece).

It is difficult then to see what it would have availed the Aeginetans to have reduced the standard which they are supposed to have got from the Phoenicians.

Their island state was of diminutive proportions; they devoted themselves almost entirely to traffick by sea, their island was an emporium where strangers resorted. In all dealings with the Phoenicians they would have to pay a drawback on their debased coin; for the cunning Phoenician or Ionian was not likely to be beguiled into taking staters of 200 grs. as equivalent to 230 grs. It is plain therefore that when we find divergencies of standard these are not due to mere *degradation*, but to some far more practical consideration, and this will be seen all the more clearly when we shall find that whilst we have divergencies in *silver* standards, the gold standard which was in use in Greece from Homeric times down to the Roman Conquest remains almost absolutely without variation. But there are other and stronger objections against the Phoenician origin of the Aeginetic standard.

Now if we accept the doctrine that the Greeks received their coin-standards across the sea from Asia, the *Aeginetic* from the Phoenician traders whose commerce lay with Aegina and Peloponnesus, the *Euboic* on the other hand from Lydia by way of the great Ionian cities on the coast of Asia Minor, we become involved in a serious difficulty. At the time represented in the Homeric Poems, there is not as yet a single Greek colony on the coast of Asia Minor[1]. Miletus, destined to be in after years the Queen of Ionia, and to be one of the greatest centres of Hellenic commerce and culture, is as yet known only as the city of the barbarous-speaking Carians[2]. Yet we find the Greeks represented in these self-same poems as already in possession of a standard for gold identical with the light Babylonian or Lydian gold shekel (130 grs.). But again we find from the same source that the Greeks were

[1] Mr D. B. Monro, *Historical Review*, January, 1886.
[2] *Il.* II. 867.

already in full commercial intercourse with one Asiatic people, but not a people who could serve as a bridge between Lydia and Euboea. Everywhere in the Homeric Poems we meet the shipmen of Tyre, who are represented as bringing the products of the skilled artists of Sidon, beautiful cloths, and cunningly wrought vessels of silver, articles of jewellery, necklaces[1] set with amber (perhaps brought from the coasts of the Baltic), and now and then as chance arose, kidnapping women and children to sell as slaves in the marts of the Mediterranean[2].

If the Hellenes had got their standard from an Asiatic source, it must have been the heavy gold shekel of 260 grains, which the Phoenicians employed, and consequently the Homeric Talent would have weighed 260 instead of 130 grains, or on the other hand if it be supposed that the Greeks might borrow and use for their own *gold* a standard used only for *silver* in Asia, the Homeric Talent ought to have weighed 225 grains, that is the Phoenician silver standard, which, as we have seen, it certainly did not.

A further difficulty arises in reference to the *Euboic* standard. No one who reflects for a moment could venture to assert that Phoenician trade and influence were limited to Southern Greece. Yet that virtually is the tacit assumption made by those who derive the standard from Asia. There is evidence to show that the Phoenicians from a very early period frequented Euboea, doubtless attracted by its copper mines (from which perhaps the famous city of Chalcis derived its name)[3]. Round no spot in Hellas do more legends cluster which connect it with Phoenician colonists than Boeotia. It was here that Cadmus settled, and introduced the Phoenician alphabet, it was here according to Greek tradition that Herakles, who is so strongly identified with the Phoenician Melkarth, had his birth. Why then should the Euboeans have been behind the rest of Hellas in receiving the Phoenician standard, which, according to Mr Head, as we saw above, did

[1] *Od.* xv. 460. [2] *Od.* xv. 470.
[3] It is more probable however that *Chalkos* copper got its name from the place (Chalcis) where it was first found in Greece. The name Chalcis may itself be connected with χαλκίς, an *owl*.

influence so powerfully the Ionic cities of the Asiatic seaboard, with which their commerce was so largely connected?

From these considerations it follows that before the Greeks came into contact with either Phoenicians or Lydians they had a weight standard of their own, the *Talanton* of the Homeric Poems, based on the *cow*, which was as yet only employed for the weighing of gold.

This standard we have found to be identical with one of the two chief standards employed in historical times for *silver*, and which from first to last was the *only* standard employed for gold in all parts of Hellas Proper.

As we have seen that gold was to silver in that region as $15:1$, there was not much difficulty in regarding fifteen *weights* or staters of silver as equivalent to one of gold of like weight. Hence there was not the same need in Greece to devise a separate silver standard as there was in Asia, where the relation of the precious metals stood as $13\cdot3:1$, a fact which made simple exchange very difficult. On the other hand we have seen that for the Aeginetans and Greeks, who used the so-called Aeginetic standard, the decimal system, the simplest and most primitive method of reckoning, had a powerful attraction.

Primitive peoples perform all their calculations by means of counters, using for such purposes their fingers and toes or seeds or pebbles.

Nature herself has supplied man with the simplest and most convenient of counters in his ten fingers. Hence naturally arises a preference amongst primitive peoples for counting by tens, and this method, although it has at times been supplanted partially (seldom altogether) by the duodecimal and sexagesimal systems, which are superior by possessing a greater number of submultiples than the decimal (*e.g.* $12 = 6 \times 2, 4 \times 3$, whilst $10 = 5 \times 2$ only), was adhered to by the Egyptians all through their history down to the latest Pharaohs. It may then perhaps be argued that it was through Egyptian influence with Greece that a large part of Greece adopted for their silver a standard based on the decimal system, especially as certain traces of Egyptian influence in very early times have been discovered of late. But as I have already pointed out above when discussing the theory of an Egyptian origin for the Aeginetan standard, because standards of like weight are found in two different regions, it by no means follows that one has borrowed from the other. If we can point out that in both Egypt and Greece there was a standard for gold almost identical in weight, it is at once apparent that there was no need for the Greeks to borrow from the Egyptians the idea of making ten silver ingots or wedges equal to one gold; especially as the decimal idea was next to that of five the simplest and most rudimentary form of calculation known to mankind. It is certainly preposterous to suppose that the Greeks were too barbarous at the time when they had attained a knowledge of silver to devise such a simple process as that of taking the fifteen ingots of silver, which from the natural laws of supply and demand they regarded as the equivalent of one gold ingot of like weight, and redividing them into ten new ingots of silver. This surely will not seem an incredible feat for the early Hellenes to perform when we recall to mind the extraordinary skill in arithmetic which is found among some barbarous peoples. " In West Africa a lively and continual habit of bargaining has developed a great power of arithmetic, and little children already do feats of computation with their heaps of cowries[1]." To imagine that the Greeks could not perform so simple a feat as that which I propose is to assume that they were in a far lower condition of culture and intelligence than the negroes of West Africa, rather resembling the lowest known tribes of men, such as the aborigines of Australia and the savages of the South American forests. To make such an assumption respecting a race which has shewn such an unrivalled potentiality of progress and development as the Greeks is absurd.

At this point it will be convenient to take a general survey of our results so far. We found in the Homeric Poems a twofold system of currency, the gold Talanton, and the cow or ox, the latter alone being employed to express values: we next found that the *Talanton* was the equivalent of the cow, the metallic unit being clearly the later in origin, and being based

---

[1] Tylor, *Primitive Culture*, Vol. I. p. 219.

on or equated to the older unit of barter. Through the sacerdotal tradition of Delos we were enabled to fix the value of the Homeric Talanton at 2 gold Attic drachms, or a Daric (135—130 grains Troy). Next came the standards used in historical Greece. (1) The Euboic (135 grains Troy) used for *silver* in the great Euboic towns, in Corinth, in Athens from the time of Solon, and as a matter of course in the Chalcidian and Corinthian colonies, and employed as the *sole* unit for *gold* in all parts of Greece Proper at all periods; (2) the Aeginetic (200—195 grains) employed in Peloponnesus, in Boeotia and Central Greece. We learned that the Euboic standard coincided with the Homeric *Talanton*, thus finding the Greeks of historical times using the same standard universally for *gold* which they had employed long before the introduction of the art of coining from Asia, and partly using this same standard for silver, whilst in other states they employed a standard for the latter metal, which was based on the gold unit, simply dividing the amount of silver equivalent to it into ten parts instead of fifteen.

We then put the question, "Is it rational to suppose that the Greeks borrowed in the 7th century B.C. along with the art of coining from Asia a standard which they themselves already long since possessed?"

At the time when I first put this view forward, I was unable to offer any concrete proof of the existence of such a standard on Greek soil before the introduction of coined money, although the literary evidence was of the strongest kind. Since then I have been enabled to obtain some data of considerable importance. I have already (Chap. II.) described the rings and spirals of gold and silver found at Mycenae, and shewn that they were not improbably made on a standard of 135 grs. We have thus found some definite evidence of the existence of a gold and possibly a silver standard, corresponding to the standard used for both metals in after ages under the name of the Euboic or Attic. It may of course be argued that though found on Greek soil, they are not really Greek in origin. For instance there may be certain indications of Egyptian art and influence in these pre-historic remains, such as the frieze discovered in the Palace at Tiryns of alabaster inlaid with blue glass which according to Lepsius and Helbig[1] is the mock *lapis lazuli* which the Egyptians were so fond of making in imitation of the rare and costly real stone which had to be brought from Tartary. Granting then for the sake of argument that the Homeric *Talent* was a standard introduced into Greece from Egypt at a very early period, it by no means follows that this standard has had a scientific origin. The Greeks it will be noticed found it necessary in taking over this standard to equate it to their primitive barter system. If then the process of human development is such that the Greeks, who above all people shewed the most extraordinary power of acquiring civilization, found it necessary even when presented with a ready made standard for metallic currency, to bring it into harmony with their immemorial system of appraising values by means of the cow, there is certainly a strong presumption that the people from whom they derived that metallic standard had not themselves obtained it by any mathematical process.

We can hardly doubt that mankind first obtained empirically the art of weighing, and that it was only at a later period that mathematics were called in to fix scientifically the standards obtained by the older and cruder method. Such is the function of mathematics still. Thus Professor Cayley observed (in his address at Stockport), "I said I would speak to you not of the utility of mathematics in any of the questions of common life or of physical science, but rather of the obligations of mathematics to these different subjects. The consideration which thus presents itself is in a great measure that of the history of the development of the different branches of mathematical science in connection with the older physical sciences, Astronomy and Mechanics. The mathematical theory is in the first instance suggested by some question of common life or of physical science, is pursued and studied quite independently thereof, and perhaps after a long interval comes in contact with it or with quite a different question[2]."

If such then is the part played by mathematics in an age when even the mathematician has come to the aid of the hang-

---

[1] Schliemann, *Tiryns*, pl. II. Helbig, *Das homerisches Epos*[2], p. 79.
[2] *Report of the British Association*, 1883, p. 21.

man, and the wretch meets a well-deserved doom in strict accordance with a mathematical formula, *a fortiori* must empirical discovery have preceded mathematical theory in the second millennium before the Christian era. Just as countless malefactors were successfully executed by empirical Jack Ketches before ever the mathematician turned executioner, so we may be certain that untold sums of gold had been weighed by means of natural seeds and according to a standard empirically obtained before ever the sages of Thebes or Chaldaea had dreamed of applying to metrology the results of their first gropings in Geometry or Astronomy.

PART II.

CHAPTER X.

THE SYSTEMS OF EGYPT, BABYLON, AND PALESTINE.

WE are now in a position to approach the last stage in our task, that which deals with the growth and development of various weight-standards, all of which start from a common unit. Of necessity Egypt, Babylon, Greece and Italy will claim a chief share of our attention. The question now is, Shall we deal with these regions according to the priority of their civilization, that is, in the order in which I have just named them, or shall we rather adhere to the principle which has hitherto guided us, of working back from that which is better known to that which is less known?

On the whole the former is perhaps the better for our present purpose. As we believe that we have discovered by the inductive method the common unit which lies at the base of all these systems, there is no longer the same necessity for always starting with that which is the less ancient. Besides, if we were nominally to pursue this course, it by no means follows that we would be starting from that which is the best known. *Prima facie* we ought to start with the Roman system, the tradition of which has remained unbroken down to our own days. We could work back through the system of the Middle Ages to the time of Constantine the Great, from Constantine to the early Empire, and from the Empire to the Republic. Moreover no weight-unit is more accurately known than the Roman pound. But the early history of Rome is so obscure that we have absolutely no records of a time, when Greece had already a literature of a venerable antiquity. Rome has no literary remains and even not more than a very few meagre inscriptions dating from before the first Punic War (263—241 B.C.), the very time when Hellas was already far advanced in the autumn of her life. Then Italy had borrowed so much from Hellas that the enquirer must be cautious as to how far he may be dealing with material of true Italian or merely adventitious origin. As we are concerned rather with the *origin* than with the later developments of weight-systems, it is plain that for dealing with our principal objects the Italian systems present us with no special aid. The late period (268 B.C.) at which the Romans struck silver coins places us at a still further disadvantage if we start with their system. Greece on the other hand presents us not only with abundant literary records of great antiquity, some of them descending from an age which knew not the uses of coined money, but also with thousands of inscriptions cut in marble or bronze, many of which contain data of great value for dealing with the history of currency and weight, and finally presents us with vast series of coins from which we can learn empirically the coin standards employed in various times and places. But it is the very wealth of material that is in some degree here our difficulty. The special feature of Greek national life was its numerous autonomous states. There was no central authority with a mint which issued coins for a whole empire as was virtually the case in the great Persian kingdom, and at a later period in the Macedonian empire of Alexander the Great. In the palmy days of Hellas each petty state issued its own coinage, following in its silver and copper mintages whatever standard or module it pleased.

To commence our constructive part with a country where we are confronted with such an array of separate coinages and of diverse standards would be unwise if it were possible to start from some region where there was a single central authority, and consequently less diversity of standards. We are thus led to choose either Egypt or Babylonia as our starting point. The former presents to us a system less developed and more simple than the latter. In fact we are tolerably well justified, in view of recent discussion, in regarding all that is more complex in the system of Egypt as borrowed from Babylonia. Yet it must not be supposed that we escape all difficulties in thus starting with Egypt. If in Hellas we found ourselves embarrassed by the wealth of coinages, in Egypt on the other hand we have no native coinage to guide us, for it was only after the conquest of Egypt by Alexander that under the Greek dynasty founded by Ptolemy Lagos the essentially Greek art of coining was introduced into Egypt. We depend therefore for our knowledge of Egyptian standards upon the actual weighings of weight-pieces and such information as can be gleaned from the ancient Egyptian documents. The same holds good likewise on the whole for the Assyrian system, where however the actual weight-pieces and statements derived from cuneiform inscriptions can in some degree be supported by collateral evidence. At the same time we must be careful not to assign as much importance to the literary evidence supplied to us by Egyptian hieroglyphic or Assyrian cuneiform as we do to the records of Greece or Rome. The keys to the former have only been obtained within the present century, and many of the translations of such documents given us by that brilliant band of savants who have opened to us the portals of a Past far exceeding in antiquity the most remote epoch of which the literatures of Greece and Rome contain even any tradition, must at the best in many cases be considered only as tentative.

Furthermore although the knowledge gained from actually existing weights, which have been gleaned from the ruins of Nineveh, Khorsabad, or Naucratis, may be regarded as positive and more or less exact, we are met by the difficulty that in the case of Egypt and Assyria, where there was no coined money, we have no means of deciding what class of weight was used for certain kinds of commodities. In Greece and in the countries which formed the Persian empire we can be sure at all events of the standards which were employed in the weighing of gold and silver: the absence of this test is a serious hindrance in the study of Egyptian and Assyrian metrology. It is easy to illustrate by a supposed example the element of uncertainty introduced. Let us suppose that in ages to come the ruins of some English ironmonger's shop were excavated, and a series of weights was found therein, a set of Avoirdupois weights ranging from a one-hundredweight to half an ounce; a set of Troy weights ranging from one pound to half a grain, and one of Apothecaries' weights consisting of ounces, drachms, scruples, and grains. Suppose likewise that some ardent metrologist of that age, in addition to this splendid find, should be able to add to his material from elsewhere one or two sovereign and half-sovereign weights, a guinea, half-guinea, quarter-guinea, and seven-shilling-piece weight, perhaps even a noble, or a half-noble weight, and then without consulting literary sources, or previously studying the standards on which the English coinage had been struck at different periods, proceeded to reconstruct the metrological system of England. It is needless to say that his conclusions would be indeed widely aberrant from the truth.

Having thus sketched however roughly some of the difficulties which beset our path, and after warning the reader that in metrology if anywhere the maxim of the old Sicilian poet is to be observed,

> Sober keep, to doubt inclined be;
> Hinges these are of the mind[1],

I shall now proceed to set forth the method in which I conceive the various systems gradually rose and expanded. Let us bear in mind the fact already proved that gold was the first of all commodities to be weighed, and that consequently the standards employed for weighing that metal are the most archaic.

EGYPT.

As has been previously remarked, we are not concerned with the long battle still raging between Assyriologists and Egyptologists as regards the respective claims of Egypt and Babylonia to the invention of measure and weight-standards.

---

[1] Νᾶφε καὶ μέμνασ' ἀπιστεῖν, ἄρθρα ταῦτα τῶν φρενῶν, Epicharmus.

Boeckh himself seems instinctively to have felt this difficulty. For whilst he took Babylonia as the birthplace and home of all the ancient systems, nevertheless he held that contemporaneously there must have existed a connection between Egypt and Babylonia in remote antiquity, from which alone certain agreements and relations between the measures and weights of Egypt and Babylonia were capable of explanation[1]. The primitive measures of length are undoubtedly by the consensus of mankind based upon the parts of the body, such as the finger, the thumb, the foot, the arm, or both arms fully extended, standards common to Egyptians and Chaldaeans alike. Whilst at a later stage in the history of all civilized peoples efforts have been made to obtain more accuracy in these standards, which of necessity have produced certain local and national divergencies, yet inasmuch as all alike started from these standards which have been supplied by nature, it is obvious that many striking similarities and relations will always be found when any comparative study of different systems is attempted. The same principle of course holds good for weight-standards. According to our argument there was a common animal unit existing in Assyria and Egypt, which was represented by a metal unit, prevailing alike in both regions possibly with certain modifications. Egypt and Assyria starting with this common unit, each in their own fashion constructed their distinctive national systems, and we need not be surprised if at a later period under certain political conditions certain parts of the system of one of these regions are found exercising some influence upon that of the other.

We shall now briefly state the Egyptian weight-system. In the oldest Egyptian documents two weights continually occur, the Kat (*Ket* or *Kite*) and the Uten (*Ten* or *Outen*). Already in the third millennium before Christ the precious metals were in full use in Egypt, and copper likewise was employed in the purchase of articles of small value. Although very large amounts are recorded, yet they had devised no larger unit than those mentioned.

[1] Boeckh, *Metrol. Untersuch.* p. 32.

To M. Chabas belongs the honour of being the first to clear up the relations between the uten and kat. The history of this discovery is an interesting proof of the fruitlessness of the purely empirical form of metrology which confines itself to the measuring of buildings, and weighing of ancient weight-pieces and coins, unless its path is made clear by means of the light derived from ancient records. The names uten and kat had been long known, as both of them recur frequently on the walls of the temple of Karnak (*Temp.* Thothmes III. 1700—1600 B.C.), and Egyptian weights were in the museums of Europe, but nevertheless "the exact relation of the one to the other remained unknown until it was fortunately disclosed by a passage in the Harris papyrus, which contains the annals of Rameses III. (circ. 1300 B.C.). From this it appears that the Uten contained ten Kats[1]." The uten therefore is the tenfold of the kat: Nissen[2] thinks that the latter was perhaps originally a gold weight (*vielleicht ursprunglich ein Goldgewicht*). These two units served for the weighing of gold, silver and copper, and there seems to be no difference noted in the documents between the units used for each purpose. In the lists of booty we read of such sums as 3144 utens of gold and 36692 utens of electrum. In lists of prices of commodities kats and utens of silver and copper are frequently mentioned. The weight of the kat has been fixed by Lepsius at 9·096 grammes (142·1 grains) and that of the uten at 90·959 grammes (1421·2 grains). But as it often happens in the case of coins that one well-preserved specimen is a better index of the normal standard than any that can be attained by taking the average of 100 bad specimens, so in the case of weights, one good specimen, made of some hard and imperishable substance, will give us a truer representation of the standard unit than the average of a large number of weights made of some less durable material, and carelessly executed, and meant merely for traffic in goods of little value. If such a weight as we have supposed is inscribed with its name, and we can also get some indication

---

[1] Head, *op. cit.* xxviii.
[2] "Griech. und röm. Metrologie" (in Iwan Müller's *Handbuch der klass. Altertumswissenschaft*, Vol. I. p. 684).

that it has all the authority that belongs to a weight used for official purposes, its value becomes still greater. Such a piece fortunately exists in the Harris Collection. It is a beautifully preserved serpentine weight, and weighs 698 grs. Troy. Allowing for its extremely slight loss we may suppose its original

FIG. 22. Egyptian Five-Kat weight (Harris Collection).

weight to have been about 700 grs. It bears the inscription, *Five Kats of the Treasury of On*. This gives 140 grains Troy as the weight of the kat[1]. This inscription also proves that the kat was the unit. For if as is commonly stated the uten is the unit, of which the kat is simply the one-tenth, we must naturally expect to find this weight described as ½ uten rather than as 5 kats. This is confirmed by a statement of the grammarian Horapollo (or Horus, who although writing about 400 A.D. nevertheless preserves much valuable information) that "with the Egyptians the didrachm is the monad. But the monad is the source of production of all numeration." As two drachms were 135 grs., it is evident that it is the kat of 140 grs., and not the uten of 1400 grs. which the Egyptians themselves regarded as the basis of their system[2]. Mr Flinders Petrie from the weights of 158 specimens found in the ruins of Naucratis, which range from 136·8 grains to 153 grains, concludes that there were two distinct kat units, one weighing

---

[1] Head, *op. cit.* XXIX. Madden's *Jewish Coinage*, p. 277.

[2] Horapollo I. 11, παρ' Αἰγυπτίοις μονάς ἐστιν αἱ δύο δραχμαί. μονὰς δὲ παντὸς ἀριθμοῦ γένεσις. εὐλόγως οὖν τὰς δύο δραχμὰς βουλόμενοι δηλῶσαι γύπα γράφουσι, ἐπεὶ μήτηρ δοκεῖ καὶ γένεσις εἶναι, καθάπερ καὶ ἡ μονάς.

142 grs., the other 152 grs. But until some literary evidence is forthcoming for the existence of this second and heavier kat[1], we must suspend our judgment. It is perfectly possible that such existed, being used for some purpose different from that of the kat of 140 grains. For instance it might have been used specially for copper owing to a desire to make certain adjustments between silver and copper, but this is of course mere conjecture.

It is worth while here to see the method by which those who believe in a scientific system of Egyptian origin obtain their unit.

Signor Bortolotti (*Del primitivo cubito Egizio*) thinks that the uten of 1400 grains is exactly the $\frac{1}{1000}$ part of the weight of a cubic cubit of Nile water, the cubit in question being not the ordinary royal cubit of 20·66 inches, but a measure which he

[1] W. M. Flinders Petrie, *Naukratis*, p. 75. It is with extreme reluctance that I must refuse to follow Mr Petrie, who for careful accuracy and scientific method stands at the head not only of metrologists but of archaeologists in general. But it seems to me that in his method of arriving at his weight-units from the weighing of weight-pieces he has overlooked one very important factor. False weights and balances have prevailed in all ages and countries, and we can hardly wrong the ancient Egyptians if we suppose that a certain number of their nation were not as honest as they might have been in their dealings. The variations in the weights of his specimens given by Mr Petrie may very well be due to false weights. And it must be carefully noted that frauds were not only perpetrated by means of light but also by means of too heavy weights. Whether the Jews learned to cheat when they sojourned in the land of Goshen or not, we cannot say, but that they used too heavy as well as too light weights is plain from the denunciations of the prophets: thus Amos (viii. 5), "When will the new moon be gone that we may sell corn? and the sabbath that we may set forth wheat, making the ephah small, and the shekel great, and falsifying the balances by deceit?" See also Ezekiel xlv. 10. But the practice of cheating with too heavy as well as with too light weights is best seen in Deuteronomy xxv. 13; "Thou shalt not have in thy bag divers weights, a great and a small; thou shalt not have in thine house divers measures, a great and a small. Thou shalt have a perfect and just weight, a perfect and just measure shalt thou have." It seems hardly likely that of the 516 weights found by Mr Petrie at Naukratis all were "perfect and just" weights. It is thus quite possible that the variations from what there is evidence to suppose is the normal standard, whether they be those of excess or deficiency, may be accounted for, at least in part, by this consideration. Mr Petrie's method, if applied to natural products such as certain kinds of seeds, will of course give the truest possible result, but when the factor of human knavery enters, his method is at once open to serious drawbacks.

calls the primitive Egyptian cubit of 19·71 inches in length. Signor Bortolotti also suggests that the standard uten of Mr Petrie's heavy system was 1486 grains, being the $\frac{1}{1500}$ part of the weight of a cubic *royal* cubit (20·66 inches) in Nile water. But as I have just pointed out the evidence is in favour of the kat being the original unit rather than the uten. Besides if the Egyptians obtained their system for the first time by the scientific process, we ought naturally to find some of those larger units such as the talent and mina, which are found in Egypt at a later epoch. But as we have seen in the case of Greeks, Hebrews, Chinese and Hindus, everywhere weight systems begin with a weight for gold, and this is naturally a small unit.

There is still one element in this matter which we must not overlook. A certain number of gold rings have been found in Egypt. Their unit is fixed by Lenormant at 8·1 grammes (128 grains). Brandis regarded them as Syrian in origin, and thus got rid of all difficulty. Others regard the rings as evidently of Egyptian manufacture, and from finding as they think a corresponding mina appearing in Egypt in Ptolemaic times regard this unit as a genuine ancient Egyptian standard in use long anterior to the Persian conquest. It may thus be very probable that the standard employed in early days in Egypt for gold (and also electrum and silver) was this unit of 128 grains, which is of course almost identical with an ox-unit. Silver, according to Erman[1], was in the time of the oldest Egyptian records more valuable than gold, for in enumeration it is always named before gold, whereas under the later dynasties it is named as with us always after gold, shewing that a great change had taken place in the relations between these metals. It is then clearly conceivable that at the outset one and the same unit of about 128—30 grains, under the name of kat, served as the unit for both gold and silver (which explains perfectly the fact that an ox is valued at a kat of silver), but that in after days when the change in the relative values of the metals came, there was found a need for a new silver unit, just as the Greeks in certain places found it necessary to

[1] Erman, *Aegypten und Aegypt. Leben*, p. 611.

form the Aeginetan and other standards, and the Babylonians found themselves compelled to form that standard which alone can with truth be termed *the Babylonian*, the silver unit of 172 grains.

We have now before us the data for the early Egyptian weight system[1]. It is simple; the unit is the kat probably based on the ox as we have seen already. The fact that weights formed in the shape of cows and cows' heads are represented in Egyptian paintings as employed in the weighing of rings, indicates that in the mind of the first manufacturer of such weights there was a distinct connection between the shape given to the weight and the object whose value in gold (or silver) it expressed. Specimens of such weights are known, and are always of small size, a sure indication that the commodity for which they were employed was very precious. The fact that we find weights in the shape of lions can be readily accounted for by the supposition that in the course of time when the connection between the ox and the original weight-unit became forgotten, and different standards had been evolved, some distinctive animal form was adopted to distinguish the weights of a particular standard. The original unit being thus obtained, the higher unit, the uten, was formed by the method most familiar to all races of men. The fingers of one hand suggested to mankind a simple means of counting; and the combined fingers of both hands gave them the decimal system. The Egyptians accordingly simply took the tenfold of the ox-unit as their highest unit. As weighing in the earliest stage was confined to the precious metals, this unit was sufficient for all practical needs[2]. It will be noticed that the process employed in forming this weight-system is exactly that which we have found in the Chinese and its related systems. The Chinese *liang* (*tael* or ounce) corresponds to the Egyptian

[1] We also find mention of a weight called the *pek*, which weighed ·71 grammes (11 grains), and was the $\frac{1}{130}$ part of the uten. Hultsch, *Metrol.*² p. 37, regards it as a provincial Ethiopian weight. Its awkward relation to the kat and uten seem to show that it did not form part of the genuine Egyptian system.

[2] The large copper coins of the Ptolemies of 1450—1350 grs. Troy (the *flans* of which were turned in a lathe) were almost certainly struck on the native uten.

kat (or shekel). Under its name of *tical* or *bat* we found it as the unit of gold in South-Eastern Asia, and for the weighing of precious metals we found that the highest unit employed was the *nén*, the tenfold of the original unit, (the *tael*) itself still the only unit in use in China for the precious metals. In process of time when ordinary commodities of life began to be reckoned by weight, the Chinese made use of the *pical* (which originally simply meant a man's load) as their highest commercial unit. Much the same process seems to have taken place in Egypt, for in later times we find *talents* of various kinds in use. Thus the Alexandrine talent which was employed for wood contained 360 utens. Was this talent originally nothing more than a man's load, which in a later and more scientific age was adjusted to the weight standard time out of mind employed for metals? In this talent of 360 utens we can see the influence of the *sexagesimal* systems of Asia Minor, which, as we shall presently see, was really a commercial standard of comparatively late development and never at any time was employed for the precious metals. The Alexandrine talent of 360 utens contained 3600 kats, just as the *royal* Babylonian talent contained 3600 shekels.

The Assyrio-Babylonian System.

Much has been written in the last thirty years concerning what is known as the *Assyrio-Babylonian* system: in fact so much has been written that it is difficult to find out the data amidst the masses of theory. What then are the facts which we have to go upon? Whence do we get the name *Babylonian*? Herodotus[1] tells us that when Darius imposed on his subjects a fixed quota of tribute instead of the occasional gifts and contributions which were brought to the king's treasury under the reigns of his predecessors Cyrus and Cambyses, those "who brought silver got orders to bring a talent of Babylonian weight whilst those who brought gold one of Euboic weight. But the

[1] III. 89, τοῖσι μὲν αὐτῶν ἀργύριον ἀπαγινέουσι εἴρητο Βαβυλώνιον σταθμὸν τάλαντον ἀπαγινέειν, τοῖσι δὲ χρυσίον ἀπαγινέουσι Εὐβοϊκόν· τὸ δὲ Βαβυλώνιον τάλαντον δύναται Εὐβοΐδας ἑβδομήκοντα μνέας.

THE SYSTEMS OF EGYPT, BABYLON, AND PALESTINE. 245

Babylonian talent amounts to seventy Euboic minas." Properly speaking then according to the ancients, the only specific Babylonian talent was one employed for silver and which was one-sixth heavier than the Euboic talent. It is to be noted carefully that the standard employed for the weighing

FIG. 23. Lion weight.

of gold is not regarded by Herodotus as peculiar to Babylon or Persia, but is treated as identical with the common Euboic standard which was used for silver in many parts of Greece, and the stater of which was the only standard employed for gold in Greece, even in those states where the Aeginetic system was in use for their silver currency. Thus in the system employed for gold in the empire of the Great King the mina contained 50 staters, and the talent 60 minas. But the discovery

A

B

FIG. 24. Assyrian half-shekel weight of the so-called Duck type[1].
    A. Side view showing cuneiform symbol = ½.
    B. View from above.

of the weights known as the Lion and the Duck weights by Sir A. H. Layard at Nineveh whilst from one point of view most fortunate, from another may be regarded as the reverse.

[1] This weight (in my own possession) said to have come from India, and almost perfect, weighed 4·29 grammes.

The large size of many of the weights caused scholars to fix their attention entirely on the larger units, and ever since then all the various efforts to reconstruct the Assyrio-Babylonian weight system have had if nothing else in common at least this that they have all commenced to build the pyramid from the top downwards. They all took the highest units, the talent or mina, as their starting-point, and proceeded to evolve from thence the small unit or *shekel*. Yet all the evidence of antiquity pointed in the opposite direction. In the Greek system, which those scholars held to be borrowed from the East, it was the small unit which was called the *stater* or "weigher," indicating clearly that it was regarded as the real basis of the standard.

Again the Phoenicians and Hebrews who from the earliest times were in constant contact with Mesopotamia ought certainly to exhibit traces in their earliest extant records of the *mina* and *talent*, if it was from these units that the weight-system started. Yet that is not the evidence afforded by the Old Testament. There is no mention of a *mina* except in Kings, Chronicles, Ezra, and Ezekiel, all books of late date. In the Book of Genesis where sums of money are mentioned, they are reckoned by shekels and nothing else. So when Abraham bought the cave of Machpelah for 600 pieces of silver, what could have been more convenient than to describe the purchase money as consisting of 12 *manahs* (*minas*)[1]? Thus, as we shall see later on, the conclusion to be drawn from the ancient Hebrew writings is the same as that which we draw from the Homeric Poems, that it is the shekel (or stater), the small unit, which was the first to be employed, and that it was only in the course of time that the higher units, the *mina* and the *talent*, make their appearance. If according to the common theory the weight standards were the actual creations of either Chaldaeans or Egyptians and only borrowed from them by other peoples, why do we not find the higher units appearing from the first amongst those supposed borrowers, if the other part of the theory is true, that they started from a high unit?

[1] If, as is held by some of the best critics, this is a late passage, there is an *a fortiori* argument against the early use of the *mina*.

Now for the evidence of the monuments themselves.

The weights found by Sir A. H. Layard fall into two classes, (*a*) those in the shape of Lions, which are made of bronze, and (*b*) those in the shape of Ducks, which are of stone[1]. "The bronze Lions are for the most part furnished with a handle on the back of the animal, and are generally inscribed with a double legend, one in cuneiform characters, the other in Aramaic." The Ducks which are inscribed have a legend in cuneiform characters only. These inscriptions contain not only the name of the king of Babylon or Assyria in whose reign they were made, but likewise a statement of the number of the minas or fractions of a mina which each weight originally represented. As these weights were found in the ancient palace some have thought that they were possibly official standards of weight deposited from time to time in the royal palaces[2]. This seems at least to be implied by the inscriptions on some of them, such as those of the largest and most ancient of the Duck weights, which run as follows:

(1) 'The palace of Irta-Merodach, King of Babylon [circ. B.C. 1050], 30 Manahs[3].'

Wt., 15060·5 grammes, yielding a Mina of 502 gram.

(2) 'Thirty Manahs of Nabu-suma-libur, King of Assyria,' [date unknown].

Wt., 14589 gram.

A small portion of this weight is broken off; if this is

---

[1] Is it possible that the so-called *Ducks* are only degraded forms of bull-head weights? The ears and horns were dropped as being inconvenient (see bull-head weight, p. 283), and at a later time when the tradition of their origin had been lost, the shapeless lump was adorned with a bird's head to serve as a handle. All the large weights from Nineveh are without any head; and it is but very rarely even on the small haematite weights that the duck's head is found fully formed.

[2] As no better selection of these weights could be made than that of Mr Head, I have followed his description. Cf. R. S. Poole, in Madden's *Jewish Coinage*, p. 261 seqq., and the Report of the Warden of the Standards, 1874—5, for a full account of these weights.

[3] The *Manah* is of course the *Meneh* so familiar from Belshazzar's vision, *mene, mene tekel upharsin* (Daniel v. 25), which the best scholars follow M. Clermont-Ganneau (*Journal Asiatique*, 1886) in interpreting as *a mina, a mina, a shekel, and the parts of a shekel*.

allowed for it will yield a Mina of about the same weight as No. 1.

(3) 'Ten Manahs' (somewhat injured); bears the name of 'Dungi,' according to George Smith, King of Babylon circ. B.C. 2000.

Wt., 4986 gram., yielding a Mina of 498·6 gram.

On three of the Lions we read as follows:

(1) 'The Palace of Shalmaneser [circ. B.C. 850] King of the Country, two manahs of the King,' in cuneiform characters, and 'Two Manahs' weight of the country' in Aramaic characters.

Wt., 1992 gram., yielding a Mina of 996 gram.

(2) 'The Palace of Tiglath-Pileser [circ. B.C. 747], King of the Country, two Manehs' in cuneiform characters.

Wt., 946 gram., yielding a Mina of 473 gram.

(3) 'Five Manahs of the King' in cuneiform characters, and 'Five Manahs' weight of the country' in Aramaic characters.

Wt., 5042 gram., yielding a Mina of 1008 gram.

The results which we obtain from these weights are that there were evidently two standards used side by side in the Assyrio-Babylonian empire, the Mina of one being about 1010 gram., that of the other about 505 gram. In other words one standard was simply the double of the other; also the weights on which Aramaic legends appear are those which belong to the double standard. Again, there is no evidence that the Talent was as yet conceived, as all the weights are Minae or fractions (or multiples) of Minae. Might we not equally well expect fractions of the Talent, as for instance to find the weight of 30 Manahs described as half a Talent, if the Talent already at this period formed part of the system[1]?

But there is one most important point to be noticed. The single mina of 505 gram. is plainly different from the mina

---

[1] Prof. Sayce (*Academy*, Dec. 19th, 1891) publishes a weight from Babylonia inscribed "One maneh standard weight, the property of Merodach-sar-ilani, a duplicate of the weight which Nebuchadrezzar, king of Babylon, the son of Nabopolassar, king of Babylon, made in exact accordance with the weight [prescribed] by the deified Dungi, a former king." This confirms my contention that the *mina* is prior in *date* to the talent.

of gold, (the Euboic mina of Herodotus) which contained 50 shekels, staters (Darics) of 130 grains (8·4 gram.) each. For it would require 50 shekels of 10·5 gram. (164 grains) each to make a mina of 505 gram. On the other hand it will be found that if we take 60 shekels of the Daric or ox-unit weight they will exactly make up the mina of 505 gram. Neither can this mina be the Babylonian silver mina of 50 shekels of 172 grains (11·2 gram.) each. For the Babylonian silver mina consists of 50 shekels of 11·2 gram., whereas the mina of 505 gram. would give 50 shekels of only 10·1 gram. each. The obvious conclusion is that this mina of 505 gram. is neither the gold nor the silver standard. It is a mina composed of 60 shekels of the weight of the gold unit (Daric or ox-unit). And its talent was composed when the system was completed, of 60 minae, as was the case with all other talents. From the weights just described it may reasonably be assumed that both the heavy and light systems were employed contemporaneously in the Assyrio-Babylonian empire. Some have suggested that whilst the light system was employed in Babylon, its double, or the heavy one, was employed in the northern part of the empire. But the fact that it is on the weights of the latter standard that we find the double legends, the second being in Aramaic characters, seems to point irresistibly to the conclusion that the heavy standard (no matter what it may have been employed for) was especially used in Syria.

It is of great significance that it is in this very quarter we find in use as the gold unit not our usual Daric or ox-unit, but its double, which is commonly known as the heavy gold shekel of 260 grains. I have suggested elsewhere that the explanation of this may be due to the fact that among certain peoples, especially those who dwelt after the fashion of the Sidonians, quiet and full of riches, and who had passed from the life pastoral into the settled agricultural stage, the yoke or pair of oxen would readily be regarded as the unit instead of the single ox of primitive days. The fact that a *zeugos* or yoke of oxen was taken as the unit of assessment by Solon for the third of the Athenian classes lends some support to this view[1]. We have

[1] Cf. Plautus, *Persa*.

likewise seen how the ancient Irish, after borrowing the Roman ounce, and equating an ounce of silver to the cow, made for their silver a higher unit by taking three ounces, which represented three cows, the ordinary price of a female slave (*cumhal*).

The Phoenicians employed the double shekel as their unit, but there is evidence to show that the light shekel was the original unit. We have seen that in Egypt, Palestine and Greece, from the remotest time, gold circulated in the form of rings made of a fixed amount of gold, and also that the unit on which they were made was our ox unit, or light shekel (130—5 grains). From the practice of using gold rings in currency as well as for ornament, we may safely conclude that the standard of 130 grains upon which these were probably made was far anterior to the use of the double shekel in Syria and Phoenicia.

The standards which we have learned from the weights found at Nineveh and Khorsabad are now generally known as the light royal talent, and the heavy royal talent, because on specimens of both standards the inscriptions describe them as weights "of the king."

It is evident that as gold and silver had each a separate standard, the "royal" standards were not employed for the precious metals. It is then most probable that they were employed for the weighing of the inferior metals such as copper, which of course played a most important part in the daily life of both Babylonians and Assyrians. We may rest assured that corn was not weighed but continued to be bought and sold by dry measure, as it was with the Hebrews in the days of the Prophets, when the *Homer* and the *Ephah* were employed to measure it.

I shall now give a tabular view of the three standards used by the peoples of Mesopotamia and their neighbours, treating the *heavy royal talent* as merely the double of the light one.

GOLD.

1 Stater = 130 grs. Troy (8·4 gram.).
50 Staters = 1 Mina = 6500 grs. (420·0 gram.).
60 Minae = 1 Talent = 390000 grs.

THE SYSTEMS OF EGYPT, BABYLON, AND PALESTINE. 251

SILVER.

1 Shekel = 172 grs.
50 Shekels = 1 Manah = 8600 grs.
60 Manahs = 1 Talent = 516000 grs.

ROYAL STANDARD.

1 Shekel = 130 grs. (8·4 gram.).
60 Shekels = 1 Manah = 7800 grs.
60 Manahs = 1 Talent = 468000 grs.

Let us now examine for a moment the current explanation of the origin and inter-relations of these standards and we shall find that they all start at the wrong end, assuming as earliest that which can be proved to be later, and deducing what are really the earliest stages from those which were in fact the historical outcome of the others.

"The proficiency of the Chaldaeans in the cognate sciences of Arithmetic and Astronomy is well known"[1,2]. The broad and monotonous plains of lower Mesopotamia had nothing to attract the eye, and impelled their inhabitants to fix their attention upon the overarching skies studded with stars that shone with exceptional clearness and lustre in the dry pellucid atmosphere of that region. There were no dark mountains looming in the distance to hinder the eye from watching down to the very horizon the heavenly bodies in their periodic movements. Thus as Geometry may be regarded as the special offspring of the Egyptian mind, so Astronomy and Astrology were the children of Babylonia. The results of their astronomical observations were duly recorded on clay tablets in the cuneiform characters, and these tablets were then baked hard, and stored up in the great libraries in their chief cities. It is recorded that when Alexander the Great captured Babylon, he obtained and forwarded to his tutor Aristotle a series of astronomical records extending back as far as the year B.C. 2234, according to our reckoning."

Certain investigations into these tablets, primarily suggested by a fragment of Berosus which described the method of divi-

[1] Brandis, 20—38. [2] Head, xxix.

ding time employed by the Babylonians, have led scholars to conclude that upon these observations "rests the entire structure of the metric system of the Babylonians[1]."

Thus was obtained the famous Babylonian Sexagesimal system. Although the French metric system of modern days has returned to the decimal system, which was the first employed by primitive men, being probably suggested to them by those natural counters, the fingers, the sexagesimal had a considerable superiority over the older decimal system (which the Egyptians had clung to) for certain practical purposes, as the number on which it was based could be resolved into fractions far more conveniently than the number 10. Dr Hultsch (*Metrologie*[2], p. 393) arrives at the Babylonian weight-unit thus: the Babylonian *maris* is equal to one-fifth of the cube of the Royal Babylonian Ell, which is itself obtained from the sun's apparent diameter. The weight in water corresponding to this measure of capacity gave the *light* Royal Babylonian Talent; this Talent was divided into 60 Minae, and each Mina into sixty parts or *Shekels*. Their *gold* Talent was derived from the *sixtieth* of this Royal Mina, with the modification that now *fifty* sixtieths of the Royal Mina made a *Mina of gold* and sixty Minae made a Talent[2].

It seems strange that the framers of this theory did not consider that just as undoubtedly the Chaldaeans must have reckoned their time by the primitive methods of sunrise, noon and sunset, "full market," or ox-loosing time for centuries before they arrived at their scientific division of time, and just as the Chaldaean artificer employed his fingers or palm, or span or foot, as a measure of length ages before the Royal Cubit was equated to the sun's apparent diameter, so in all probability they employed as measures of capacity, gourds or eggshells (as did the Hebrews) and for weights the seeds of plants.

---

[1] Berosus. Synkellos 30, 6 (Eusebii chronic. ed. Alfr. Schoene vol. ι. col. 8): ἀλλ' ὁ μὲν Βηρωσσὸς διὰ σάρων καὶ νήρων καὶ σώσσων ἀνεγράψατο· ὧν ὁ μὲν σάρος τρισχιλίων καὶ ἑξακοσίων ἐτῶν χρόνον σημαίνει, ὁ δὲ νῆρος ἐτῶν ἑξακοσίων, ὁ δὲ σῶσσος ἑξήκοντα. Fragm. Script. Hist. Graec.

[2] Hultsch, *op. cit.* p. 407.

But since, after what we have already seen, it is perfectly clear that the first of articles to be weighed is gold, and that the unit of weight is consequently small, we at once join issue with several points in the theory of Brandis and his school. First they start with the Talent as the unit, and only arrive at the shekel (the *weight* par excellence) by a twofold process of subdivision; secondly, it is assumed that the Royal Talent which we have had reason to believe was a purely commercial Talent, seeing that it was employed neither for gold or silver, was the first to be invented, and that it was only at a later stage that the mina and talent specially employed for gold were developed, not out of the primal unit obtained originally from the one-fifth of the cube of the *maris*, but from the sixtieth of the mina of that Royal Talent; thirdly one asks in wonder why did the Chaldaeans, who only achieved their famous Sexagesimal system after gazing at the stars through unnumbered generations, abandon this precious discovery the very moment they set about the construction of a weight-unit for gold, for instead of taking one-sixth of the cube of the *maris*, they are represented as following their old decimal system with invincible obstinacy by taking one-*fifth* of the *maris* as their point of departure; lastly, it is astonishing that the Chaldaeans did not employ their new discovery in the weighing of the precious metals, the thing which above all others ought to have called for the most scientific accuracy.

The fact is, that just as children find some difficulty in realising that their parents were ever children, so when we stand in the presence of the remains of the great cities of Egypt and Babylonia, those ancients of the earth, we are too prone to forget that Thebes, Babylon or Nineveh had ever their day of small things. The familiar tale of Romulus and Remus with their band of outlaws dwelling in their hovels beside the Tiber has kept people in mind that "Rome was not built in a day." If we can but just approach the question of the first beginnings of Egyptian or Chaldaean civilization with the same idea, it will be far easier to project ourselves into the past of those great races, and thus to realize far better the conditions under which they grew and lived.

There can be little doubt that the unit of the Babylonian system was the light shekel (Daric or ox-unit) of 130—5 grs. Troy. But I have shown that the Chaldaeans were aware of and made use of the method of fixing weight-units by means of grains of corn such as we have found to be the universal practice from Ireland to China, and we have at once removed all need for supposing that it was only when they had discovered a scientific method of metrology that the Chaldaeans constructed their weight-unit.

After what we have shown upon p. 115 concerning the methods employed in the buying and selling of corn, where it has been made clear that of all commodities corn is one of the very last to be weighed because of its bulkiness in proportion to its cheapness, I think no one will readily accept M. Aurè's ingenious hypothesis[1].

Are we not now justified in supposing that, just as the peoples of Mesopotamia had marked their seasons and time by primitive methods, and used their fingers and hands and feet as measures long before they dreamed of scientific methods, so that likewise they had employed for weighing their gold the natural weight-unit which lay ready to their hands in the wheat-ears that crowned their plains.

Let us now start with the light shekel as our unit. According to our argument it was nothing more than the amount of gold which represented the value of the cow, the unit of barter throughout all Europe, Asia and Africa, as it still is over considerable areas of both the latter continents. There is no reason for not believing that as among other people, all articles of property, utensils, weapons, clothes, ornaments and the various kinds of animals stand to one another in well-known relations of value, so the same principle was in full force among the Semites of Mesopotamia. We found that the wild tribes of Laos had a regular scale commencing with a hoe as their lowest unit, leading up through kettles and porcelain jars to the buffalo, their main unit; we also found

---

[1] *Recueil des travaux relatifs à la Philologie et l'Archéologie Égyptiennes et Assyriennes*, Vol. x. fasc. 4, p. 157.

that the weight of a grain of corn in gold was equated to a hoe, and that thus by a simple process of multiplication it was easy to ascertain the value of a buffalo in gold. The unit thus attained was kept from fluctuating, as it was known to every one how many grains of corn gave the true weight of the unit. The practical accuracy of this method of fixing monetary units has been demonstrated from the case of the Early English and Mediaeval English silver penny (p. 180). There is complete evidence to show that the light shekel system was older than the heavy system. Firstly the so-called Duck weights with their cuneiform inscriptions point to the fact that Babylonia was the special home of this system, whilst the Lion weights with their Aramaic inscriptions point to a later period, when the Assyrian Empire was in immediate touch with the merchants of Phoenicia. But, in the next place, a far more powerful argument can be drawn from the Hebrew system. In later times the heavy shekel system prevailed in Palestine, in accordance with which the maneh contained 50 heavy or double shekels of 200 grs. each. But that this maneh was simply imposed on the older light shekel system is demonstrated from the fact that when in two parallel passages articles of a certain weight of gold are mentioned, in the one the weight is given at three manehs, in the other at 300 shekels, the maneh thus being counted at 100 shekels. These 100 shekels are equal to the 50 heavy shekels of the heavy Assyrian or Aramaic maneh. Now it is evident that if the heavy system had been the original one employed by the Hebrews, the maneh would simply have been reckoned at 50 (heavy) shekels. As the matter stands it is evident that on the contrary, the heavy mina was introduced into a system where the unit was simply the light shekel, and the Hebrews therefore clinging to their old unit, described the maneh as consisting of 100 shekels instead of 50. Further evidence to the same effect will be adduced later on. Finding thus the light shekel in Babylonia, in Palestine and in Egypt, and current even under the Assyrian Empire side by side with the heavy system even amongst people who used the Aramaic system of writing, we may without any hesitation regard it as the older.

The process by which the gold Talent was arrived at was somewhat thus:

The ox-unit of 130—135 grs. is the basis.

Next the fivefold of this was taken, whether from five being the simplest multiple, since it was suggested from the primitive method of counting by the fingers of one hand, or far less likely from a slave being estimated at 5 oxen, somewhat as we find among the Homeric Greeks an ordinary slave-woman estimated at four cows, and in ancient Ireland at three cows. This weight is known as the Assyrian five-shekel standard, and from it Mr Petrie derives the 80-grain standard which he detects as the unit of a certain number of weights found at Naucratis (*Naukratis*, p. 86). Whilst the Egyptians contented themselves with the 5 ket and 10 ket, or uten, as their highest unit, the Chaldaeans advanced to the fifty-fold (5 × 10), and thus obtained that which probably for a long time formed their highest unit.

What was this *Maneh?* Is it a Semitic word or is it rather an Aryan, as the present writer has argued elsewhere [1]? At all events it is interesting to find the appearance of a similar word in the Rig Veda and that too in connection with gold: this has been regarded by some as a loan word from Babylon [2]. But it is equally possible, that it is a "loan word" from India to Babylon. The maneh evidently belongs to a period anterior to the development of the sexagesimal system, for if it had come into use along with or subsequent to that system, we should certainly find 60 instead of 50 shekels in the mina of gold and the mina of silver: hence it cannot in any wise be regarded as a distinctive feature of the Babylonian scientific system, as it plainly existed at the time when the decimal system was still dominant. As the latter was the system which prevailed among the Indians of the Vedic period there was no reason why they should borrow the Chaldaean term. On the contrary there is rather a reason why the Chaldaeans would have borrowed the term from India. Gold did not pass into India from Babylonia, for as we have already

---

[1] Kaeji in Fleckeisen's *Jahrbücher*, 1880, first calls attention to this word.
[2] Hultsch, *Metrol.*[2], p. 131.

seen there are no auriferous strata in Mesopotamia, but it passed from the rich surface deposit of the valley of the Oxus and Central Asia into Chaldaea. Now if the same term intimately associated with the same commodity is found among two different peoples, and it is known as a matter of certainty that one of these countries supplies the other with this particular article, there is a considerable probability that the peculiar term connected with the commodity has passed along with it from the source of its production into the country which imports it.

We saw above that there was no native gold in Chaldaea and therefore it must have been imported by those Chaldaean merchantmen from India by way of the Persian Gulf. But was there no gold in Chaldaea until the shipmen of Ur were able to construct vessels capable of a voyage, even albeit only a coasting voyage, to the mouths of the Indus? Working in metals must have been far advanced when such ships were built. That gold came from India we can have little doubt. But it probably came overland for ages before anything in the form of a ship larger than a 'dug-out' had ever floated on the Indian Seas.

The first voyage undertaken to the ancient El Dorado may have been to search for the region from whence came the gold, somewhat in the fashion that in after-times Pytheas of Massalia sallied forth to investigate the sources of the tin and amber which reached Marseilles overland from Britain and the Baltic. After weighing these considerations we shall be careful to avoid any dogmatic declarations as to the origin of the word *mana*. One thing however is clear, and that is that the ancient Hindus were employing certain lumps of gold probably of uniform size in Vedic times, as we saw[1]. The Indians of the Vedic times had thus a gold unit of their own (and as we have shown above probably based on the value of a cow) before they as yet knew the use of silver or had as yet reached the sea in their downward advance into the peninsula of Hindustan. Even granting that they borrowed the *Manā* from Babylonia, it is plain that they had already their own gold unit, for otherwise instead of em-

[1] Rig Veda, *Mandala*, VI. 47, 23—4.

ploying *hiranya pinda*, a most primitive term meaning only *gold-lump*, they would certainly have borrowed the term *shekel* along with the *maneh*. But the fact of most importance for us at present is that, whether *maneh* be Semitic or Aryan, in either case it seems to mean not a *weight* but a *measure*. It will be remembered that we found the *catty* or pound of Further Asia was in origin a natural unit of capacity, as was shown by its Cambodian name *neal*, which simply means a cocoa-nut, and that we found in China the joints of the bamboo of certain sizes serving as their measures of capacity, and both cocoa-nuts and bamboo joints among the Malays of the Indian Isles. This will naturally suggest the question, Is it possible that the *maneh* had a somewhat similar origin? Was some natural object, such as the gourd, which is at the present moment the ordinary unit of capacity at Zanzibar, taken to serve as a measure of liquids or of corn? It is probable that the Greek *cyathus* ($\kappa\acute{\upsilon}\alpha\theta o\varsigma$) like its Latin congener *cucurbita* meant originally some kind of gourd. But there is a certain amount of probability that the Semitic peoples used gourds in primitive times for vessels, not simply from *à priori* considerations, but from the fact that the most archaic pottery obtained by Mr Petrie from his excavations on the site of the ancient city of Lachish in 1890 show unmistakable signs of being modelled after the shape of a gourd. Although the Chinese never have employed their *ching* (catty) for the precious metals, yet the Cambodians have advanced to counting silver not only by the *catty* but also by the *picul*. Did then the Babylonians make 50 shekels of gold or silver roundly equal to their *maneh* or measure of capacity? This is of course pure speculation, but it is at least supported by the comparison of what has actually taken place elsewhere; and even from the empire of the Great King himself can we get an insight into the method by which the *maneh* (and likewise the Talent) may have been brought into the weight system. Herodotus[1] tells us that when the tribute of gold (largely in gold dust) and silver was brought to the King he stored it thus: "he melts it and pours it into earthenware jars, and when he has filled the vessels he strips off the earthenware, and when-

[1] Herod. III. 96.

ever he wants money, he cuts off as much as he needs on each occasion." We saw above that the Cambodian *catty* of silver is twice the weight of the catty of rice, the Cambodian *catty* being simply the cocoanut, the ordinary unit of capacity, which after being filled with rice or silver and then weighed has given two different *catties*. The Great King no doubt poured his gold into jars of known capacity, and the weight of such a jar when filled with gold was well known. It seems then not unlikely that in this way from either a jar, or from the gourd which preceded the jar, the mina was derived. However the *maneh* may have been determined, it is fairly certain that the Babylonians fixed upon 50 as a convenient multiple of the gold unit when silver first came into use; as we have seen above it was probably equal if not superior in value to gold and it was naturally weighed by the same unit. But in the course of time as it became more plentiful, and at the same time if likewise the art of weighing began to be employed by merchants in the traffic in the costly spices and balsams of the east, a necessity would be specially felt among traders for a somewhat heavier unit than the original shekel. Possibly then the Aramaean merchants adopted the double shekel (based on the double ox-unit) for the purpose of weighing silver (when that metal had now become much more plentiful than gold), and for trade in precious gums and spices. Such a procedure can be well paralleled by the old English pound of silk, which is simply two pounds Troy weight. Silk was of course of great value, and was accordingly weighed after the same system as the precious metals; but when it became less costly and more abundant the weight unit was simply doubled. We may therefore regard the doubling of the original shekel as an early step towards the development of a commercial standard. It is not difficult to understand how in the course of time a nation of traders like the Phoenicians preferred this double standard even for their gold, and made it perhaps, as we shall shortly see, the basis of their silver standard.

We saw above that there is every reason to believe that when silver first became known to mankind, they esteemed it as highly as gold, if not more so. It would naturally, therefore, be weighed on the same standard as gold. This would continue until, in the course of years, a time came when the relation between gold and silver had become fairly fixed over all Asia Minor. We know that in the beginning of the 5th cent. B.C. gold was to silver as 13 : 1 (or rather 13·3 : 1). Herodotus, in the celebrated passage in which he describes the organisation of the Persian empire into satrapies, and details the amount of tribute appointed by Darius for each, tells us that the gold was reckoned at thirteen times the value of silver. Now for ordinary purposes of exchange this relation would be extremely inconvenient, and the more accurate relation of 13·3 : to 1 would be still more so. It became thus desirable to fix some separate standard for silver by which a convenient number, such as 10, of silver ingots would be equal to the gold ingot of the ox-unit standard. Metrologists are wont to speak of the desirability of being able to exchange a round number of talents of silver for a talent of gold. But not even in the palmiest days of the wealthy Orient lands was the ordinary individual so rich that he felt any inconvenience in the way of exchanging *talents* of gold and silver. The Great King might deal out talents as he pleased, but his subjects were chiefly concerned with the exchange of silver and gold shekels. I have made this remark because it appears to me that many of the misconceptions connected with this whole subject have arisen from scholars concentrating all their attention on the talent, and taking it as their point of departure.

The Babylonians arrived at their silver standard as follows :

1 gold shekel of 130 grs. was worth 1730 grs. of silver (130 × 13·3), since gold was to silver as 13·3 : 1.

130 grs. gold = 1730 grs. silver.

They divided this amount of silver by 10, and thus :

1 gold shekel of 130 grs. = 10 silver shekels of 173 grs.

As we stated already, Herodotus says that the Babylonian talent was equal to 70 Euboic minas, that is, one-sixth more than the Euboic talent. The latter contained 390,000 grs. Troy, therefore the Babylonian ought to give 455,000 grs. If we multiply our silver shekel by 50 and then by 60, we shall obtain a total amount for the talent of silver of 519,000 grs. Unfortunately several inaccuracies have crept into the text of Herodotus, numerals always being especially liable to corruption in MSS. He seems, however, to have regarded the relation of the Euboic to the Babylonian talent as about that of 5 : 6, and also to have estimated the current weight of the Persian silver piece at about 162 grs. Troy. But there can be little doubt that the full standard weight of the Babylonian silver shekel was 169 grs. (or, according to Mr Head, 172·9 grs.).

From this it is easy to construct the Babylonian *silver* system, which was employed in Lydia and in the Persian empire.

  1 shekel = 169 grs.
  50 shekels = 1 mina = 7450,
    60 minae = 1 talent 447000.

From the double gold shekel was formed another silver standard known as the *Phoenician*.

Gold being to silver as 13 : 1,
  1 double shekel of 260 grs. = 3380 grs. silver,
    3380 grs. silver = 15 shekels of 225·3 grs.

As this silver standard is found in the same area as the double gold shekel, I have thought it best to follow the usual derivation, but at the same time it is worth pointing out that it may have been gained directly from the light shekel.

The light shekel (which in the form of coined money appears either as the gold of Croesus, or the Daric), in the case of the Babylonian system was made equal to ten silver didrachms, or 20 drachms known under the name of Sigli; it likewise is equal in value to 15 Phoenician didrachms of 112·6 grs. Thus, whilst in one region they obtained a silver unit, ten of which would be an equivalent to the gold unit, in another they formed a silver unit, 15 of which would be equivalent to the same gold unit of 130 grs. In each case a number convenient for purposes of exchange was substituted for the extremely unmanageable number 13 (or still more intractable 13·3) of the older system, according to which silver was made into ingots of the same size as those of gold.

These now are the systems on which depended all traffic and currency of the precious metals throughout Western Asia for many centuries. I have been compelled in the statement of the two silver systems to anticipate one step in the growth of the fully developed weight system by speaking of the *Talent*. We have seen that the mina of silver, like that of gold, contains only 50 shekels, thus evidently having likewise been developed before the full elaboration of the Chaldaean system of numeration, or at least before the application of that system to their metric standards. But when we come to deal with the talent we find that in every case alike, whether it be the gold, silver, or royal talent of commerce, the talent invariably consists of *sixty* minae. From this we may with safety infer that it was at a period posterior to the invention of the sexagesimal method that the *Talent* was added to the gold and silver systems. When we turn to the royal system (both light and heavy), we find that the mina consists of *sixty* shekels, just as the talent consists of 60 minae, and consequently we are constrained to believe that this royal system was fixed at a date long after the growth of the gold and silver *minae*, and when the sexagesimal system had now complete sway. We have already seen good reason for considering the *royal* talent to be essentially a mercantile unit. It certainly was not used for gold or silver. Corn was not sold by weight, and so in all probability it was meant for copper, iron, lead, and merchandise of value. We have learned from our studies in the metal trade of primitive peoples that copper and iron are not weighed but are sold by measurement, being wrought into bars or plates of a well defined size. It is only when communities are well advanced in culture that they begin to employ the scales for the buying and selling of the common metals. We argued above that the double shekel system arose from a desire amongst a nation of traders like the Phoenicians for a heavier standard, more serviceable for such goods as were less valuable than gold. It was probably the same desire which found its complete realization in the royal system. Whilst gold and silver had only the mina as their highest unit, there was a new system developed scientifically from the ancient shekel or ox-unit. The sixty-fold of this unit was taken to form a mina considerably heavier than the old gold mina, and now a new higher unit, the sixty-fold of the mina, was introduced. This we know under its Greek name of *talent*, but it was called *kikkar* in the Semitic languages. Now are we to suppose that this *kikkar* or talent was purely and simply nothing more than a higher unit formed by taking a convenient multiple of the lower unit, just as in the French metric system the kilogram is 1000 times the gramme; or was it rather some ancient natural unit, originally formed empirically, and at a later epoch, when science had advanced, fitted into the system of commercial weight by being made exactly the sixty-fold of the *mina*? Comparison with other systems in various lands will incline us to the latter alternative. If we enquire for a moment in what manner the highest unit of weight for merchandise is fixed among barbarous and semi-civilized nationalities, we shall find that the *load*, that is, the amount that a man of average size and strength can carry, is the universal unit. Readers of the various recent books of African travel frequently meet in their dreary and monotonous pages allusions to so many *loads* for which porters have to be supplied. The amount of the *load* seems to vary in different parts. Thus amongst the Madi or Moru tribe of Central Africa, a pure negro race, according to that admirable observer Mr Felkin, the *load* is about 50 lbs. in weight, whilst according to Major Barttelot, the *load* carried by the Zanzibaris on the Emin Pacha Relief expedition was 65 lbs. (besides the man's own rations for several days). We have already had occasion to refer to the *picul* of Eastern Asia, which we found was simply the Malay word for a *load*; and we also found that the load varied in different places. Finally, we found that the Chinese had introduced the *picul* into their system of commercial weight, fixing it at 100 *chings* (catties), but at the same time excluded it from their silver and gold system, where the *tael* (ounce) has remained always the highest unit. Yet in Cambodia we find that the further step has been made, and that the commercial system of the catty and *picul* has been called into service for the weighing of silver. In

Java, whilst gold and silver are weighed by units of small size, copper is sold by the *picul*.

It seems to me not unreasonable to suppose that the origin of the talent has been analogous to that of the *picul*. There is certainly nothing in either the Hebrew *kikkar* or the Greek *talanton* to imply in the slightest degree that they represented a numerical multiple of the mina. The Greek word means simply a *weight*, whilst the Hebrew seems to mean nothing more than a *round mass* or *cake* of anything, whether applied to a tract of country, as the region round the Jordan (as in Nehemiah vii. 28), or a loaf of bread (Exodus xxix. 23; 1 Samuel ii. 36). For as the talent was only introduced into the Hebrew system at a late period the term was probably applied to a *cake* or *pig* of copper or iron the weight of the ordinary *load*. That there was a direct connection between the kikkar and a man's *load* seems implied by the fact that Naaman " bound *two* talents of silver in *two* bags, with two changes of garments, and laid *them* upon *two* of his servants; and they bare *them* before him" (2 Kings v. 23). As we find Naaman asking Elisha for "two mules' burden of earth" (v. 17) it is at least certain that the Semites regularly estimated bulky weights by some kind of *load*. We saw above that in Assyrian the same ideogram stands for *tribute* and *talent*. If a *load* of corn was the regular unit for tribute, the use of a single ideogram may be explained. In the case of *talanton* we have no difficulty in directly regarding it as a *load*, whilst with *kikkar* it is not difficult to see how easy it was for the meaning of a *load* of a certain weight to spring from the earlier meaning of the word. Its use as a loaf is interesting in connection with the fact noted on p. 159 that in Annam the largest unit in use for gold and silver is called a *loaf*.

When under a strong central government a metric system more or less scientific was introduced at Babylon, it was natural that an accurate adjustment of the old empirical unit of merchandise, the *load*, to the mina and shekel should be carefully carried out, just as in China the Mathematical Board have fixed the *picul* of commerce as the hundred fold of the *ching* (*catty*), giving it a value equal to $133\frac{1}{3}$ lbs. avoirdupois.

Such scientific adjustments take place in all countries with the advance of civilization and commerce, and above all under the influence of a strong central government. Let us reflect how long it has taken for the English Statute Acre to conquer the local ancient acres in use in various parts of the United Kingdom, such as the Irish, the Scotch or the Winchester acre. In like fashion, although the standards of weight and capacity were regulated by Act of Parliament in 1824, local usage still held on, and units of weight unknown to the Statute still survive in the usage of provincial places. Now it is not unreasonable to suppose that the name *royal* or *king's weight* was given to the Babylonian commercial system, which was constructed on purely sexagesimal lines, because it was enforced by royal proclamation and power throughout the whole of the empire, and that in like manner the *royal cubit* mentioned by Herodotus (I. 178) owes its origin to the establishment of one uniform standard for the dominions of the Great King. In fact no better illustration of what took place can be found than that afforded by our own terms such as *imperial pint*, or *imperial* gallon, or in a less degree by the *statute* acre, as contrasted with the older customary pints, or gallons, or acres. The mistake made by metrologists, in regarding the scientifically constructed Babylonian system as the first beginning of the art of weighing, is just as great as if a person writing a manual of English Metrology were to start with the metric legislation of 1824 as the first beginning of our metrology, and were to try and explain all traces of an earlier system or systems by forcing the facts into some sort of conformity with our modern standards. Undoubtedly in such an effort great facility would be found inasmuch as the present scientific standards are simply the ancient units of the realm accurately defined. But the reader will best understand the relations which probably existed between the Babylonian *royal* standard (both single and double) by having a short account of the adjustment of our standards laid before him. Great inconvenience having been felt in the United Kingdom for a long time from the want of uniformity in the system of weights and measures, which were in use in different parts of it, an Act of Parliament was passed in 1824 and came into force on January the 1st 1826, by which certain measures and weights therein specified were declared to be the only lawful ones in this realm under the name of *imperial weights and measures*. It was settled by this Act (1) that a certain yard-measure, made by an order of Parliament in 1760 by a comparison of the yards then in common use, should henceforward be the *imperial yard* and the standard of *length* for the kingdom: and that, in case this standard should be lost or injured, it might be recovered from a knowledge of the fact that the length of a pendulum, oscillating in a second *in vacuo* in the latitude of London and at the level of the sea (which can always be accurately obtained by certain scientific processes), was 39·13929 inches of this yard: (2) that the half of a double pound Troy, made at the same time (1760), should be the *Imperial Pound Troy* and the standard of *weight*; and that of the 5760 grains which this pound contains, the pound *Avoirdupois* should contain 7000; and that, in case this standard should be lost or injured, it might be recovered from the knowledge of the fact that a *cubic inch* of distilled water at the temperature of 62° Fahrenheit, and when the barometer is at 30 in., weighs 252·458 grains: (3) that the *imperial gallon* and standard of *capacity* should contain 277·274 *cubic inches* (the *inch* being above defined), which size was selected from its being nearly that of the gallons already in use, and from the fact that 10 lbs. Avoirdupois of distilled water weighed in air at a temperature of 62° Fahrenheit, and when the barometer stands at 30 in., will just fill this space. On p. 180 we saw that the standard gallon in the Tudor period ultimately depended on the pennyweight, which was, as we found, fixed by being the weight of 32 grains of wheat, dry and taken from the midst of the ear of wheat after the ancient laws of the realm. It was from the descendants of this gallon that the *imperial gallon* of 1824 was fixed, with a slight modification so as to make it contain 10 lbs. of distilled water weighed in air at a temperature of 62° and when the barometer stands at 30 in. The double pound Troy made in 1760 depended in like fashion for its ultimate origin on the wheat-grains, and it also affords us an interesting illustration of the doubling of the original single unit, such as we find in the heavy *royal* Babylonian system. We may find further analogies between our own system and that of the Babylonians. Whilst at the Mint gold and silver are weighed for coinage by Troy weight, the copper coinage on the other hand is regulated by the lb. Avoirdupois, the ordinary commercial standard. As already remarked, it is almost certain from the method of elimination that copper was the principal article for which the *royal* Babylonian system was employed, as gold and silver had separate standards of their own, and corn was sold by measure and not by weight.

To sum up then the results of our enquiry into the Assyrio-Babylonian system, we started with the so-called light shekel or ox-unit as the basis of the system; and found that gold and silver were weighed by it and by its fifty-fold, the *maneh*, which may have been itself a natural measure of capacity, such as the catty used in Eastern Asia, where we know for certain that this weight was originally a measure of capacity obtained from the joints of bamboos or the cocoa-nut; that in a certain part of the empire a need was felt for a slightly heavier unit for the weighing of silver and precious commodities such as gums and spices, and that accordingly the great trading Aramaic peoples used the two-fold of the ox-unit (260 grains Troy); that at the earliest period copper would not be sold by weight but would be sold by bars or plates of fixed dimensions, as is still the practice with iron and copper among the barbarous peoples of Further Asia and Africa; that with the advance of culture the art of weighing was extended to copper and other articles of small value in proportion to their bulk, and that, as the maneh, or contents of a gourd, and the *load* or amount that a man could carry on his back, had been most probably in general use as units for common merchandise, the time came when under the all-mastering authority of the Great King a standard based on the ancient ox-unit, but framed on the new scientific sexagesimal system, was established for copper and certain other kinds of merchandise; that in this system 60 shekels made the maneh, and the *load* (the *kikkar* or talent) was adjusted to the new system as the sixty-fold of the maneh; and that in the course of time this higher unit of the *kikkar* or talent was added to the gold and silver systems, sixty manehs in each case making the *kikkar* as in the case of the royal or commercial system; that in the case of silver, which on its first discovery and employment was as valuable as gold, and was therefore weighed on the same standard, when in course of time it became about thirteen times less valuable than gold, and there was a difficulty experienced in exchanging the units of gold and silver, a separate standard was created by dividing into ten new parts or shekels the amount of silver which was the equivalent of the gold shekel (ox-unit); that this was probably developed before the royal commercial mina of 60 shekels had been formed, as in that case the silver mina would have contained 60 shekels likewise; we were able to give an explanation of the name *royal* as applied to the commercial standard by regarding it as of late origin, created by a supreme central authority for the regulation of the commerce of a great empire made up of a heterogeneous mass of races, just as in the present century our own *imperial* standards have been fixed for the whole kingdom, being based, as was the Babylonian, on an ancient unit empirically obtained; and just as the royal arms are stamped on our imperial standards, so the weights of the Assyrian *royal* system were shaped in the form of a lion, the symbol of royalty throughout the East. Finally we found that at the base of the Assyrio-Babylonian system lay, as the determinant of the ox-unit or shekel, the grain of wheat, which we have already traced all across Europe into Asia. We can therefore now come to a very reasonable conclusion that the Assyrio-Babylonian weight system was in its origin empirical, and that it was only at a comparatively late date in its history, just as in the case of our own standards, that a certain uniformity between the standards of measures and weights was brought about by the (not complete) application of the sexagesimal system of numeration, the invention of which is their eternal glory.

Having now dealt with Egypt, and the systems which prevailed in the Assyrio-Babylonian empire, it will be best to treat of the region which lay between them. In both the former countries we found the light shekel or ox-unit in use from the earliest times; and it will also be remembered that at an earlier stage we found that Abraham was able to traverse all the wide country that lay between Mesopotamia and the ancient kingdom of the Nile with his flocks and herds, and that he dwelt in the land of Canaan in close neighbourhood and on friendly terms with the sons of Heth, or Hittites, who were then the possessors of that land; and that furthermore monetary transactions were then carried on by means of certain small ingots of silver, as we see from the purchase of the Cave of Machpelah. These ingots, translated *shekels* in the English version and called *didrachms* in the Septuagint, are termed in Hebrew *Keseph* (כֶּסֶף), simply *pieces of silver*, or *silverlings*. In the old Hebrew literature values in silver and gold are expressed either in *shekels* or by a simple numeral with the words "of silver," "of gold" added (where the latter method is followed the English version supplies *pieces* or substitutes "a thousand silverlings" for "a thousand of silver" (Isa. vii. 23). The Septuagint renders the skekel by the Greek *didrachm*). There are several inferences to be drawn from this. It is evident that pieces of silver (and no doubt of gold also) of a certain quality and weight were employed as currency in Palestine, and we may likewise suppose with some probability that these pieces of silver were according to the standard in common use in Egypt and Chaldaea. Again, since we have already shown that gold in the form of rings and other articles for personal adornment was exchanged according to the ox-unit of 130–5 grs., as evidenced by the story of the ring given to Rebekah, it follows that there was but one and the same standard for gold from the Euphrates to the Nile. This is confirmed by the story of the sale of Joseph by his brethren to the company of Ishmaelites "who came from Gilead with their camels bearing spices and balm and myrrh going to carry it down to Egypt"; to these Ishmaelites or Midianites Joseph was sold for twenty pieces of silver[1]. Here we have evidence that the same silver unit was current from Gilead to Egypt. There are various other large sums of silver mentioned both in Genesis and also in the Book of Judges and in Joshua. Thus Abimelech, King of Gerar, is said to have given Abraham

[1] For 20 pieces of *gold* (εἴκοσι χρυσῶν) LXX.

a thousand [pieces] of silver[1], whilst the lords of the Philistines persuaded Delilah to beguile Samson into telling her wherein lay his great strength by the promise of eleven hundred [pieces] of silver, which money she afterwards received[2]. Abimelech the son of Jerubbaal (Gideon) was enabled to form his conspiracy by hiring 'vain and light persons'[3] with the three-score and ten [pieces] of silver taken by his mother's brother from the house of Baal-berith. Finally, we have a sum of eleven hundred [pieces] of silver which were stolen by that "man of Mount Ephraim whose name was Micah" from his mother, of which his mother took (when he had restored the money) two hundred [shekels] and gave them to the founder, who "made thereof a graven image and a molten image[4]." Now although all these are considerable sums, all exceeding a *mina*, yet there is no mention whatever made of the latter unit of account in any of these passages. The story of another theft shows that gold as well as silver was reckoned originally only by the shekel and not by the mina. Thus Achan "saw among the spoils a goodly Babylonish garment and two hundred shekels of silver and a wedge of gold of fifty shekels weight[5]." As fifty shekels were a mina, here if anywhere we ought to have found the latter term. From this we infer without hesitation that the shekel was the original unit.

But there is another word besides *keseph* which is translated *piece of money* or piece of silver. This is the term *qesitah* (קְשִׂיטָה) which occurs in three passages of the Old Testament. Thus Jacob bought the parcel of ground where he had spread his tent at the hand of the children of Hamor, Shechem's father, "for an hundred pieces of money" (Gen. xxxiii. 19); and the same word is used in the parallel passage in Joshua (xxiv. 32) where the children of Israel buried Joseph's bones in Shechem in the parcel of ground which Jacob bought for an hundred pieces of money. Lastly, Job's kinsfolk and acquaintances gave him every man a *piece of money*, and every one a ring of gold (xlii. 11). It has been always a matter of doubt what this piece of money really was. The Septuagint translates

---

[1] Gen. xx. 16. [2] Judges xvi. 5. [3] Judges ix. 4.
[4] Judges xvii. 2—4. [5] Joshua vii. 21.

qesitah in these three passages by ἑκατὸν ἀμνῶν, ἑκατὸν ἀμνάδων, and ἀμνάδα μίαν, thus in every case regarding it as a *lamb*. The most ancient interpreters all agree in this, whilst some of the later Rabbis regarded it as signifying a coin stamped with the form of a lamb: one of them says that he found such a coin in Africa[1].

Long ago Prof. R. S. Poole, speaking of this word, said: "The sanction of the LXX, and the use of weights bearing the forms of lions, bulls, and geese by the Egyptians, Assyrians, and probably Persians, must make us hesitate before we abandon a rendering [lamb] so singularly confirmed by the relation of the Latin *pecunia* and *pecus*[2]." The connection between

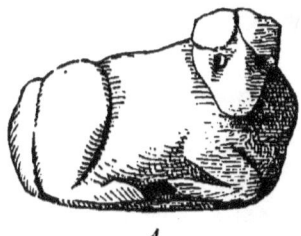 

A. *B.*

Fig. 25. Weights in the form of Sheep[3].

weights and units of currency is especially close at a time when coined money is as yet unknown, and hence when we find weights in the form of sheep coming from Syria, and also recollect that sheep were employed as a regular unit in Palestine for the paying of tribute, and with the light obtained from primitive systems of currency, we may well conclude that the *qesitah* was an old unit of barter, like the Homeric ox, and

---

[1] Cf. Buxtorf and Gesenius *sub voce*.
[2] Madden's *Jewish Coinage*, p. 7.
[3] *A* is from Beirut, in the Greville Chester Collection in the Ashmolean Museum, of white and yellow crystalline stone; wt. 32·160 gram. (a very slight chip from the base); on the base is engraved a rude ibex and another figure. *B* is from Persia, slightly chipped on side of head, yellowish white stone, veined with red, like jasper; wt. 22·450 gram.; on the base are two ibexes. I am indebted for this information to Mr A. J. Evans, Keeper of the Ashmolean Museum, by whose kindness I am likewise enabled to give representations of the weights.

as the latter was transformed into a gold unit, so the former was superseded by an equivalent of silver. We read (2 Kings iii. 4) that Mesha, king of Moab (now so famous from the inscription which bears his name), was a sheep-master, and he rendered unto the king of Israel one hundred thousand lambs, and one hundred thousand rams with the wool. When payment in metal came more and more into use silver served as the sub-multiple of gold, just as sheep formed that of the ox, and it is not surprising that in later times when coins were struck

FIG. 26. Coin of Salamis in Cyprus.

by the Phoenicians, as at Salamis in Cyprus and many other places, bearing a sheep or a sheep's head, there arose some doubt as to whether the *qesitah* was a *sheep*, a piece of uncoined silver, or a coin stamped with a sheep. The very fact of the Phoenicians having such a predilection for this type is in itself an indication that the silver coin in its origin represented the value of a sheep. At a later stage, when we come to deal with the early Greek coin types, we shall develope this principle more completely. The mere fact that the sheep on the Phoenician coins is sometimes found accompanying a divinity does not militate against our doctrine, as I shall explain when I deal with the coins of Messana and Thasos.

But then comes the question, which was the shekel employed by the Hebrews? It must have been either (1) the ox-unit of 130 grs., used alike for gold and silver in early days both in Egypt and Mesopotamia and Greece, or (2) the double of this, or heavy shekel of 260 grs., used for gold only in parts of Asia Minor, or (3) the Phoenician shekel of 225 grs., used only for silver and electrum along the coast of Asia Minor, and never employed for gold, or (4) the Babylonian or Persic standard of 172 grs., used only for *silver*. In later times the silver shekel in

THE SYSTEMS OF EGYPT, BABYLON, AND PALESTINE. 273

use amongst the Jews was most undoubtedly the Phoenician shekel, obtained, as we saw above, by dividing the amount of silver equivalent to the double gold shekel into 15 parts. But it may be reasonably doubted whether the silver piece or shekel (called always a *didrachmon* in the Septuagint) mentioned in Genesis and Judges is the Phoenician shekel. It is used without any distinctive epithet, as if it were the weight *par excellence*, and is employed for *gold* as well as silver. But when we turn to certain other passages we find mention made of a shekel called the *Shekel of the Sanctuary*[1]. This shekel is frequently mentioned, generally in connection with silver, and in reference to such things as the contribution of the half-shekel to the Tabernacle, the redemption of the firstborn, the sacrifice of animals, and the payment of the seer. Yet we find this shekel likewise employed in the estimation of *gold*, a fact which at once shews that it is neither the Phoenician shekel of 220 grs. nor the Persic of 172 grs., both of which were confined to *silver*. It must then have been either the ox-unit of 130 grs. or the heavy shekel of 260 grs. As the latter was confined in use to *gold* it follows that the ox-unit of 130 grs. alone fits the conditions required. If then we can discover what in the case of either silver or gold was the weight of this shekel, we shall have determined it for both metals, for it will hardly be maintained that there was one shekel of the Sanctuary for gold and one of different weight for silver.

Now we read in Exodus (xxxviii. 24 *seqq.*) that "all the gold that was occupied for the work in all the work of the holy [place], even the gold of the offering, was twenty and nine talents and seven hundred and thirty shekels, after the shekel of the Sanctuary. And the silver of them that were numbered of the congregation was an hundred talents and a thousand seven hundred and three-score and fifteen shekels, after the shekel of the Sanctuary; a bekah for every man, that is, half a shekel after the shekel of the Sanctuary, for every one that went to be numbered from twenty years old and upward, for six hundred thousand and three thousand and five hundred and fifty men. And the brass of the offering was seventy

[1] Exod. xxx. 13. Levit. v. 15, etc.

talents and two thousand and four hundred shekels." From this passage we learn that, whilst the gold and silver were estimated on the shekel of the Sanctuary (or Holy Shekel), the brass was probably reckoned by some other standard.

It is also of importance to note that it is the shekel which is regarded as the *unit* of the system, for we never hear of a talent or mina of the Sanctuary. From this passage likewise we readily discover that the talent of silver contained 3000 shekels (603,550 ÷ 2 = 301,775 shekels − 1775 = 300,000 ÷ 100 = 3000 shekels).

Now when king Solomon made three hundred shields of beaten gold, three minas (translated *pounds* in the Authorized Version) went to one shield (1 Kings x. 17). But in the parallel passage (1 Chron. ix. 1) we read that "three hundred shields made he of beaten gold, three hundred shekels went to one shield," from which it is evident that a maneh of gold contained 100 shekels[1]. A very important conclusion follows from these facts, for it is plain that when the Hebrews adopted the heavy or double maneh from the Phoenicians they did not adopt for *gold* and silver at the same time the double shekel, of which that maneh was the fifty-fold, but on the contrary they retained their own old unit of the light shekel, and made one hundred of them equivalent to the Phoenician or heavy Assyrian mina. Since this light shekel was employed in the estimation of the gold and silver dedicated by King Solomon for the adornment of the Temple, this shekel can hardly be any other than the Holy Shekel of the Sanctuary.

We are thus led to conclude that the shekel was the same both for gold and silver, and was simply the time-honoured immemorial unit of 130—5 grs.

It is natural on other grounds that this should be the unit employed by the Israelites for the precious metals, since it was the unit employed both for silver and gold in Egypt, the land of their bondage.

[1] The question of the date at which certain documents were written or took their final shape is of course important. But it does not at all follow that a document written at a later period cannot contain traditions of real historical value. Thus here we find Chronicles, placed quite late by the critics, gives the weight in *shekels*, whilst Kings, supposed to be far earlier, gives it in *minas*.

The question next suggests itself, Why was the shekel called by a distinctive name? It is only when there are two or more examples or individuals of the same kind that any need arises for a distinctive appellation: again, as we have already observed, in such cases the older institution continues to prevail in all matters religious or legal. It is important to note that in Exodus xxi. 32, a passage which the best critics consider of great antiquity, the penalties are expressed in shekels simply without any distinctive appellation. At that period there was probably only one shekel (the ox-unit of 130—5 grs.) as yet in use, and so there was no need to distinguish the shekel in which fines were paid. This shekel was then described in the later part of Exodus, where there was a second standard in use, as the holy shekel. As a matter of fact we have another weight mentioned in 2 Samuel (xiv. 26), where it is related of Absalom that "when he polled his head (for it was at every year's end that he polled it: because the hair was heavy on him, therefore he polled it) he weighed the hair of his head at two hundred shekels after the king's weight[1]."

Now it will be observed that in the passage from Exodus quoted above, whilst the shekel of the Sanctuary is carefully mentioned when amounts of gold and silver are enumerated, no such addition is made in reference to the "seventy talents and two thousand and four hundred shekels of brass." If then the heavy or double shekel and its corresponding mina and talent, known to us hitherto as the royal Assyrio-Babylonian heavy standard, had already been introduced among the Hebrews (and we have just seen that according to the First Book of Kings it was in use, at least a mina of 50 double shekels (100 light) was employed for gold), nothing is more likely than that this standard would bear a title similar to that which it enjoyed in Babylonia and Syria, and be known as the king's weight or *stone*. As I have observed in the case of

[1] The mere question as to whether the 200 shekels is far more than the average crop of hair can weigh, does not concern us. If the writer wished to exaggerate the amount of Absalom's hair he would naturally make the shekel as heavy as possible, and say that the weight was in the *heavy* or *royal* shekels, employed for merchandize.

the royal Assyrian standards that they were employed for copper, lead, and commodities sufficiently costly to be sold by weight, so we may with considerable probability conjecture that this king's weight was employed regularly among the Semites for the weighing of the less precious metals, and other merchandise. Hence it is that there was no need to add any explanation of the nature of the standard by which the 70 talents of brass were weighed, and it was only because in the case of Absalom's hair we have an article not commonly weighed, that it was thought necessary by the writer to make clear to us by which of the two standards usually employed the estimate of the weight of the year's growth of hair was made. We may therefore conclude with probability that "the king's shekel" was no other than the double shekel (260 grains). It will have been noted that in Genesis and Judges, admittedly two of the oldest books, there is mention made of only one kind of shekel, and that it is only in Exodus, Numbers and Leviticus, all of late date, that we find the shekel distinguished as that of the Sanctuary, and that it is only in Samuel that we find reference made to the *royal shekel*. It is also worthy of notice that neither in Genesis nor Judges is there any mention made of a maneh or talent, although there was full opportunity for the appearance of the former if it had been then in use, as we find such sums as 400 shekels (4 manehs), 1100 shekels (11 manehs) and 1700 shekels (17 manehs), whilst in the other series of books named we find both the maneh and the talent. It is not unreasonable therefore to suppose, that with the advent of the *maneh* and *kikkar* or talent from their powerful kinsfolk and neighbours came also the practice of employing the double shekel, the fiftieth part of the mina of gold and mina of silver, which was employed in that part of the Assyrio-Babylonian empire, where the use of the heavy Assyrian shekel was in vogue. Besides gold and silver, spices were likewise weighed according to the shekel of the Sanctuary. "Take thee also unto thee principal spices, of pure myrrh five hundred [shekels], and of sweet cinnamon half as much [even] two hundred and fifty [shekels], and of sweet calamus two hundred and fifty [shekels], and of cassia five hundred [shekels], after the shekel of the Sanctuary[1]." If we had any doubt as to whether it was not possible that there were two separate shekels of the Sanctuary, one for gold, and one of different standard for silver, our misgivings are at once dispelled by finding spices weighed after the holy shekel. It is certainly incredible that there could have been a separate standard of the Sanctuary for the weighing of spices. There seems then no reasonable doubt that there was only one shekel of the Sanctuary, and that the unit of 130 grains. In support of this we may adduce Josephus[2], who made the Jewish gold shekel a Daric (which as we have already seen is our unit of 130 grains). This in turn derives support from the fact that the Septuagint, which regularly renders the Hebrew *sheqel* (which like the Greek *Talanton* means simply *weight*) by both *siklos* and *didrachmon*, not unfrequently renders *shekel of gold* by chrysûs[3], which means of course nothing more than gold *stater*, that is a didrachm of gold, such as those struck by the Athenians, by Philip of Macedon, Alexander and the successors of the latter, including the Ptolemies of Egypt, under whom was made the Septuagint Version. We have thus found the earliest Hebrew weight unit to be that standard which we have found universally diffused, and which we have called the ox-unit.

Next let us see how from this unit grew their system. In several passages the shekel of the Sanctuary is said to consist of 20 *gerahs*[4], a word rendered simply by *obolos* in the Septuagint. As before observed, the Hebrew metric system was essentially decimal, like that of Egypt; in fact had Tacitus been a metrologist he might have quoted this as an additional proof that the Jews were Egyptian outcasts, expelled by their countrymen because they were afflicted with a plague, perhaps the *scabies*[5], which so frequently affects swine. The measures of capacity, both dry and liquid, are decimal, and so accordingly we find a decimal division applied to the shekel. The latter is divided into two *bekahs* (בֶּקַע, "a division," "a half"), and each *bekah* is

---

[1] Exod. xxx. 23—4. [2] *Antiq.* III. 8, 10.
[3] Pollux, IX. 59, observes that when χρυσοῦς stands alone, στατήρ is always to be understood.
[4] Exod. xxx. 13. [5] *Hist.* v. 3.

divided into 10 *gerahs* (גֵּרָה). The latter signifies "a grain" or "bean." The Hebrew literature does not state what kind of seed or grain it was, although it is defined by Rabbinical writers as equal to 16 barleycorns. But the fact is that, as we see from the Septuagint rendering, the name in the course of time came to be considered simply as that of one-twentieth of the shekel, whether that shekel was the shekel of the Sanctuary, the Phoenician silver shekel of 220 grains, or the king's shekel of 260 grains used for copper and lead. The *gerah* of the gold shekel or shekel of the Sanctuary was probably the most ancient and came closest to the natural seed from which it derived its name; this *gerah* would be about 6½ grains (130 ÷ 20 = 6·5). On an earlier page (p. 194) we gave the weights of a number of grains and seeds of plants, and amongst them that of the lupin, called by the Greeks *thermos*. According to the ancient tables the *thermos* is equal to two *keratia*, or *siliquae* (the seeds of the carob tree); but since each *siliqua* = 4 wheat grains, the *thermos* = 8 wheat grains, or 6 barleycorns, or 6 Troy grains. If the wheat grain in Palestine was as heavy as that of Egypt or Africa (·051 gram. instead of ·047 gram.), the 8 wheat grains, would = 6·4 grains troy. Again, the Roman metrologists estimated the *lupin* as the third part of the *scripulum*, which weighed 24 grains of wheat[1]; thus the Roman *lupin* also = 8 wheat grains. We may therefore have little doubt that the *gerah* was simply the *lupin*[2]. But what about the Rabbinical *gerah* of 16 barleycorns? In the first place let us recall the confusion which exists in the Arab metrologists respecting the *habba*, some making three habbas, some four equal to the *karat*. This arose, as we saw, from confounding the wheat and barley grain. If the 16 grains assigned to the *gerah* by the Rabbis are really wheat grains, all is at once clear. The *gerah* to which they refer is that of the royal or double shekel (260 grs.), or in other words it is a double *gerah*. We have just found the *gerah* of the Sanctuary shekel to be the lupin, and equal to 8 wheat grains, accordingly its double will contain 16 wheat grains.

[1] Hultsch, *Metr. Scrip. s. v.* Lupinus.
[2] In Gesenius' *Lexicon*, II. 88; II. 144, it is suggested that the *gerah* is the lupin.

Nothing is more common than a change in the value of a natural weight unit, when in the course of time its real origin has been forgotten, and it has been adjusted to meet the requirements of newer systems. Thus the value of the Greek *thermos* and its Roman equivalent the *lupin* both suffered in later days, and were regarded as only equal to 6 wheat grains instead of the original 8 owing to a like confusion between wheat grains and barleycorns. Finally there is a further reason why the authors of the Septuagint Version would translate *gerah* by *obolos*. Writing at Alexandria under Ptolemaic rule, at a time when the Ptolemaic silver stater of 220 grains contained exactly 20 obols of the Attic or ordinary Greek standard of 11 grains, they would all the more readily adopt a rendering, which harmonized so well with the monetary system of their own day; at the same time the Greek habit of dividing all staters into 12 *obols*, no matter on what standard the stater was struck, naturally would incline them all the more to regard the *gerah* not as an actual weight, but simply as the twentieth of the shekel, be the shekel what it might.

The Hebrew gold standard accordingly consisted of a shekel of 130 grains, subdivided into 2 *bekahs* or *halves*; each of which in turn contained 10 *gerahs* or lupins: 100 such shekels made a maneh, and according to Josephus[1] 100 manehs made a *kikkar* or talent. It would thus appear that, just as in the time of Solomon the heavy mina had been introduced which was equal to 100 shekels of the Sanctuary, so the Hebrews carried out consistently this principle by making 100 minae go to the talent. It is however most probable that before that time they had employed a maneh of their own of 50 light shekels, for we have seen above that the talent of silver mentioned in Exodus consisted of only 3000 shekels, just as in all the other gold and silver systems of Asia Minor and Greece: and since we have proved that the silver shekel of the Sanctuary was the ordinary light shekel of 130 grains, it is evident that the silver talent is not made up of 3000 double shekels, but is really nothing more

---

[1] *Antiq.* III. 6, § 7, λυχνία ἐκ χρυσοῦ...σταθμὸν ἔχουσα μνᾶς ἑκατόν, ἃς Ἑβραῖοι μὲν καλοῦσι κίγχαρες, εἰς δὲ τὴν Ἑλληνικὴν μεταβαλλόμενον γλῶσσαν σημαίνει τάλαντον.

than the sixty-fold of a mina which contained 50 shekels of the ox-unit standard. If gold was weighed at all by any higher standard than the shekel, it is almost certain that it must have been weighed by this mina and talent[1]. However, by the time of the monarchy it is most probable that the double or heavy mina had been introduced for silver as well as for gold. In fact the probabilities are that it was applied for the weighing of silver before that of gold. Thus when Naaman the leper set out to go to the Hebrew prophet, "he took with him ten talents of silver, and six thousand [pieces] of gold, and ten changes of raiment[2]." Here the 6000 gold pieces are perhaps the 6000 light shekels which would make a talent of the heavy Assyrian standard after the ordinary Phoenician system of 50 shekels = 1 mina, and 60 minae = 1 talent: and doubtless Naaman counted these 6000 gold pieces as a talent of gold; but inasmuch as the Hebrews had a peculiar system of their own, by which 100 minae, and 10,000 light shekels went to the *kikkar*, these 6000 are not described as a talent by the Hebrew writer. We may thus regard the silver talent as consisting of 3000 light shekels, at the earliest period, and later on as of 3000 heavy shekels: finally, when coinage was introduced and money was struck under the Maccabees on the Phoenician silver standard, it consisted of 3000 shekels of 220 grs. each. But there is one period about which we find great difficulty in coming to any conclusion. After the return from the Babylonian captivity what standards were employed for gold and silver? As Judaea formed part of the dominions of the Great King, we would naturally expect to find in Nehemiah and Ezra traces of the standard then employed throughout the Persian Empire for the precious metals. As we have found that the light shekel formed the unit for gold from first to last, and as it was also the gold unit of the Babylonians and Assyrians, we may unhesi-

[1] Even granting that the parts of Exodus (the priestly Code) took their present form in post-Exile times it is perfectly possible that the metrological data contained therein are based on a genuine old tradition, just as Homer, although in its present shape differing much in linguistic forms from what must have been its original, gives us an archaic talent quite different from those in use when it took its final shape.

[2] 2 Kings v. 5.

tatingly assume that it formed the basis of the Jewish system
in the days of Nehemiah (446 B.C.). As regards the silver
standard we have fortunately one piece of evidence, which may
give us the right solution. We found that in Exodus each
male Israelite contributed a *bekah*, or half a shekel (of the
Sanctuary) to defray the cost of the tabernacle: this half-shekel
was a drachm of about 65 grs. Troy. Now after the Return
from Captivity, we find Nehemiah (x. 32) writing: "We made
ordinances for us, to charge ourselves yearly with the third part
of a shekel[1] for the service of the house of our God." Why the
third of a shekel instead of the half of earlier days? When we
read of the generous and self-sacrificing efforts made by the
Jews to restore the ancient glories of the Temple worship, we
can hardly believe that it was through any desire to reduce the
annual contribution. The solution is not far to seek when we
recollect that the Babylonian silver stater of that age weighed
about 172·8 grs. This formed the standard of the empire, and
doubtless the Jews of the Captivity employed it like the rest
of the subjects of the Great King. The third part of this stater
or shekel weighed about 58 grains; so that practically the third
part of the Babylonian silver shekel was the same as the half of
the ancient light shekel, or shekel of the Sanctuary. From this
we may not unreasonably infer that after the Return the Jews
employed the Babylonian silver shekel as their silver unit, and
this probably continued in use until Alexander by the victories
of Issus and Arbela overthrew the Persian Empire, and erected
his own on its ruins. But although the Babylonian shekel was
the official standard of the empire there can be no doubt that
the old local standards lingered on, or rather held their ground
stubbornly in not a few cases. We saw above that the
Aramaean peoples had especially preferred the double shekel,
and from it they developed the so-called Phoenician or Graeco-
Asiatic silver standard. Gold being to silver as 13·3 : 1, one
double shekel of 260 grains of gold was equal to fifteen reduced
double shekels of silver of 225 grains each. Now it is important to note that the Phoenician shekel or stater was always
considered not as a didrachm but as a tetradrachm; a fact which

[1] LXX. τρίτον τοῦ διδράχμου.

is explained by its development from the old double shekel, which of course was regarded as containing four drachms, and which at the same time explains why it is that in the New Testament the Temple-tax of the half shekel is called a *didrachm*, the term applied to the shekel itself in the Septuagint. When the Jews coined money under the Maccabees, they struck their silver coins on this Phoenician standard, and their shekel was always regarded as a tetradrachm. For the ancient half shekel of the Sanctuary they soon substituted the half of their shekel coins, that is about 110 instead of 65 grains of silver. This change probably took place under the Maccabees; silver had then probably become much more plentiful in Judaea as shown by the fact that they were able to issue a silver coinage. When those who collected the Temple-tax asked Christ for his didrachm, he bade Simon Peter go to the sea and catch a fish, in the mouth of which he would find a *stater*, "that give him, said he, for both me and thee." As the stater evidently sufficed to pay a didrachm for each, there can be no doubt that the shekel or stater was considered by the Jews to be a tetradrachm.

It is very uncertain whether the Hebrews at any time employed a *maneh* of 60 shekels. They most certainly did not do so for gold and silver, and probably not even for copper and other cheap commodities. Very unfortunately the famous passage in Ezekiel (xlv. 12), which deals with weights and measures, is so confused in the description of the maneh that we cannot employ it as evidence. The one element of certainty is that the gold shekel never varied from first to last. It is likewise probable that, whilst the heavy maneh was introduced for gold silver and copper alike, the shekel always remained the same, 100 shekels being counted to the mina of gold and silver in the royal system, whilst 50 shekels always continued to be regarded as composing the maneh of the Sanctuary, such as we found it in the Book of Exodus. To confirm this view of the shekel we can cite the Bull's-head weight (fig. 27), which came from Jerusalem, and weighs 36·800 grammes, which represents the amount of 5 light shekels (making allowance for a small fracture), the light shekel being 8·4 grams. (130 grs.). It is plain that this is a multiple of the light and not of the heavy shekel, for it is not likely that such a multiple as 2½ would be employed. On the other hand, we found the five-fold multiple of the light shekel appearing in the Assyrian system, and also the Egyptian.

FIG. 27. Bull's-head Five-shekel Weight.

The Hebrew systems, as we have tentatively set them forth, may be seen in the following tables.

I. Earliest period. Shekel of 130-5 grs. alone employed for gold and probably silver.

II. Mosaic period. *Gold and Silver.* (The old light shekel or ox-unit is now called shekel of the Sanctuary to distinguish it from its double.)

   50 light Shekels = 1 Maneh
   3000 light Shekels = 60 Manehs = 1 Kikkar (*talent*).

III. Regal period. *Gold.*

100 light (= 50 double) shekels = 1 heavy Maneh
5000 heavy (= 10,000 light) „ = 100 heavy Manehs = 1 talent.

The same system was probably employed for *silver and copper*, but instead of counting 100 light shekels to the Maneh as in the case of gold, they reckoned silver and copper by the double shekel, probably called the king's shekel in contradistinction to that of the Sanctuary.

IV. After the Return. The light shekel still retained for *gold*, and the Babylonian, or Phoenician silver standard, employed for *silver*.

V. **Maccabean Period.** *Gold* on the old standard, and *silver* (now first coined) struck on the Phoenician silver standard of 220 grains.

Copper was estimated most probably on the the old double shekel system; and most likely the royal Assyrian heavy system of 60 shekels to the maneh and 60 manehs to the talent was adopted in its entirety for copper and other articles of no great value in proportion to their bulk[1].

[1] We are unfortunately unable to gain any definite knowledge from Ezekiel xlv., as v. 12, which gives the weight system, is confused, and there is a great discrepancy between the Hebrew and Greek texts. Though it is a prophetic passage, there is no reason for supposing that the prophet did not clearly understand the standard weight system of his time (600 B.C.), for his account of the metric system is singularly clear. It is best to give the whole passage as it appears in the Revised Version: " Thus saith the Lord God: Let it suffice you, O princes of Israel: remove violence and spoil, and execute judgment and justice; take away your exactions from my people, saith the Lord God. Ye shall have just balances, and a just ephah, and a just bath. The ephah and the bath shall be of one measure, that the bath may contain the tenth part of an homer, and the ephah the tenth part of an homer: the measure thereof shall be after the homer. And the shekel shall be twenty gerahs; twenty shekels, five and twenty shekels, fifteen shekels shall be your maneh." (vv. 9—12.) One thing is clear at least, and that is that the passage is a protest against over-exaction, and we may infer that the weight system here mentioned is for precious metals, seeing that there is no mention made of the talent. The shekel is to be 20 gerahs, that is, the shekel of the Sanctuary. If the princes had sought to exact payment in *royal* shekels instead of the old shekel, and also to make the maneh of silver contain 60 shekels instead of 50, we can see every reason for the cry of the oppressed being loud.

The confusion in the Hebrew text may be due to the fact that there were two manehs in use, that of 50 shekels for gold and silver, and that of 60 shekels for other commodities. The Septuagint version is perfectly capable of explanation on the principles which I have indicated. The LXX. runs thus: καὶ τὰ στάθμια εἴκοσι ὀβολοί, πέντε σίκλοι, πέντε καὶ σίκλοι, δέκα καὶ πεντήκοντα σίκλοι ἡ μνᾶ ἔσται ὑμῖν. So Tischendorf.

There is a MS. (Cod. Al.) reading οἱ πέντε σίκλοι, καὶ πέντε καὶ οἱ δέκα σίκλοι. Tischendorf's text can hardly be right, πέντε καὶ σίκλοι, δέκα καὶ πεντήκοντα contain two most unnatural collocations. δέκα καὶ πεντήκοντα is absolutely absurd as a way of expressing 60. εἷς καὶ πεντήκοντα up to ἐννέα καὶ πεντήκοντα to express 51 to 59 are reasonable and found universally, but to add on 10 to one of the main multiples of 10 in the decimal system is a method unknown, and is just as absurd in Greek as it would be if in English we were to say 10 and 50, meaning thereby 60. Again in the previous clause, the words πέντε καὶ point to some other numeral such as 10, or 20, as necessarily following. This is obtained by taking the MS. reading πέντε καὶ δέκα σίκλοι, καὶ πεντήκοντα, κ.τ.λ.

PHOENICIAN STANDARD.

The total loss of the literature and records of the Phoenicians, and the fact that neither in their own country nor in the greatest of their colonies, Carthage, did they employ coined money until a comparatively late period, make the task of restoring their weight system very difficult if not hopeless. The *silver* standard called Phoenician or Graeco-Asiatic is the sole evidence to show that they employed as their unit for gold the heavy Babylonian shekel of 260 grs. On the other hand we have just seen that their close neighbours, the Hebrews, from first to last, and the ancient people of the Nile with whom the Phoenicians were in the closest trade relations (having large trading communities settled in the Delta, and from whom

Now the LXX. gives the plural στάθμα for "*shekel*": στάθμα means the actual weights employed in weighing the amounts of gold or silver so weighed. Ezekiel is describing the various weight-units to be employed: "And the weights are 20 gerahs (lupins), *the* five shekel weight, *the* fifteen shekel weight, and fifty shekels shall be your maneh." The article οἱ is very rightly used before πέντε, for it refers to the well known multiple of the shekel, of which we spoke above when dealing with the Bull's-head weight. The same explanation may probably be given of *the* fifteen shekel weight. The maneh of 50 shekels of 20 gerahs each is the old maneh of the Sanctuary (Period II.), not the royal maneh which contained 100 light shekels.

Now turning to the Hebrew version we find "twenty shekels, five and twenty shekels and fifteen shekels," the sum of which makes a maneh of 60 shekels, or the royal Assyrian and Hebrew *commercial* maneh. It is also to be observed that the position of *fifteen* is unnatural; it ought to come in the series before "twenty" and "five and twenty." Fifty stands in the corresponding place in LXX. Has the Hebrew text altered 50 into 15 so as to obtain a total of 60? But there is another question; Why do we find "five" and "fifteen" stand first in LXX., and "twenty" and "twenty five" in Hebrew? On the theory, that of the Septuagint translators, that the prophet is describing a series of weight-pieces, it is quite simple. Combine the numbers of both versions, and place them in order thus: 1 shekel, 5 shekels, 15 shekels, 20 shekels, 25 shekels (½ maneh), 50 shekels (maneh). This gives a rational explanation of how the discrepancy arose. The LXX. translated from a text which probably ran thus, 5 shekels, 10 shekels, 15 shekels, and went no further with the series. For it is not at all improbable that the reading οἱ δέκα is due to the fact that after οἱ πέντε σίκλοι stood οἱ δέκα, which was followed by οἱ πεντεκαίδεκα σίκλοι. The Jews of a later date, knowing only of the commercial mina of 60 shekels, left out some of the numerals, and altered 50 into 15 to make up 60 shekels.

they had borrowed the hieroglyphic syllabic symbols, which with them became the Alphabet), had employed the light shekel, the only *gold* unit that likewise from first to last prevailed throughout the vast regions of Central Asia Minor, and as we have seen, was the unit of Greece even in the early days when the great cities of Mycenae and Tiryns were in direct contact with, and deriving their arts and civilization from Asia or from Egypt.

The derivation of the Phoenician *silver* standard of about 225 grs. (14·58 gram.) according to the hitherto received doctrine is as follows. As the Babylonians formed their silver standard by making into *ten* pieces the amount of silver equivalent to the "light gold shekel," so the Phoenicians and Syrians are supposed to have divided the amount of silver equivalent to "the heavy shekel" into *fifteen* pieces, gold being to silver in each case as 13·3 : 1. But we ask why did the Phoenicians adopt so awkward a scale as the quindecimal when it was possible for them to employ the decimal or duodecimal? In the next place by the supposed system $7\frac{1}{2}$ silver shekels were equal to one light shekel, that is the gold unit which was universally employed amongst all the peoples with whom they traded: and what number could be more awkward for purposes of exchange than $7\frac{1}{2}$? If therefore we can show that it is probable that at one period silver was exceedingly abundant in Phoenicia compared with gold, and that consequently gold was worth considerably more than 13 times its weight in silver, the sole support for the heavy shekel being the Phoenician unit is removed, and the theory of the *fifteen stater* system falls to the ground. It is well known that the Phoenicians had much of the trade of Cilicia and the other coast regions of Asia Minor in their hands. It was Cilicia that produced the chief supplies of silver for Western Asia[1]. From this land therefore the Phoenicians obtained vast quantities of silver, and it was from them almost certainly the Egyptians, who had no native silver, obtained a supply of that metal. But this was not all. About 1000 B.C. the Phoenicians, in their quest after new and unexhausted regions, made their way westward and reached

[1] Herod. III. 89, *seqq.*

Spain. I have already related the ancient stories which embody the account of the marvellous amount of silver which the first bold explorers brought back. We need not wonder then if in the days of king Solomon, "silver was nothing accounted of" in Syria and Palestine. We also saw that the relative value of gold and silver was just as liable to fluctuate in ancient, as in modern times, according to the supply of either metal, and when we come to deal with the Greek system we shall find many instances of this. If we then suppose that gold was to silver as $17:1$ in Phoenicia, the gold shekel of 130 grs. would be worth ten silver pieces of 220 grs. each. ($130 \times 17 = 2210$; $\frac{2210}{10} = 221$). This is in reality far closer to the actual weight of the coins than the result obtained by the old hypothesis: $260 \times 13\cdot3 = 3466 \div 15 = 231$ grs. Troy, which is about 10 grs. higher than the actual coin weights.

The approximation gained by our conjectural relation of $17:1$, is far closer than that obtained by that of $13\cdot3:1$. The conclusion is probable that silver was far cheaper in Phoenicia and the contiguous coasts than elsewhere in Asia Minor, and that it was natural that the weight of the silver unit was increased in order to preserve the relation in value between one gold unit, and ten silver units. Lastly we may point out that at no place on the coast of Phoenicia or Asia Minor, the region especially in contact with the Phoenicians, do we find *gold* pieces struck on the heavy shekel. *Electrum* certainly was coined on this foot; but of this we shall be able to give a satisfactory explanation. We have (with the exception of some Lydian pieces) to go as far north as Thasos or Thrace before we find a gold coin of such a nature, which is of course nothing more than a double stater.

The Phoenician gold mina was probably like the Hebrew, which was most likely borrowed from it, the fifty-fold of the heavy shekel, 100 gold shekels and 100 silver shekels constituting a maneh, as amongst the Hebrews in the time of Solomon. But we can conjecture with some probability that at an earlier stage they weighed their gold and silver according to the old common ox-unit, which we found in use among the Hebrews under the name of the Holy Shekel or shekel of the

Sanctuary. No doubt the mina for gold always contained 100 light or 50 heavy shekels, and when their own peculiar shekel of 220 grs. came into vogue for silver, 50 such shekels made a mina. Finally, there can be little doubt that 60 minas invariably went to the talent.

In the case of commercial weights, it is most probable that 60 heavy shekels made a mina: this is rendered almost perfectly certain by the Lion weights with Phoenician as well as cuneiform inscriptions found at Nineveh, 60 heavy minas forming a heavy talent.

THE PHOENICIAN COLONIES.

It is worth while before going further to enquire whether we can gain any light from the systems of weight employed by the famous daughter-cities of Phoenicia, such as Gades and Carthage. A weight bearing in Punic characters the name of the Agoranomos and the numeral 100 has been found at Jol (Julia Caesarea) in North Africa, but unfortunately it has suffered so much by corrosion from water and the loss of its handle that it is impossible to make any tolerable approximation to its original weight. Hultsch[1] conjectures with some probability that, making allowance for its loss, it represents 100 *drachms*, and deduces from this that the Carthaginians treated the drachm as their *shekel*, but for this latter hypothesis there seems no sufficient evidence. If this supposition were true, the weight would represent a half-mina of the Phoenician *silver* standard. But there is one thing which this weight does prove, and that is that, whether it be a mina or half-mina, it is the drachm or shekel, which was evidently regarded as the unit of the system, not the mina. Thus once more we get a confirmation of our general thesis that the mina and talent are the multiples, and that it is the shekel or stater which is the basis. Nor does the coinage of Carthage furnish us with all the information that could be desired, for it was only after 410 B.C. that that great "mart of merchants" began to strike coins, and even then it was only in her Sicilian pos-

[1] *Metrol.*², p. 420.

sessions that she did so, no doubt induced to adopt the practice by constant contact with her Greek enemies: for not only the type (of Persephone) was borrowed from Syracusan coins, but the very dies were engraved by the hands of Greek artists. The gold coins are struck on a standard of about 120 grs. Troy, whilst the silver issue consists of tetradrachms of the so-called Attic (or more simply light shekel or ox-unit) standard of 130—135 grs. Since during the same period (405—347 B.C.) Syracuse[1] was issuing gold pieces on the Attic standard, it is most probable that it is only through the want of heavier specimens that we are compelled to set the Siculo-Punic coins issued at Panormus (Palermo) and other places in Italy so low as 120 grs. It was not until about the time of Timoleon (340 B.C.) that money was coined at Carthage itself. This coinage consists wholly of gold, electrum and bronze, down to the time of the acquisition of the rich silver mines of Spain, and the foundation of New Carthage in that country by Hasdrubal, the son-in-law of Hamilkar Barca and brother-in-law of Hannibal, in the interval between the First and Second Punic wars (241—218 B.C.), when large silver coins both Carthaginian and Hispano-Carthaginian seem to have been first struck[2].

The gold and electrum coins of the first period are of the following weights: *gold* 145 and 73 grs.; *electrum* 118, 58 and 27 grains. The gold unit is thus some 10 grains higher than the normal value of the ox-unit. If these coins belonged to an earlier period we might with some confidence affirm that the variation was due to the plentiful supply of gold derived by the Carthaginians from the still unexhausted gold deposits of Western Africa. This is perhaps the true explanation even at the late period when the coins were issued, but there may have been a desire to adjust the three metals, gold, electrum and silver, so that they might be conveniently exchanged. It will be observed that the electrum coins are struck on a unit of 118 grs., and it is not at all improbable that silver was reckoned by the same unit, even though not yet coined; for when the silver coins appear they are struck on a standard of 118 or 236 grs. It will be at once noticed that

[1] *Metrol.*², p. 153.　　　[2] Head, *op. cit.* p. 789.

this standard is considerably higher than the Phoenician silver standard found along the coasts of Asia Minor. It may thus have been found convenient to raise by a few grains the weight of the gold unit so as to harmonize the relations between the three metals. Further speculation is vain, as we do not know the proportion of gold contained in the electrum coins[1]. From what we shall shortly learn about the electrum of Cyzicus, it is not impossible that the gold piece of 73 grs. was worth an electrum stater of 118 grs.

Coming to the Phoenicians of Spain we find that Gades, which did not begin her coinage until about 250 B.C., employed a standard for her silver of 78 grains, and that the island of Ebusus (*Iviza*) struck didrachms of 154 grs., a half-drachm of 39 grs. and a quarter-drachm. This coincides closely with the 78 grain drachm of Gades. It is palpable that there is no connection between this standard and the Phoenician standard of 220 grs. As the same system is found in the cities of Emporiae and Rhoda (*Ampurias* and *Rosas*) in the north-east of Spain, and in the earliest drachms of Massilia (*Marseilles*)[2], it is far more reasonable to suppose that the relations between gold and silver throughout Spain were such that, in order to make a certain fixed number of silver pieces equivalent to the gold ox-unit, it was found necessary to make the silver didrachm of about 156 grs. and the drachm of 78 grs.

It would thus seem that the principle which we shall seek to establish for the Greek silver standards held true of the Phoenician likewise,—that whilst the gold unit, the basis of all weight, remains unchanged or was but very slightly modified even at a late period (when the idea of the original ox-unit must have become dimmed by time), in order to effect a more complete harmonizing of a threefold system of gold, electrum and silver, the silver units shew every kind of variety, which can only be accounted for by supposing that owing to the different relations between gold and silver in various regions

---

[1] The amount of gold in electrum varies greatly. Pliny, *H. N.* xxxiii. 4. 23, ubicumque quinta argenti portio est, et electrum uocatur. The Carthaginian electrum probably came from Spain (cp. p. 94).

[2] Head, *op. cit.* p. 2.

and at various periods in the same regions, it was found necessary from time to time to increase or diminish the weight of the silver unit. Thus if gold was to silver as 12 : 1 in the 3rd century B.C., we find a ready explanation for the standard of Gades and Emporiae. The gold unit of 130 grs. would be worth ten silver units of 156 grs. each (130 × 12 = 1560 ÷ 10 = 156). So too the 118 gr. standard of Carthage may be explained by supposing that gold was to silver as 11 : 1; for then 1 gold unit of 130 grs. = 12 silver of 118 grs. each (130 × 11 = 1430 ÷ 12 = 119 grs.), duodecimal division perhaps being preferred to the decimal owing to the relations between electrum and silver, the former perhaps being as in Lydia[1] counted at 10 times the value of the latter. If gold was to silver as 12 : 1, and electrum to silver as 8 : 1, electrum being thus nearly two-thirds gold, one gold piece of 75 grs. = 1 piece of electrum of 118 grains, and 8 pieces of silver of 116 grs. each (75 × 12 = 900; 116 × 8 = 928), and 1 piece of electrum of 118 was worth 8 pieces of silver of 116 grs. each. All this is, be it remembered, purely conjectural, as we know nothing of the actual relations existing between any pair of the metals.

However, when we come to deal with the electrum of Cyzicus we shall be able to produce some data, which will at least show that our suggested explanation of the relations existing between gold, electrum and silver at Carthage is not purely chimerical.

Lastly comes the question of the commercial weight-system. We have already spoken of the badly preserved weight from Jol, but we could not say whether it was used for the precious metals, or more ordinary merchandize. However, the great Phoenician inscription of Marseilles, already referred to, makes it plain that even in the weighing of meat they reckoned by the shekel and not by the mina; for we find in it mention of 300 [shekels] and 150 [shekels] of flesh from the victims. This completely accords with the 20 shekels of food mentioned by Ezekiel (iv. 10), and clearly indicates that even in what we may well believe to be the heavy commercial shekel, the ancient decimal system had not been superseded by the sexa-

[1] Pliny, *H. N.* xxxiv.

gesimal; and, further, that the mina had not succeeded in supplanting the more ancient fashion of counting by shekels; for had such been the case, the weight of the meat would have been expressed in 6 manehs, or 3 manehs. This piece of evidence confirms the results which we arrived at in the case of the Hebrews—that it was only at a later period that reckoning by manehs came into use. The Phoenician colonies of the West, including Carthage herself, had probably been planted before the influences of the Chaldaean system had obtained a solid footing in Palestine. We may however not unreasonably believe that the Carthaginians employed some such form of talent as we find in the Book of Exodus, 3000 shekels ($50 \times 60 = 3000$) going to the talent, though as yet no record has revealed to us the actual existence of either *talent* or *mina*.

CHAPTER XI.

THE LYDIAN AND PERSIAN SYSTEMS.

"The Lydians," says Herodotus, "were the first of all nations we know who struck gold and silver coin[1]," a tradition also attested by Xenophanes of Colophon, according to Julius Pollux[2]. These statements of the ancient writers are confirmed by an examination of the earliest essays made in Asia in the art of coining; from which the best numismatists have been led to ascribe it to the seventh century B.C. and probably to the reign of Gyges, who from being a shepherd, by means of the "virtuous ring" became the founder of the great dynasty of the Mermnadae, and of the new Lydian empire as distinguished from the Lydia of a more remote antiquity. The first issues of the Lydian mint were rudely executed coins of electrum, being staters and smaller coins of the standards usually known as the Babylonian and Phoenician, of which the earliest staters weigh about 167 and 220 grs. respectively[3]. It is most likely that the Babylonian standard was intended for commerce with the interior of Asia Minor, and the Phoenician for transactions with the cities of the western seaboard, to coincide with the silver standards in use in these respective regions. The proportion of gold and silver in electrum is ex-

---

[1] Herod. I. 94, πρῶτοι δὲ ἀνθρώπων, τῶν ἡμεῖς ἴδμεν, νόμισμα χρυσοῦ καὶ ἀργύρου κοψάμενοι ἐχρήσαντο.
[2] Julius Pollux, IX. 83.
[3] Head, op. cit. p. 544.

ceedingly variable: according to Pliny[1] any gold alloyed with one-fifth of silver (and by implication any containing any higher proportion of silver) was called electrum. We shall soon find that the electrum staters of Cyzicus contained about an equal amount of either metal; but the analysis of Lydian electrum gives a proportion of 73 per cent. of gold to 27 per cent. of silver, or practically 3 to 1. As gold in the central parts of Asia Minor stood to silver as $13\cdot3:1$ in the reign of Darius and probably long before, we may not unreasonably assume that such also was the relation between them in the reign of Gyges, at least in the interior. In this case electrum would stand to silver as $10:1$, a proportion exceedingly convenient for exchange, as a single standard served for both metals, one electrum ingot of 168 grs. being equal to 10 silver ingots of like weight. We have already seen that one gold unit of 130 grs. was equivalent to 10 silver units of 168 grs., therefore the gold ox-unit was exactly represented in value by the electrum ingot of 168 grs., for, according to our statement of the composition of the Lydian electrum, 168 grs. of that alloy would contain 126 grs. of pure gold. If we were certain that on the coast of Asia Minor the relation between gold and silver was $13\cdot3:1$, we should be compelled to follow Brandis and the rest in making the double gold shekel of 260 grs. equal to 15 silver shekels of 220 grs. each; again, if we accept as universal the relation of gold to electrum as $4:3$, and accordingly make one piece of electrum of 220 grs. equal to 10 silver pieces of the same standard, we shall find it impossible to obtain any convenient relation between the gold stater of 130 grs. and the electrum stater of 220 grs. But from this difficulty it is not hard to find an escape: 224 grs. of electrum = 168 grs. of gold; that is exactly $1\frac{1}{3}$ gold shekels ($\frac{129}{3} = 43 \times 4 = 172$). The division into thirds and sixths is of course a well-known feature in the coinage of the Asiatic coast-towns. Thus there would be no practical difficulty in the ordinary monetary transactions, for three Phoenician drachms of electrum (= 168 grs.) would = 1 gold shekel; and 4 gold Thirds (*Tritae*), or 8 gold Sixths (*Hectae*), would equal one electrum stater of 224—220 grs.

[1] *H. N.* xxxiii. 4. 23, ubicumque quinta argenti portio est, et electrum uocatur.

If on the other hand silver held a lower value in relation to gold on the coasts of the Aegean, and the electrum employed in that quarter was alloyed to a greater extent with silver, two disturbing elements are introduced. The probabilities are in favour of silver being cheaper in Cilicia and the contiguous region, and most certainly at Cyzicus the electrum was half silver, whilst the Phocaic electrum had a bad name in antiquity, since according to Hesychius Phocaic gold was synonymous with bad gold. Is it then possible that 220 grains of electrum were equivalent to 130 grs. of pure gold? This gives about 60 per cent. of gold. If gold was to silver as 13·5 : 1, the gold unit of 130 grs. is equal to 8 silver pieces of 220 grs. ($130 \times 13·3 = 1765 \div 8 = 220·6$). In our present state of knowledge it is impossible to decide in favour of either view, but it is at least evident that some such relation and adjustment must have existed between the three metals. In fact the problem which the Lydians tried to solve was not merely that of *Bimetallism*, but of *Trimetallism*.

These early electrum coins are simply bullet-shaped lumps of metal, like the so-called *bean* money formerly employed by

FIG. 28. Lydian electrum coin.

the Japanese, having what is termed the obverse plain or rather striated, as a series of lines in relief run across the coin, whilst the reverse has three incuse depressions, that in the centre oblong, the others square. The coin here figured (from the British Museum specimen) is on the Babylonian silver standard (166·8 grs.), but it is on the staters of Phoenician standard that we first find any attempt at types or symbols. The idea of engraving some symbol on the punches used for stamping the incuse depressions was in truth the grand step towards the creation of a real coin. Thus a stater of 219 grs. which bears in the central incuse a running fox, in the upper square a stag's head, and the lower an X-like device, may be regarded as the first complete coin as yet known. It would seem from this, therefore, that it was on the coast-region, where the Lydians came into contact with the artistic genius of the Greeks, that the real start in the art of striking money took place. Electrum was employed because it was found native in great quantities in the whole district which lay around Sardis, in the valleys of Tmolus, and the sands of Pactolus. The ancients found considerable difficulty in freeing the gold from the associated silver (p. 97).

Once known, Miletus and other important Ionian cities were not long in improving on the Lydian invention. The advantages of a metallic currency were so obvious that an intelligent and progressive race hastened to avail themselves of it. "Only those," says Captain Gill (speaking of the borders of Thibet and China), "who have gone through the weary process of cutting up and weighing out lumps of silver, disputing over the scale, and asserting the quality of the metal, can appreciate our feelings of satisfaction at being once more able to make payments in coin[1]." No sooner had the Ionians commenced coining than they appear to have adorned the face of the ingot with a symbol, probably both as a guarantee of weight and purity, and perhaps as a preventive of fraudulent abrasion. During this period it is not improbable that the arts of Ionia had made their influence felt in Lydia, and hence "it is impossible to distinguish with absolute certainty the Lydian issues from those of the Greek towns, but there is one type which seems to be especially characteristic of Lydia as it occurs in a modified form on the coinage attributed to the Sardian mint and to the reign of Croesus; this is the Lion and the Bull. These coins have on the obverse the forefronts of a lion and a bull turned away from one another and joined by their necks[2]," whilst the reverse shows three incuse depressions. This is Phoenician in weight (215·4 grs.). There

[1] *River of Golden Sand*, II. p. 78.
[2] Head, *op. cit.* p. 545.

are other coins, often attributed to Miletus, which may be assigned to Lydia; some with a recumbent lion on the obverse, and a reverse exhibiting the fox, stag's head, and X of the coin already described. To these may be added a series of coins bearing a lion's head with open mouth, and with what is commonly regarded as a star above it, but which is more probably part of the lion's hair, and on the reverse incuse sinkings, in some cases containing an ornamental star[1]. These coins have now with great probability been assigned by the eminent numismatist, Mr J. P. Six, to the Lydian king, Alyattes, the father of Croesus.

When Croesus ascended the throne in 568 B.C., one of his earliest acts seems to have been an attempt to propitiate the Greeks both of Asia and Hellas proper by sending offerings of equal value to the two most famous shrines of Apollo, Delphi and Branchidae. In the course of some fourteen years he reduced under the sway of Lydia all the regions that lay between the river Halys and the sea. "It seems probable (says Mr Head) that the introduction of a double currency of pure gold and silver, in place of the primitive electrum, may have been due to the commercial genius of Croesus." If this be so, the monarch seems to have acted with thrift in his offerings, for according to Herodotus his dedications at Delphi were all of *white gold*, *i.e.* electrum. Perhaps then he got no more than he deserved when, induced by the declaration of the Delphic prophetess that he would destroy a mighty kingdom, he made war upon Cyrus with disastrous issue. There however can be no doubt that Croesus made some important monetary change, for in after years there still remained a clear tradition of Croesus' stater (Κροίσειος στατήρ), just as the famous gold stater of Philip of Macedon was known as the *Philippean* or *Philippus*[2]. In his monetary reform Croesus seems to have had regard to the weights of the two old electrum staters, each of

---

[1] *Ibid.* p. 503.

[2] Pollux, III. 87, εὐδόκιμος δὲ καὶ ὁ Γυγάδας χρυσὸς καὶ οἱ Κροίσειοι στατῆρες: ix. 84 *sq.*, ἴσως δὲ ὀνομάτων καταλόγῳ προσήκουσιν οἱ Κροίσειοι στατῆρες καὶ Φιλίππειοι, καὶ Δαρεικοί, καὶ τὸ Βερενίκειον νόμισμα καὶ Ἀλεξανδρεῖον, καὶ Πτολεμαϊκὸν καὶ Δημαρέτειον, κ.τ.λ.

which was now represented by an equal value, though not of course by an equal weight, of pure gold.

Thus the old Phoenician electrum stater of 220 grs. was replaced by a pure gold coin of 168 grs., equivalent like its predecessor in electrum to 10 silver staters of 220 grs. each, and the old Babylonian electrum stater of 168 grs. was replaced by a new pure gold stater of 126 grs., equal in value like it to 10 silver staters of 168 grs. each, "as now for the first time coined." These gold coins bear as obverse the foreparts of a lion and a bull facing each other, and on the reverse an oblong incuse

FIG. 29. Coin of Croesus.

divided into two parts (Fig. 29). Of the Babylonian standard we find:

Stater	168 grs.
Trite	56 ,,
Hecte	28 ,,
Hemihecton	14 ,,

And of the light shekel:

Stater	126 grs.
Trite	42 ,,
Hecte	21 ,,
Hemihecton	11 ,,

Of Babylonian standard *silver*:

Stater	168 grs.
½ stater	84 ,,
⅓ stater	56 ,,
1/12 stater	14 ,,

This double standard for gold is at first sight somewhat strange until we observe that the two systems are in complete harmony. For the gold piece of 168 grs. is nothing more than $1\frac{1}{3}$ of the light shekel ($168 \div \frac{4}{3} = 126$ grs.). The third of the light shekel (42 grs.) is the fourth of the Babylonian of 168 grs. There can be no doubt that the coins of 168 grs. were simply an experiment suggested by the coincidence that the number of grains (168) in the Babylonian silver shekel was exactly one-quarter more than those in the *light* gold shekel, in the hope doubtless of obtaining a single standard for gold electrum and silver. The division of the silver stater into thirds would facilitate the process of exchange, as 13 silver staters and one-third would be equivalent to the gold piece of the same Babylonian standard, whilst 10 silver staters would be equivalent to one of the old electrum pieces of 168 grs. It is at all events certain that the standard of 168 grs. was not a regular gold unit, for it simply makes its appearance for a brief space, there being no trace of it at any earlier period, nor does it afterwards appear save in its own legitimate province of silver. A perfectly analogous case is that of the gold pieces struck by the Ptolemaic kings, who, starting with the gold stater of Philip and Alexander and the Phoenician standard for silver (after the founder of the dynasty had for a short time used the so-called Rhodian standard), presently struck gold pieces on the same standard as their silver. But the experiment of Croesus, if such it was, did not succeed. For the eastern mind was still too much impressed with the necessity of cleaving fast to the original weight unit obtained from the ancient unit of barter. For whether the attempt had failed before the reign of Croesus was brought to a sudden end by the conquests of the great Cyrus, or whether he continued up to the very hour of the Persian conquest to coin, at least for one part of his dominions, the gold pieces of the Babylonian silver standard, it matters little. As we have no evidence on the point, we cannot say whether there were two gold minae and two gold talents in use, one being of course the ordinary gold talent (called Euboic) of 3000 light shekels of 130 grs., the other containing 3000 shekels of 168 grs. each. The probability I think is that only the former existed. As 50 of the latter shekels made 1⅙ minae, there was no practical difficulty in making any calculations; on the other hand, if there had been two separate minae, and two separate talents, it would have led to great complications. The fact that we hear nothing about any such second gold system existing in Asia, and that when Darius fixed the tribute from each region he did not make it the basis of his payment, which he would probably have done as he would thus have made a considerable gain, by causing the payments in gold as well as those in silver to be made on the Babylonian standard, seems to put beyond all doubt that the 168 grain gold piece was not a real unit, but was simply regarded as 1⅙ shekels, and was nothing more than a temporary effort to simplify the trimetallic monetary system of Lydia.

What system the Lydians employed for commercial purposes we have no means of knowing, but we may conjecture plausibly that the light royal mina of 60 shekels was the standard employed.

The Persian Standard.

We may adopt the generally received belief that the Persians, like the Medes and Babylonians, did not coin money (although they were probably acquainted with the Lydian stater) until after the conquest of Asia Minor and Egypt by Cyrus and Cambyses, and the reorganization of the empire by Darius the son of Hystaspes (522—485 B.C.). For although the learned *savants* MM. Oppert and Révillout[1] hold that Daric (Δαρεικός) is unconnected with the name Darius (Δαρεῖος), an opinion supported by Dr Hoffmann[2], and rather regard it as derived from the Assyrian *darag mana*, "degree (i.e. $\frac{1}{60}$) of a mina," and although Mr G. Bertin has read the word *dariku* on a Babylonian contract, dated in the twelfth year of Nabonidas, five years before the conquest of Babylon by Cyrus[3], it does not at all follow that either *darag* or *dariku* refers to a

[1] *Annuaire de Numismatique*, 1884, p. 119.
[2] *Zeitschr. für Assyriologie.* Vol. II. 48 (1887).
[3] *Proceedings of the Society of Biblical Archaeology*, 1883—4, p. 87.

coin. That the unit was employed for gold ages before the Persians ever descended from the mountains there can be little doubt But whether we adopt or reject the Greek tradition that the Daric (Δαρεικός) was named from Darius, as the Philippean and Croesean staters were called after the sovereigns who first struck them, it is perfectly certain that Darius organized the whole numbering system of the great empire to which he had succeeded, and that he coined gold pieces of the first quality: for Herodotus tells us that Darius, having refined gold to the greatest extent possible, had coin struck[1]. This would be very analogous to the course pursued by Croesus and Philip; gold in some form was current in the dominions of both these princes before their reigns, but it was owing to certain reforms introduced and to the issue of a gold coin of a certain pattern, that the names of both became associated with particular kinds of gold coins. By the time of Xerxes the son of Darius vast quantities of these Darics were circulating through Asia Minor, for Herodotus relates that the Lydian Pythius had in his own possession as many as 3,993,000 of them, a sum afterwards increased by Xerxes to 4,000,000. They became the gold currency of all the Greek towns not only of Asia Minor, but also of the islands, and made their way in considerable quantities into the great cities of the mainland of Hellas, and wrought as much harm in disuniting the various states of Greece as did the gold staters of Philip at a period a little later. Darics formed a regular part of the wealth of a well-to-do Athenian at the time of the Peloponnesian war. Thus Lysias[2] relates that when his house was entered and plundered by the minions of the Thirty, his money chest contained 100 Darics, 400 Cyzicenes, and 3 talents of silver. It is only necessary to enumerate some of the passages in the Greek authors, where mention is made of their coins, to show how wide an influence they exercised in the eastern Mediterranean. Besides Herodotus and Lysias already mentioned, Thucydides, Aristo-

---

[1] IV. 166, Δαρεῖος μὲν γὰρ χρυσίον καθαρώτατον ἀπεψήσας ἐς τὸ δυνατώτατον νόμισμα ἐκόψατο.

[2] *Or.* XII. 70 τρία τάλαντα ἀργυρίου καὶ τετρακοσίους κυζικηνοὺς καὶ ἑκατὸν δαρεικοὺς καὶ φιάλας ἀργυρίου τέσσαρας.

phanes, Xenophon, Demosthenes, Arrian, Diodorus and many others all make mention of these famous coins[1]. No classification of them according to the reigns of the monarchs by whom they were issued is possible, for this is precluded by the absence of all inscriptions, and the great uniformity of style. They bear on the obverse the king of Persia bearded crowned and clad in a long robe; he kneels towards the right on one knee; on his back is a quiver, in his right hand is a long spear, and in his outstretched left a bow (from which came the familiar Greek name of Archers for these pieces). The reverse is simply marked by an oblong incuse.

Their weight may be set at 130 grs., which of course is the light shekel or ox-unit. We have no difficulty in fixing the gold mina or talent. In fact we have already seen on p. 260 that the Persian talent of gold was the same as the Euboic-Attic talent. Hence

$$1 \text{ Daric} = 130 \text{ grs.}$$
$$50 \text{ Darics} = 1 \text{ mina} = 6{,}500 \text{ grs.}$$
$$3000 \text{ Darics} = 60 \text{ minas} = 1 \text{ talent} = 390{,}000 \text{ grs.}$$

For silver currency the Persians employed half of the Babylonian silver stater of 168 grs., its usual weight being about 84 grs. This coin was in every way similar to the Daric and in fact is sometimes called by the same name by writers of a later age[2], but the more usual appellation in the classical writers was the *Median* siglos (Μηδικὸς σίγλος) or simply *siglos*. Twenty of these sigli were equivalent to one gold Daric, for Xenophon appears to count 3000 Darics as equal to 10 talents of silver, or in other words to 60,000 sigli (6000 × 10 = 60,000). The siglos may therefore be regarded as the Persian drachm or half-stater. As 130 grains of gold are thus made equal to 1680 grs. of silver (84 × 20), gold held to silver the old ratio of 13 : 1.

---

[1] Thuc. VIII. 28; Xen. *An.* I. 1. 9; I. 3. 21; I. 7. 18; V. 6. 18; VII. 6. 1; *Cyrop.* V. 27; Dem. XXIV. 129; Aristoph. *Eccl.* 602; Arrian *Anab.* IV. 18. 7; Diod. XVII. 66, etc.

[2] Plutarch, *Cimon*, x. 11, φιάλας δύο, τὴν μὲν ἀργυρείων ἐμπλησάμενον Δαρεικῶν, τὴν δὲ χρυσῶν.

THE LYDIAN AND PERSIAN SYSTEMS.

The Persian silver standard was formed thus:

1 siglos = 84 grs.
100 sigli = 50 staters = 1 mina = 8400 grs.
6000 sigli = 3000 staters = 60 minae = 1 talent = 504,000 grs.

As regards commercial weight we may fairly assume that the old light and heavy *royal* systems continued in use in the respective regions where they had been employed in early days.

CHAPTER XII.

The Greek System.

We are now come to the most important portion of our task, the development of the Greek and Italic systems. In the Homeric Poems we found the Talanton (or value of a cow in gold) the sole unit of weight, and that only employed for gold. This Talanton has been shown to be the same in weight as the light gold shekel of Asia Minor, which, under the form of coin, we have just been discussing as the Croesean stater and Persian Daric. It was therefore nothing else than the Euboic or Attic stater of historical times, which at all periods and at all places that fall within our knowledge formed the sole unit for the weighing of gold.

Besides the Talanton based on the ox, there was in all probability another higher unit in occasional use in Greece Proper. This was the threefold of the ox-unit. We have already had occasion to notice the small gold talent, called by some writers the Macedonian, which was equal to three Attic staters. The same weight under the name of the Sicilian talent was employed likewise for gold only in the Greek colonies of Sicily and Southern Italy. The conservatism of colonists is too well known to need illustration, and we may with high probability infer that the Greek settlers in Magna Graecia brought the small talent from their original homes. What was the origin of this weight? We have seen that everywhere all over our area the slave is the occasional higher unit. Thus the Irish slave (*cumhal*) was a unit of account equal to three cows. The slave in the Welsh Laws is equal to 4 cows, whilst in Homer we found a slave woman valued at 4 cows also. From the way in which this notice of her price occurs, it is probable that Achilles did not give a woman of the most ordinary kind as a prize, for had she been the ordinary slave-woman of account, there would have been no need to mention the price, as any one would have known how many cows exactly she was worth. It is then not improbable that three cows were commonly reckoned as the value of a slave, and accordingly the small gold talent, which is the multiple of the ox-unit, is simply the metallic representative of the slave, just as the Homeric Talanton itself is that of the cow.

What the exact weight of this unit was on Greek soil we are now enabled to ascertain by the aid of the treatise on the Constitution of the Athenians known to the ancients as the work of Aristotle, and the brilliant discovery and identification of which by the officials of the British Museum reflects much credit on British scholarship.

We had previously known from Plutarch (who ascribed the first coinage of Athens to Theseus[1]) that amongst his other reforms Solon caused drachms to be coined of lighter weight than those previously in currency, so that 100 of the new ones would be equal in value to 73 old ones. Some scholars have inferred that this was an expedient for relieving debtors, who would be allowed to pay in the new coin debts contracted in the older currency. The newly discovered Constitution dispels this assumption, and also affords us some most valuable additional matter[2]: "In his Laws then he appears to have made these enactments in favour of the people, but before his legislation he appears to have wrought the cancelling of debts, and afterwards the augmentation of the measures and weights, and the augmentation of the currency. For in his

[1] *Thes.* xxv., ἔκοψε δὲ νόμισμα βοῦν ἐγχαράξας.

[2] p. 27 (ch. 10) (Kenyon's ed.), ἐν μὲν οὖν τοῖς νόμοις ταῦτα δοκεῖ θεῖναι δημοτικά, πρὸ δὲ τῆς νομοθεσίας ποιησάσθαι τὴν χρεῶν ἀποκοπήν, καὶ μετὰ ταῦτα τήν τε τῶν μέτρων καὶ τῶν σταθμῶν καὶ τὴν τοῦ νομίσματος αὔξησιν. ἐπ' ἐκείνου γὰρ ἐγένετο καὶ τὰ μέτρα μείζω τῶν Φειδωνείων, καὶ ἡ μνᾶ πρότερον ἔχουσα παραπλήσιον ἑβδομήκοντα δραχμὰς ἀνεπληρώθη ταῖς ἑκατόν. ἦν δ' ὁ ἀρχαῖος χαρακτὴρ δίδραχμον. ἐποίησε δὲ καὶ σταθμὸν πρὸς τὸ νόμισμα τρεῖς καὶ ἑξήκοντα μνᾶς τὸ τάλαντον ἀγούσας, καὶ ἐπιδιενεμήθησαν αἱ μναῖ τῷ στατῆρι καὶ τοῖς ἄλλοις σταθμοῖς.

day the measures likewise were made larger than those of Pheidon, and the mina, which previously had almost seventy drachms, was filled up by a hundred drachms[1]. But the ancient type was the didrachm[2], and he also made as a standard[3] for his coinage 63 minas weighing the talent, and the minae were apportioned out by the stater, and the other weights."

The first point to engage our attention is the formation of a new standard for the *silver* coin (for no gold was coined for nearly two centuries): sixty-three old minas were taken to form a new talent, which of course was divided henceforward into 60 new minas. As the weight of the Attic talent in post-Solonian times is most accurately known, we can at once discover the weight of the ancient mina by dividing the ordinary weight of the talent (405,000 grs.) by 63 : $405,000 \div 63 = 6428$ grs., that is 322 grs. less than the post-Solonian mina of 6750 grs. As there are 50 staters in the mina, the ancient stater weighed 128·56 grs., or just a grain lighter than the Daric (129·6 grs.). The old mina of 6428 grs. had been equal to 70 drachms; each of these then must have weighed 92 grs. nearly, that is, the ordinary weight of an Aeginetic drachm. There can be no doubt that the coins of Aegina were used as

Fig. 80.   Coin of Eretria.

currency at Athens before Solon's time, where they circulated side by side in all probability with the coins of Euboea which bore the bull's head, whence arose the tradition of the earliest

---

[1] I have translated the παρά .[μικρὸν] of Kaibel and Wilamowitz instead of Kenyon's παραπλήσιον. According to Plutarch (Solon. 15) the old (silver) mina contained 73 drachms. The apparent discrepancy is easily explained. In the prae-Solonian mina there were 70 drachms of 92 grs. each. Plutarch writing at a later time took the number of drachms of 92 grs. in the post-Solonian mina of 6750, which is just 73. The information supplied by the *Polity* is evidently older and better.

[2] Tho. Reinsch needlessly regards ἦν δὲ ὁ ἀρχαῖος κ.τ.λ. as an interpolation.

[3] Kaibel and Wilamowitz read σταθμὰ instead of σταθμὸν.

coinage of Athens consisting of didrachms stamped with an ox. The old mina (63 of which went to the new *silver* talent) was of course the ancient standard used for weighing *gold* and *silver* before coined money was employed. It was that known as the Euboic, based on the ox-unit. The Aeginetic standard was only used for *silver*, *gold* at all times being weighed by the Euboic standard even where the Aeginetic was in use for silver. This standard was of course in full use for gold and evidently likewise for silver in prae-Solonian times, even though the Aeginetic drachms passed as currency at Athens. For if they had adopted the Aeginetic *standard*, 100 Aeginetic drachms would have been reckoned to the mina, but as only 70 drachms went to the mina it is evident that the old ox-unit (so-called Euboic) standard of unit 130 grs. with its corresponding mina was always the national Athenian standard.

We showed at an earlier stage that in the age when the art of coining was first introduced into Greece by Pheidon of Argos, it was probable that gold stood to silver in the proportion of $15:1$. For convenience, then, in Peloponnesus and in Central Greece a system was adopted by which 10 pieces of silver were equivalent to one piece or ingot of gold. This system, known as the Aeginetic, was thus otained.

Gold being to silver as $15:1$,

1 gold ingot (Talanton) of 130 grs. $\times 15 = 1950$ grs. of silver, 1950 grs. $\div 10 = 195$ grs.

Therefore 1 gold Talanton of 130 grs. $= 10$ pieces of silver of 195 grs. each.

It is possible that this method of making 10 silver pieces equal to one gold unit was developed at the time of the introduction of coined money, but it is more likely that it may have been in use even before that time.

Now it is worth observing that all through the classical period of Greek history the term stater is generally confined in use to gold pieces. Thus silver coins, unless they weighed 135 grs., are not described as silver *staters*, but are regularly termed didrachms. So general evidently was this practice that the adjective *chrysous* (χρυσοῦς) was regularly employed to express the gold unit, the masculine gender showing that the noun understood is *stater* (στατήρ). Thus Pollux says: "Some were termed staters of Darius, some Philippeans, other Alexandrians, all being of gold, and if you say *gold piece*, *stater* is understood: but if you should say *stater*, gold is not absolutely to be understood[1]." From the fact that Pollux draws attention to the exceptional use of *stater* to express a silver coin, on the principle that *exceptio probat regulam*, it is evident that stater regularly represents a gold piece of two Attic drachms. The familiar practice in Attic Greek, when speaking of a considerable sum of silver without employing either the term mina or talent, is to say 1000 drachms, 2000 drachms and the like, but not 1000 staters or 2000 staters, etc., whilst on the other hand, under like conditions, the practice is to enumerate gold not by drachms, but by *staters*. Thus in a fragment from the *Demi* of Eupolis quoted by Pollux[2] a man is described as possessing 3000 *staters* of gold. We certainly hear of an Aeginean stater and a Corinthian[3] stater (both of silver), but both are found in writers of comparatively late date, when usage was getting less exact, and besides, as the Aeginetic system had a separate individuality of its own, its unit being perfectly different from the Euboic Attic, might with justice be termed a stater. We are thus justified in considering the gold stater the legitimate descendant of the Homeric Talanton, the stater or *weigher* representing the Talanton or *weight* of the older time. As long as no other unit than the ox-unit or Talanton was employed, the Talanton or weight *par excellence* was sufficient to describe it, but when under Asiatic influences the higher unit of the *mina* (μνᾶ) and *talent* were introduced, a term was substituted which indicates clearly that the gold unit of 130 grs. was *the weigher* or basis of the whole system. Starting then with our ox-unit, we find already in Homer definite traces of a decimal, but nothing to indicate the existence of a sexagesimal system. *Ten* talents of gold are mentioned in several passages.

[1] Pollux ix. 59.
[2] Pollux ix. 58 ἔχων στατῆρας χρυσίου τρισχιλίους.
[3] Thuc. (i. 27) speaks of Corinthian drachms not *staters;* and (v. 47) of Aeginetic *drachms.*

Starting then with the ox-unit of 130 grs. we can thus arrive at the fully elaborated Greek systems. The term mina ($\mu\nu\hat{a}$) is beyond doubt a borrowing from the East. How far it was ever much employed in the reckoning of gold it is hard to say, but it is at least remarkable that, when we hear so frequently of *minae* of silver in the Attic writers, no instance of a mina of gold is quoted in our books of reference. From this one is led to infer that it was for the purpose of measuring the less precious metal, silver, that the term *mina* was brought into use in Greece. In fact, as stater is essentially a term which clings to gold, so *mina* is especially a term used of silver. With the mina the Greeks borrowed likewise the highest Asiatic unit (the *kikkar* of the Hebrews), which became the Talanton or talent of historical Greece. But it is remarkable that the Greeks did not borrow its Asiatic name along with the unit itself. They simply gave it their own name *weight* (literally, 'that which can be lifted,' cp. $\tau\lambda\acute{a}\omega$, *tollo*, etc.). This fact can be explained readily if we suppose that the Greeks, like all those other primitive peoples whom we have mentioned, had a rough and ready unit for estimating bulky wares, the standard of *the load*, or as much as a man could conveniently carry on his back. Having already such a unit they would have no difficulty in adopting the *load* or talent, which had been fixed according to the Sexagesimal system, and which had permeated all Western Asia. In fact their position towards the Asiatic *load*, which had been accurately fixed by the mathematical skill of the Babylonians, would be exactly analagous to that of the Malays of Java and Sumatra towards the accurately adjusted Chinese *picul*. Because the Malays themselves were accustomed to use *loads* of various weights as their rough highest unit of bulk, they have with all the more readiness received the form of the same unit, which the clever Chinese have incorporated into their commercial weight system by making it equal to 100 *chings* (catties, or pounds). But it is doubtful if at any time in Greece Proper the talent of gold was ever considered as a monetary unit. We have found Eupolis speaking of "3000 staters of gold" instead of simply saying a talent of gold, and when we do find mention made of talents of gold, as in a famous passage of Thucydides, where he describes the amount of gold employed by Pheidias in the making of the world-renowned chryselephantine statue of Athena for the Parthenon, whilst the computations in silver are expressed simply by talents, the gold is enumerated as talents *in weight*. We may assume that gold was weighed throughout Greece in historical times on the following system:

$$\begin{aligned}
1 \text{ stater} &= 130 \text{ grs.} \\
50 \text{ staters} = 1 \text{ mina} &= 6500 \text{ grs.} \\
3000 \quad ,, \quad = 60 \text{ minae} = 1 \text{ talent} &= 390{,}000 \text{ grs.}
\end{aligned}$$

When silver came into use it was probably weighed all through Hellas, as in Asia and Egypt, on the same standard as gold. This continued always to be the practice amongst the great trading communities of Euboea, Chalcis and Eretria, and their colonies, and also with Corinth and her daughter states. Hence the system was commonly known as the Euboic, sometimes as the Corinthian, and in later times, for a reason to be presently given, the *Attic*. But in this silver system it is no longer the stater which represents the smaller unit, but rather the *drachm* (δραχμή). Furthermore we find in most constant use a subdivision of the *drachm* called the *obol* (ὀβολός *nail* or *spike*), six of which made a drachm. There can be no doubt that this silver obolos represented the value in silver of the ancient copper unit from which it took its name, which itself was not estimated by weight but probably, as we saw above, was simply appraised by measure, as is done by all primitive peoples in the estimation of copper and iron, nay even in the very earliest stage of gold itself (p. 43). As six of these *nails* or *obols* made a handful (δραχμή) in the ancient copper system, so when each of them was equated to a certain amount of silver, the equivalence in silver was called an *obol*, and the six silver *obols* obtained the old name of *handful* or *drachm*. In the ordinary Greek system of reckoning silver it is 100 drachms, not 50 staters, of silver which form the mina. But of course at the earlier stages of the use of silver we may with some boldness assume that silver was simply weighed by the stater (or Homeric Talanton).

It is important then to note that among the smaller weight denominations silver has virtually no term peculiarly its own: for we have seen that *stater* belongs essentially to gold, whilst *drachm* and *obol* have originated in the use of copper. This is in complete harmony with what we know of the history of the metals themselves, gold and copper being known and employed long before men had learned to utilize silver; and so too, we find the late-introduced term *mina* in especially close connection with the latest employed of the three metals. This Euboic-Attic *silver* system maybe stated as follows:

$$6 \text{ obols} = 1 \text{ drachm}$$
$$100 \text{ drachms} = 1 \text{ mina}$$
$$60 \text{ minae} = 1 \text{ talent.}$$

The Corinthians, whilst making the *obol* of the same weight as the Euboic, made a different division of the silver stater; for as Corinth occupied the very portals of Peloponnesus where the Aeginetic system was universal, she found it convenient for purposes of exchange to divide her silver stater of 135 grs. into *three* drachms of 45 grs. each, one of which was for practical purposes identical with the Aeginetan *half drachm*. Thus two Corinthian drachms of 45 grs. each were equal to one Aeginetan drachm of 90 grs.

The Aeginetan Standard.

The desire to obtain 10 silver pieces equivalent in value to the gold ox-unit induced the Aeginetans, who were famous merchantmen, to make a silver system distinct from that of gold. Gold being to silver as 15 : 1,

$$130 \times 15 = 1950 \text{ grs. of silver.}$$
$$1950 \div 10 = 195 \text{ grs.}$$

With the Aeginetans as with the Euboeans in their silver system, the ancient copper units of the *nail* and *handful* played an important part. The story of Pheidon[1] having hung up in the temple of Hera at Argos the ancient currency of nails of

---

[1] Cp. p. 214.

copper and iron as soon as he struck his first issue of silver coins, if not absolutely true in all details, at least contains a most probable statement of what did actually take place when a real silver currency was first introduced. We have seen how the Chinese, starting with a barter currency of real hoes and knives, the objects of most general demand, gradually replaced those larger and more cumbrous articles by hoes and knives of a more diminutive size, until finally they became a real currency when they had been so reduced in size as to be utterly unfit for practical use. We saw likewise how that at the present moment the real hoe is the lowest unit of barter among the wild tribes of Annam, and that small bars of iron of given size are used in Laos, and that plates of metal ready to be made into hoes, and hoes themselves, are employed by the negroes of Central Africa, whilst on the west coast axes of a size too diminutive for actual use are employed as a real currency. As the day came when the Chinese finally replaced the archaic knife by the full developed copper coin called the cash, so the Aeginetans and Argives of the days of Pheidon superseded by a real coin ancient monetary-units consisting either of real implements of iron and copper, or bars of those metals of certain definite dimensions, or possibly mere Lilliputian representatives of such, which had previously served them as a true currency. On the whole however it is safest to assume from the names *nail* (*Obol*) and *Handful* (drachme) that the form in which copper or iron served as currency in Peloponnesus and the mainland of Hellas in general was that of rods of a certain length and thickness. We have cited already many analogous forms from modern Asia and Africa, and from the ancient Kelts, to which we shall presently add the ancient Italians. But just as we found that in the Soudan, whilst the slave and ox were universally the higher units of value, each particular district had its own distinctive lower unit according to the nature of its products and requirements, so it is most likely that there were many different units of value (but all alike sub-multiples of the cow) in use among the various Greek communities. It is also probable that they must have exercised a certain effect in the formation of the units of silver currency.

Nor is evidence wanting for this. I have already maintained (p. 5) that the fact of the occurrence of the type of the cow, or cow's head, on early Greek coins is evidence that the original monetary unit was the ox. Thus we find the forepart of an ox on the early electrum staters of Samos of the Phoenician standard (217 grs.), which was probably equivalent to a pure gold ox-unit of 130 grs. The bull's head also appears on the electrum coins of Eretria and of other places in Euboea. But it is with the silver currency that we are now especially concerned. Whilst it was extremely likely that silver coins might in process of time bear the impress of an ox, the general unit of currency, it was still more natural that, as pieces of silver supplanted as units not the ox but its sub-multiples, that is the particular series of articles of barter in use in any particular district, so these silver coins should bear some traces in their types of the ancient units thus supplanted. That eminent scholar Colonel Leake many years ago remarked that the types of Greek coins generally related "to the local mythology and fortunes of the place, with *symbols referring to the principal productions* or to the protecting numina."

Modern scholars have more and more lost sight of the doctrine contained in the words which I have italicized, and directed all their efforts to giving a religious signification to everything[1]. The forepart of the Lion and the Bull on the coins of Lydia become symbols of the Sun and Moon, the Tortoise on the didrachm of Aegina is regarded as a symbol of

FIG. 31. Coin of Cyrene with Silphium plant.

Aphrodite, the Ashtaroth of the Phoenicians, in her capacity of patron divinity of traders; even the silphium plant of Cyrene,

[1] P. Gardner, *Types of Greek Coins*, passim.

which yielded a salubrious but somewhat unpleasant medicine, is regarded not as holding its place on the coins of Cyrene and its sister towns because it formed the chief staple of trade, but because forsooth it may have been the symbol of Aristaeus, "the protector of the corn-field and the vine and all growing crops, and bees and flocks and shepherds, and the averter of the scorching blasts of the Sahara." There is probably just as much evidence for this as there is for believing that the beaver on some Canadian coins and stamps is symbolical of St Lawrence, after whom the great Canadian river is named, the warm skin of the beaver indicating that the saint of the red-hot gridiron is the averter of the cruel and biting blasts that sweep down from the icy North. I do not for a moment mean that mythological and religious subjects do not play their proper part in Greek coin types. But it is just as wrong to reduce all coin types to this category as it would be to regard them all as merely symbolic of the natural and manufactured products of the various states. If however we can show that certain coins, even in historical times, were regarded as the representations of the objects of barter of more primitive times, we shall have established a firm basis from which to make further advances.

In those now famous Cretan inscriptions found at Gortyn[1] certain sums are counted by kettles (*lebetes*, λέβητες) and pots (*tripods*, τρίποδες). Some have thought that these are the same objects which are called staters in later forms of the same documents. But recently M. Svoronos[2] has advanced a very plausible hypothesis that the *lebetes* and *tripods* of the inscriptions really refer not to an actual currency in the kettles and pots of the old Homeric times, but to certain Cretan coins which are countermarked with a stamp, which he recognizes in many examples as a *lebes*, and in at least one case as a *tripod*. Whether the first hypothesis, that actual kettles and pots were indicated

---

[1] Comparetti, *Leggi antiche della città di Gortyna in Creta*, 1885; *Museo Italiano* II. 195, no. 39: *ibid.* II. 222. Roberts, *Greek Epigraphy*, p. 53.

[2] *Bulletin de Correspondance Hellénique*, 1888, p. 405 seqq. (where he gives an engraving of a stater so countermarked). Mr B. V. Head (*Numism. Chron.* 3rd ser. IX. 242) in a notice of this paper lends his great authority to the support of Svoronos' view.

in the earlier inscriptions and that they had been replaced afterwards by coins, or the hypothesis of M. Svoronos, be true, is immaterial for us. In either case there is evidence of a direct and unbroken succession which connects the silver currency of Crete with an earlier currency of manufactured articles. The very fact that a lebes or a tripod stamped upon a coin gave it currency, not merely in the town of issue but among neighbouring states, indicates that in a previous age the common unit of currency corresponding in value to the coin so marked was an actual lebes or tripod. Such is the evidence preserved for us in this remote corner of Hellas where life moved slowly, and where the archaic style of writing known as *boustrophedon* (the lines going from right to left and left to right alternately, as the plough turns up and down the field) still lingered on long after it had disappeared from every spot on the mainland of Greece. If then amongst the symbols which appear on the earliest coins of Greek communities, which began very early to strike money, we can find some which have not been identified as religious, and which we can show represent objects which actually did or may well have formed a monetary unit in such places, we shall have advanced a step further; and if we succeed in making good this fresh position, we may in turn find a non-religious explanation for certain types, which at present are regarded as mythological symbols.

The types with which we shall deal must be those found on the most archaic coins, and which therefore date from a time when barter was just being replaced by a monetary currency. Thus in the case of cities like Athens and Corinth, which began to coin at a comparatively late period and which had been long accustomed to use the issues of other states before they struck money of their own, we should hardly expect to find any trace of the old local barter-unit in their coin types, as such a unit had long since been replaced by the foreign coins.

Let us first turn to the well-known type of the tunny fish (πηλαμύς, θύννος), vast shoals of which were continually passing through the sea of Marmora (Propontis) from the Black Sea to the Mediterranean[1]. This type appears invariably upon the

[1] Head, *op. cit.* 450, who quotes Marquardt's *Cyzicus*, p. 45.

electrum coins of Cyzicus, and a tunny's head is found upon some very archaic silver coins from the Santorin 'find' which Mr Head places at the top of the whole Cyzicene series, but no one has, as far as I am aware, yet hitherto attempted to mythologize it[1], although the fecundity of this fish would make it just as suitable an emblem for Aphrodite as the "lascivious turtle," and the traders of Cyzicus might quite as well wear the badge of the goddess of the sea as the merchants of Aegina, for there is just as much or just as little evidence for Phoenician influences at Cyzicus as there is at Aegina. From what we have learned in an earlier chapter we know that the articles which form the staple commodities of a community in the age of barter virtually form its money. In a city like Cyzicus whose citizens depended for their wealth on their fisheries and trade, rather than on flocks or herds and agriculture, the tunny fish singly or in certain defined numbers, as by the score

Fig. 32. Coin of Cyzicus with tunny fish.

or hundred and the like, would naturally form a chief monetary unit, just as we found the stock fish employed in mediaeval Iceland. Are we not then justified in considering the tunny fish, which forms the invariable adjunct of the coins of Cyzicus, as an indication that these coins superseded a primitive system in which the tunny formed a monetary unit, just as the Kettle and Pot counter-marks on the coins of Crete point back to the days when real kettles formed the chief medium of exchange? But far stronger evidence is at hand to show that the tunny fish was used as a monetary unit in some parts of Hellas. We have had occasion to refer to the city of Olbia which lay on the north shore of the Black Sea. It was a Milesian colony,

---

[1] Fishermen offered to Poseidon the first tunny they caught (Athen. p. 346), but this was simply an offering of first fruits and not because the tunny was sacred.

and was the chief Greek emporium in this region. There are bronze coins of this city made in the shape of fishes, and

Fig. 33. Coins of Olbia in the form of tunny fish.

inscribed ΘΤ, which has been identified as the abbreviation of θύννος, *tunny*. Others are inscribed ΑΡΙΧΟ, which Koehler read as τάριχος, salt fish, but which the distinguished German numismatist Von Sallet[1] regards as meaning a basket (ἄρριχος). He holds those marked ΘΤ as the legal price of a tunny fish, those marked ΑΡΙΧΟ as that of a basket of fish[2]. When we recall the Chinese bronze cowries, the Burmese silver shells, the silver fish-hooks of the Indian Ocean, the little hoes and knives of China, and the miniature axes from Africa, we are constrained to believe that in these coins of Olbia, shaped like a fish, we have a distinct proof of the influence on the Greek mind of the same principle which has impelled other peoples to imitate in metal the older object of barter which a metal currency is replacing. The inhabitants of Olbia were largely intermixed with the surrounding barbarians, and may therefore have felt some difficulty in replacing their barter unit by a round piece of metal bearing merely the imprint of a fish, while the pure-blooded Greek of Cyzicus had no hesitation in mentally bridging the gulf between a real fish and a piece of metal merely stamped with a fish, and did not require the intermediate step of first shaping his metal unit into the form of a tunny. We shall find that this tendency to shape metal into the form of the object which it supplants may perhaps be traced in the coins of Aegina and Boeotia.

In the same quarter of Hellas we find another instance of a coin type which may be regarded as evidence that the silver coin which bears it was the representative of an older barter

---

[1] *Zeitschrift f. Numismatik*, x. 144 seqq.

[2] The tunny is a very large fish, usually four feet long, and is hardly likely to have been sold by the basketful.

unit. The island of Tenedos, lying off the Troad, struck at a very early date silver coins bearing for device a double-headed

FIG. 34. Coin of Tenedos with double-headed axe.

axe (the Latin *bipennis*). This "Axe of Tenedos" (Τενέδιος πέλεκυς) was explained by Aristotle[1] as a reference to a decree of a king of Tenedos which enacted that all who were convicted of adultery should be put to death. This explanation is probably a bit of mere aetiology to explain the existence of an emblem, the true origin of which had been forgotten. However, it yields one important result, for it shows that the emblem was not religious. Had that been its nature, priestly conservatism would have kept an unbroken tradition of its origin. But from another source some light may be obtained: Pausanias[2] in the 2nd century A.D. saw at Delphi axes dedicated according to tradition by Periclytus of Tenedos, and then proceeds to relate the following tale: Tennes, an old King of Tenedos about the time of the Trojan War, cut with an axe the ropes with which his father Cycnus had moored his ship to the shore, when he came to ask pardon of Tennes for having cast him and his sister in a chest into the sea, in a fit of anger caused by the false accusation of a stepmother. We may gather that according to this form of the legend the Janiform head, male and female, on the obverse of the coins of Tenedos alludes to the brother and sister. But Pausanias makes no attempt to connect Periclytus in any way with Tennes except as being a native of Tenedos. This is hardly enough to account for the dedication of the axes at Delphi. Two explanations suggest themselves. It was the custom of kings or communities to send offerings to Delphi of the best products of their land. Thus Croesus sent vast quantities of his Lydian electrum, and,

[1] *Apud Stephanum Byzant.* s. v. *Τένεδος.*
[2] x. 14. 1.

still more to the point, the people of Metapontum in South Italy, whose land was famous for its wheat, after an especially favourable harvest sent to Delphi a wheat-car ($\theta\acute{\epsilon}\rho o\varsigma$) of gold. Were the double axes in like fashion an especial product of Tenedos? Or was this dedication analogous to that of Pheidon when he hung up in the temple of the Argive Here the ancient nails and bars? The first explanation is the more probable, for there was no reason why the Tenedians should not have dedicated their cast off currency of axes in some temple at home. I have already mentioned the hoe currency of ancient China, and the axes used as such in Africa. I shall now show that such double-axes as those stamped on the coins of Tenedos formed part of the earliest Greek system of currency. I have already enumerated the various articles used in barter in the Homeric poems. The prizes offered in the Funeral games of Patroclus are of course merely the usual objects of barter and currency, slavewomen, oxen, lebetes, tripods, talents of gold and the like. "But he (Achilles) set for the archers dark iron, and he set down ten axes ($\pi\epsilon\lambda\acute{\epsilon}\kappa\epsilon\alpha\varsigma$), and ten half-axes ($\dot{\eta}\mu\iota\pi\acute{\epsilon}\lambda\epsilon\kappa\kappa\alpha$)[1]." The axe is undoubtedly of the same kind as that on the coins of Tenedos, the name (*pelekys*) being the same in each case, and the Homeric one beyond doubt is double-headed like the Tenedian, since the half-axe (*hemi-pelekkon*) must obviously mean a single-headed axe[2]. The double-axes formed the first prize, the ten half-axes the second, for "Meriones took up all the ten axes, and Teucer bore the ten half-axes to the hollow ships[3]." These axes and half-axes then seem to go in groups of ten as units of value, the half-axes representing half

---

[1] *Iliad*, XXIII. 850—1,

Αὐτὰρ ὁ τοξευτῇσι τίθει ἰόεντα σίδηρον,
κὰδ δ' ἐτίθει δέκα μὲν πελέκεας, δέκα δ' ἡμιπέλεκκα.

[2] No doubt the axe was often used as a religious emblem; double-headed axes borne in procession are seen on Hittite sculptures (Perrot et Chipiez, *Histoire de l'Art dans l'antiquité*, IV. p. 637). It was also the symbol of Dionysus at Pagasae. So amongst the Polynesians we find processional axes as well as real ones like our sword of state as contrasted with real swords.

[3] *Ib.* 882—3,

ἂν δ' ἄρα Μηριόνης πελέκεας δέκα πάντας ἄειρεν,
Τεῦκρος δ' ἡμιπέλεκκα φέρεν κοίλας ἐπὶ νῆας.

the value of the double-headed. If then the kettle and tripod of Homeric times are found as symbols on the coins of Crete, why may not the axe on those of Tenedos represent the local unit of an earlier epoch? and that such axes were evidently an important article in Tenedos is proved by the dedication at Delphi.

But could we only find a contemporary description of the type on one of the earliest coins of Asia Minor, the cradle of the art of coining, we might get our ideas on the nature of the coin types greatly cleared. Fortunately such an opportunity is afforded to us by an unique coin in the British Museum, the oldest as yet known which bears an inscription. It is an oblong electrum coin (Fig. 35), the reverse having the usual incuse, but on its obverse it bears a stag feeding, and over it runs

FIG. 35. Coin of Phanes (earliest known inscribed coin).

(retrograde) in archaic letters I AM THE MARK OF PHANES ($\Phi a\nu o\varsigma$ $\epsilon\mu\iota$ $\sigma\epsilon\mu a$ = $\Phi \acute{a}\nu o\nu\varsigma$ $\epsilon\grave{\iota}\mu\grave{\iota}$ $\sigma\hat{\eta}\mu a$). There can be no doubt that the *mark* of Phanes is the stag. If there was no inscription it would have been at once asserted that the stag was the symbol of the goddess Artemis, and who could deny it? But as it stands it is plain that the stag is nothing more than the particular badge adopted by the potentate Phanes, when and where he may have reigned, as a guarantee of the weight of the coin and perhaps the purity of the metal. The Daric itself needs no inscription to tell us that its type is not religious. The figure of the Great King with his spear and bow and quiver can hardly be allegorized even by an Origen[1]. Emboldened by these instances we may even hold up our hands against the host of Heaven, and raise doubts as to whether the foreparts of the

[1] Although Mr Frazer (*Golden Bough*, I. 8) has given abundant evidence to show that kings were in some places worshipped as gods, no one can maintain that the Persians, who were Zoroastrians, would have treated their king as a god.

lion and bull upon the coins of Lydia represent the Sun-god and the Moon-goddess. May not the lion simply be the royal emblem? I have already suggested this explanation for the lion weights of Assyria. Undoubtedly from the earliest times the king of beasts (as in *Aesop's Fables*) was regarded in the East as the true badge of royalty. "The Lion of the tribe of Judah" is familiar to us all, and it is more rational to regard the lions which guarded the steps of Solomon's throne as emblems of kingship rather than as symbols of the Sun. Is then the Lion on the coins of Lydia nothing more than the king's badge, just as the stag is the badge of Phanes? But what about the bull or cow? Shall I go too far if I regard it as indicating that the coin is the ox-unit? When the Greeks

FIG. 36. Archaic coin of Samos.

borrowed the art of coining from Lydia it is easy to understand that they would likewise borrow the type either in a complete or modified form, and hence it is that we find the lion or lion's head on the coins of Miletus[1], the lion's scalp on those of Samos (on which the cow's head also is found), the lion's head on the coins of Cnidus, of Gortyn in Crete, at Rhodes, at Mi-

FIG. 37. Coin of Cnidus.

letus, and at the Phocaean towns of Velia in Lucania, and Massalia in Gaul, and put by the Samian exiles on their coins at Zancle. If the Greeks had been barbarians they would have

[1] The electrum coins with the lion's head with open jaws formerly ascribed to Miletus are now assigned to the Lydian king Alyattes by M. J. P. Six, *Num. Chron.* N. S. Vol. x. 185 *seqq.* (1890).

slavishly copied the lion coins of Lydia, just as the Gauls copied the lion of Massalia, and at a later time the stater of Philip, and as the Himyarites of South Arabia, the "owls" of Athens[1], and as in mediaeval times the Danes of Dublin copied the coins of the Saxon kings[2]. But the artistic genius of the Greeks could submit to no such trammels, and the lion type was varied and diversified according to the fancy of each community. The same holds good of the type of the cow and cow's head. The Greek genius gave us these beautiful types such as the cow suckling her calf (Dyrrachium), the cow with the bird on her back (Eretria), the cow scratching herself (Eretria), the two calves' heads seen on the coins of Mytilene, and the magnificent charging bull on the coins of Thurii. The cow or bull's head on the early gold and

Fig. 38. Coin of Thurii.

electrum coins was the indication of the value. In later times when the connection between ox and coin was only traditional, the ox was put on coins simply as symbolical of money.

Again Phocaea, one of the very earliest Greek towns to issue coins, employed a symbol which cannot be termed religious. Her coins bear a seal (*phoca*) a *type parlant* referring to the name of the town. Many examples of the same kind

Fig. 39. Coin of Rhoda in Spain.

can be quoted, the rose (ῥόδον) on the coins of Rhodes ('Ρόδος) and also on those of Rhoda in Spain, the bee (*melitta*) on those

---

[1] Head, *Op. cit.* 6. 88.
[2] Lindsay, *Survey of the Coinage of Ireland*, p. 6 *seqq*.

of Melitaea, perhaps even the owl (χαλκίς) on coins ascribed to Chalcis in Euboea. These considerations will serve to show that we may expect many things on coins besides religious symbols. Thasos was famous for its wine, and accordingly the wine-cup is a regular adjunct of its coins, either standing alone, or held in the hands of old Silenus, who quaffs therefrom a "draught of vintage that hath been cooled a long age in the deep-delved earth." All who have read Horace remember the fame of the wines of Chios, and accordingly the wine-jar is a regular adjunct of the mintage of that island. Now there is proof that the trade in wine was of extreme antiquity, if not in the islands just mentioned, at least in Lemnos, and that that trade was carried on by barter, for we read in Homer how "many ships stood in from Lemnos bringing wine, which Euneos the son of Jason had sent forward, whom Hypsipyle had borne to Jason shepherd of the folk, but separately for the sons of Atreus, Agamemnon and Menelaus, the son of Jason gave wine to be fetched, a thousand measures. From thence used the flowing-haired Achaeans to buy their wine, some with copper, some with glittering iron, some with hides, others with the kine themselves, others again with slaves[1]." From what we have seen in an earlier chapter it is clear that a measure of wine would have a known value in relation to the various articles here enumerated. Thus in North America where the beaver skin was the unit, a gallon of brandy = 6 skins, a brass kettle = 1 skin, an ounce of vermilion = 1 skin and so on[2]. In other words, the ordinary currency with which the Lemnians would purchase wares from other people who had no wine of their own would be wine, the unit of which was the *measure* (which elsewhere I have tried to show was the cup δέπας, Smith's *Dict. Antiq. s.v.* Mensura). This measure would be the size of the vessel ordinarily employed for wine, probably much the same as the two-handled vase out of which Silenus is seen drinking on coins of Thasos.

With the introduction of silver currency nothing is more likely than that an effort would be made to equate the new

[1] *Il.* VII. 468 *seqq.*
[2] A. Dobbs, *Account of Hudson's Bay* (1744).

silver unit to that which had formed the principal unit of barter. That the earliest types should indicate the object (or its value) which the coin replaced is in complete accord with the statement of Aristotle (quoted on an earlier page) that "the stamp was put on the coin as an indication of value[1]." As no numerals appear on the early Greek coins, it is evident that Aristotle regarded the symbol, whether ox-head, or tunny, or shield, as the index of the value. If it be said that the putting of a cow, or axe, or tunny on a coin was simply a picturesque way of indicating a single unit, we may reply that it is far easier to understand why a certain people chose a particular symbol, if in their minds the object symbolized was identified with the value of the silver or gold coin. It is at all events certain that Aristotle did not regard the type as religious in origin. But we are not without actual evidence that such an equating of the silver unit to the barter-unit really took place in Greece. It is held by the best numismatists that Solon was the first to coin money at Athens. It is also well known that the highest class in his constitution, called Pentacosiomedimni (*Five-hundred-measure-men*), were rated at 500 drachms. Thus the Olympic victor received 500 drachms to qualify him to be a Five-hundred-measure-man[2]. Furthermore Plutarch distinctly tells us that Solon reckoned a drachm as equivalent to a measure[3] or a sheep. It is hardly possible to doubt that the first Attic coined silver drachm was equated to the old barter unit of a measure (either of corn or oil). The same may be said in reference to the olive sprig which from the earliest issue is found on the coins of Athens. The sacred olive-trees (μορίαι) which belonged to the state, and for the care of which special officials were appointed, and even the very stumps of which, and the spot on which they had grown, were under a taboo[4], were a source of considerable

[1] *Politics* II. 1257 D ὁ γὰρ χαρακτὴρ ἐτέθη τοῦ πόσου σημεῖον.
[2] Plutarch, *Solon* 18.
[3] *Ibid.* 23 Εἰς μέν γε τὰ τιμήματα τῶν θυσιῶν λογίζεται πρόβατον καὶ δραχμὴν ἀντὶ μεδίμνου· τῷ δ' Ἴσθμια νικήσαντι δραχμὰς ἔταξεν ἑκατὸν δίδοσθαι, τῷ δ' Ὀλύμπια πεντακοσίας· λύκον δὲ τῷ κομίσαντι πέντε δραχμὰς ἔδωκε, λυκιδέα δὲ μίαν, ὧν φησιν ὁ Φαληρεὺς Δημήτριος τὸ μὲν βοὸς εἶναι, τὸ δὲ προβάτου τιμήν.
[4] Lysias, *de Sacra oliva*, 6.

revenue to the state in the 6th century B.C. The fact that they were all supposed to be scions of the sacred olive-tree on the Acropolis, which was itself supposed to be the gift of

Fig. 40.   Tetradrachm of Athens.

Athena, and the religious care bestowed on them, puts it beyond doubt that the olive at an early date formed one of the most important products of Attica. The instances given already of the employment of various kinds of food as money are sufficient to show that there is nothing far-fetched in supposing that olives and olive-oil may have been so employed at Athens.

We have already spoken of the silphium or laserpitium plant on the coins of Cyrene, Barca, Euesperides and Teuchira, and mentioned the interpretation which makes it the symbol of the hero Aristaeus. It seems however far more reasonable to treat it on the same principle as the others just discussed. The silphium formed the most important article produced in that region, and it is perfectly in accordance with all analogy that certain quantities of this plant and of the juice extracted from it should be employed as money. We saw above that at the present moment tea is so employed on the borders of Tibet and China, and raw cotton in Darfur. But there is also some positive evidence in favour of this assumption, for Strabo[1] tell us that a traffic was carried on at the port of Charax between the Carthaginians and Cyrenaeans, the former bringing wine wherewith to purchase the silphium of the latter. There must have been a wine-unit, and also an unit for the silphium, or otherwise the barter could not have been carried on; and just as in Gaul[2] a jar of wine purchased a boy fit to serve

[1] Strabo, xvii. 836.
[2] Diodorus Siculus v. 26. 2 διδόντες γὰρ τοῦ οἴνου κεράμιον ἀντιλαμβάνουσι παῖδα κτλ.

as a cupbearer, a certain measure of wine being equated to a slave-boy, so we may conclude that some such wine-unit was equated to a packet or bale of silphium, the latter in turn having a certain amount of silver equated to it, which when coinage was introduced was stamped with the silphium device. That the silphium was packed in bales of a fixed weight is proved by a now famous vase-painting which repre-

FIG. 41. Vase from Cyrene, shewing the weighing of the Silphium.

sents the weighing (on ship board?) of the bales of silphium in the presence of Arcesilas the king of Cyrene[1]. The figure who points to the scales is marked *sliphiomachos* (σλιφιο-

[1] Baumeister, *Denkmäler*, s.v. Silphium. Studnicyna, *Kyrene*, p. 22. Birch, *Ancient Pottery* (frontispiece). The vase is in the Paris Bibliothèque.

μαχος) which is taken to mean *silphium-weigher* (σλιφιο- being either a mis-spelling of the artist, or the local form of the word, whilst the latter part is connected with the Egyptian *mach* = to *weigh*). Close to the silphium packets is the word MAEN, which has not been explained, but which may be simply a form of the word *mina* (*manah*, *meneh*) and denotes that each packet weighed that amount.

Fig. 42. Coin of Metapontum.

The ear of corn (wheat) on the coins of Metapontum[1], an old Achaean colony in Magna Graecia, is explained by modern writers as a symbol of Demeter: but the story told by Strabo of how the early settlers dedicated a golden ear at Delphi because they had amassed such great wealth from agriculture, indicates a far simpler solution, that the chief product and chief article of barter of Metapontum was naturally placed on her coins. As the tunny adorns the coins of Cyzicus, so we find the cuttle-fish on the coins of Croton and Eretria. As this creature was devoured with great gusto by the ancients, as it is at the present day at Naples and in Palestine, there is

[1] The only evidence to show that Demeter was worshipped at Metapontum is that a female head on certain of her coins is accompanied by the legend Σωτηρία. It has been inferred that this is an epithet of Demeter, but this is most unlikely, for in that case we should expect Σώτειρα, as on the coins of Hipponium, Syracuse, Agrigentum, Corcyra, Cyzicus, and Apamea, not Σωτηρία, as the adjective. Thus we always find Ζεὺς Σωτήρ, not Σωτήριος: cf. Σώτειρα Εὐνομία, Pind. *Ol.* IX. 16, Σώτειρα Τύχα, *Ol.* XII. 2, Σώτειρα Θέμις, *Ol.* VIII. 21. Σωτηρία is rather *Safety* (Lat. *Salus*), who, as my friend Mr J. G. Frazer points out to me, was worshipped at Patrac and Aegeum, two of the chief towns of Achaea (Pausan. VII. 21. 7; VII. 24. 3). We also find such names of divinities as Ὑγιεία, Ὁμόνοια and Νίκα on the coins of Metapontum. As Metapontum was an Achaean colony, it is likely that *Salus* was worshipped there also. Besides it was to Apollo, and not to Demeter, that they dedicated their golden ear as a harvest thank-offering. Θέρος is the ear cut from the stalk after the ancient way of reaping, cf. θέρη σταχύων, Plut.

no necessity to regard it as a symbol of Poseidon, or of treating it in any way different from the tunny.

Fig. 43. Coin of Croton with cuttle fish.

I now come to two most important types, the Tortoise of Aegina, and the Shield of Boeotia. I have already mentioned the

Fig. 44. 'Tortoise' of Aegina.

symbolic interpretation given by E. Curtius to the former. That various natural productions, such as gourds, cocoa-nuts, joints of bamboo, served and still serve as vessels and measures of capacity in various countries we have seen already, and we likewise found that in the ancient Chinese monetary system of shells the shell of the tortoise stood at the top as the unit of highest value, and that down to a comparatively late epoch it was still highly prized in Cochin China for making bowls of great beauty. In both Greek and Latin there is abundant evidence to show that the functions which in a later time were performed by pottery were discharged by natural shells at an earlier period. Thus, if we do not find any actual vessel called a *chelône* (tortoise) in use amongst the Greeks, we at least find one called a Sea-urchin (Echinus, ἐχῖνος): for not only was the shell of this creature used as a vessel for containing medicines and the like, but vessels of artificial construction of the same shape and name were actually employed; thus the casket in which were deposited and sealed up the documents produced at the preliminary hearing of an Athenian lawsuit was called an *Echinus*. There was likewise a small vessel called *conché* (κόγχη), after the shell-fish of that name, the Latin *concha*, whilst a cognate name, *conchylion*, was applied to the case placed over the seals of wills.

Nay, *ostrakon*, the common word for a potsherd, familiar to us from its famous derivative Ostracism, or *Voting by Potsherds*, so called because the people inscribed their votes on pieces of pottery, meant originally nothing more than an oyster shell. In Latin *testa*, the ordinary name for an earthenware vessel, means nothing more than the covering of a shell-fish, and from this word *testudo*, the Latin name for the tortoise, is simply a derivative. Such instances could be multiplied if it were necessary, but those mentioned are sufficient to show the high probability of so valuable a shell as that of the tortoise having been employed. Owing to its beauty it would probably hold its place in Greece as the choicest kind of vessel for centuries after the art of pottery was known, just as it did in Cochin China. It would be only when the art of glazing and embellishing pottery had made some progress that vessels of baked clay could compete with the lustrous, many-hued shell. Nor are we without some direct evidence for the use of tortoise shell among the Greeks. The famous story of the invention of the lyre by the god Hermes is not without significance. According to the Hymn to Hermes, "the precocious divinity on the very day of his birth sallied forth and found a tortoise feeding on the luxuriant grass in front of the palace, as it moved with straddling gait." His eye was caught by the dappled shell (αἰόλον ὄστρακον), and carrying home his spoil, he made of it a lyre. The legend which thus explains why the sounding-board of the lyre is so called points back to a time when the best form of bowl or hollow vessel for making a sounding board for a musical instrument was that afforded by the shell which was probably one of the common articles of everyday life.

But, in addition to all this indirect evidence, we are able to point to actual Greek vessels made of earthenware, fashioned in the shape of a tortoise. In the second Vase Room of the British Museum (case 48 and 49) there are two terra cotta vases from the island of Melos, wrought in the shape of this creature, and with these before us it is hardly possible to regard as other than wooden bowls carved in the shape of the same animal *the wooden tortoises* with which the Thessalian women pounded to death Lais the famous courtezan, in the temple of Aphrodite, after she had taken up her residence in their country[1]. We can parallel this development of artificial vessels of wood and earthenware from the use of the actual shell in modern times. Lady Brassey saw in the Museum at Honolulu, amongst the ancient native weapons and swords, "tortoise-shell cups and spoons, calabashes and bowls[2]." Now in the Cambridge Ethnological Museum there is a very fine wooden bowl from the South Seas, carved in the shape of a tortoise, and also earthenware vessels in the shape of tortoises from Fiji, which shows that the islanders of the Pacific not only used the real shells for vessels, but likewise imitated them in wood[3].

On an earlier page I quoted the statement of Ephorus that the Aeginetans took to commerce on account of the barrenness of their island. But they must have had something to give in exchange to other people before they could have developed a carrying trade, and as the island had been the resort of merchants from very early days, it must have had something to attract strangers as well as its position. Let us take the case of an island with barren soil in modern days, and see what it has to export. Thus Dhalac Island in the Red Sea is frequented by the Banyan merchants for the sake of its pearls, and at Massowah tortoise-shell forms an important article of commerce. Just as the Banyans come to Dhalac[4], so the Phoenicians probably came to Aegina, searching for the murex

[1] Athenaeus XIII. p. 589 ab; Schol. on Aristophanes, *Plutus*, 179; Suidas, s.v. χελώνη.

[2] *Voyage of the Sunbeam*, p. 276 (London, 1880). [L.M.R.]

[3] We learn from Strabo, 773, that the Greeks were familiar with the employment of tortoise shells, for a tribe called Tortoise-eaters on the north coast of Africa used the shells of these animals, which were of large size, for roofing purposes. Pausanias (VIII. 23. 9) tells us that there were large tortoises well suited for making lyres in Arcadia, but the people would not touch them as they were under the protection of Pan. As Pan was lord of the forest and mountain, the tortoise being especially large would naturally be regarded as his special property.

[4] Mansfield Parkyn, *Abyssinia*, Vol. I. p. 407.

(purple fish) and tortoise. No doubt tortoise-shell must have been the chief article of export from Tortoise Island, described by Strabo (773), as situated in the Arabian Gulf (Red Sea).

The foregoing considerations make it not at all improbable that the tortoise on the coins of Aegina simply indicates that the old monetary unit of that island was the shell of the sea-tortoise (ἡ θαλαττία χελώνη), which was considerably larger, and therefore more valuable for making bowls, than that of the land or "mountain" tortoise (ἡ ὀρεινὴ χελώνη). There was a well-known headland on the Coast of Peloponnesus called "Tortoise Head" (Chelonates), and this creature must have been a peculiar feature of the shores of Aegina, or it would not have been chosen as the type for her coins, whether it be a religious symbol or not. At all events we know from the story of Sciron the robber, slain by Theseus, that the sea-tortoise was a familiar feature on the shores of the Saronic Gulf, as the hapless travellers who were kicked over the rocks by the caitiff were devoured by a large sea-tortoise which frequented the strand below. This creature's picture is handed down on a well-known vase-painting which commemorates the exploits of Theseus. Finally, it may well be supposed that had not its connection with the invention of the lyre attracted to that instrument the name of "Tortoise" both in Greek and Latin, we should have found the name employed for some sort of vessel, as is the case with the Echinus.

Coming now to Central Greece, we find on the coins of all the Boeotian towns (with the exception of Orchomenus in her earliest issues) the well-known device of the Boeotian shield.

FIG. 45. Coin of Boeotia with shield.

This has been confidently pronounced to be a sacred emblem, symbolic of a common worship, conjectured to be that of Athena Itonia, whose temple near Coronea was the meeting-place of the

Boeotians[1], whilst at Coronea golden shields were preserved in the Acropolis[2]. This may be so, but it is equally possible that the shield represented a common monetary unit in ancient times. The shield of early Hellas was a simple ox-hide buckler, described in Homeric language simply as an *ox-hide*[3]. Amongst barbarous peoples, as we saw above, weapons form one of the regular commodities commonly employed as currency; the Achaeans bought wine with hides as well as with oxen from the ships that came from Lemnos, and as there can be no doubt that the hide was a regular sub-multiple of the cow, it is very probable that the ox-hide shield stood in a similar relation to the cow, the chief or most universal unit; and as we find axes and half-axes among the prizes offered by Achilles as well as kettles and caldrons, so we learn from a famous passage[4] that shields were amongst the most usual articles offered as prizes and therefore were regular units of currency: " For they strove neither for an ox to be sacrificed nor yet for an ox-hide shield which are wont to be the prizes for the feet of men, but they strove for the life of the horse-taming Hector."

When silver money was struck, it was natural that the barter-unit which came nearest in value to the silver didrachm would be equated to it, and the piece of silver would accordingly be termed *Shield* or *Tortoise*, just as the silver equivalent for the old copper rod was called the Obol, and in due course the corresponding device would be impressed on the silver coinage. The same explanation may probably be

Fig. 46. Coin of Lycia.

applied in other cases, such as that of the boar on the coins of Lycia. On the coins of the Gaulish tribe Sequani who made

---

[1] Pausan. ix. 34.
[2] Pausan. i. 25.
[3] *Iliad* xvii. 381.
[4] *Iliad* xxii. 158.

the best bacon and hams which came into the Roman market, the swine is found[1]. Doubtless this animal was their chief source of wealth, and formed a unit of barter, but we have not space for any more examples.

It is worth noting that it is quite possible that the men who issued the earliest coins of Boeotia and Aegina were influenced in the shape they gave these coins by the actual objects which they were replacing. The coins of Aegina with their high round upper side and flat under side suggest the general outline of a tortoise. As the people of Olbia, like the Chinese, Burmese and Ceylonese, had to make coins in the shape of a fish, so the Aeginetans acting under a like instinct may have wished to give a conventional representation of the tortoise. The earliest coins have the incuse on the reverse divided into *eight* triangular compartments. Are these the *eight* plates which form invariably the *plastron* or under surface of all the tortoise family? Later on the Aeginetan incuse is always in five compartments, but in the two well-known triangular depressions we perhaps find an echo of the tortoise-*plastron*[2]. The earliest coins seem to represent a sea-tortoise, for the feet are real *flippers* quite distinct in shape from the legs shown on the later coins. As the plates of the *carapace* (upper surface) are not fully represented in the archaic coins, this omission may not be merely due to rudeness of work, but rather because in the case of the sea-tortoise the *thirteen* plates of the *carapace* are not so prominent as in the land-tortoise. On the later coins where the feet are those of the land-tortoise the coins accurately represent the *thirteen* plates.

It has to be borne in mind that the shape of the incuse depressions on the reverse of coins is very constant. Thus on the Aeginetan coins we never find what is known as the mill-sail incuse which is the peculiar feature of the reverse

---

[1] Strabo 192, ὅθεν οἱ ἄρισται ταριχεῖαι τῶν ὑείων κρεῶν εἰς τὴν ῾Ρώμην κατακομίζονται. Hucher, *Art Gaulois*, Pl. 78. The swine is also found on coins of Bellovaci, Pictones and Armorican Gauls.

[2] On the plastron of the sea-tortoise eight triangular patches are made very conspicuous by pigmentation.

of the early Boeotian coins, nor on the other hand do we even find the eight-fold incuse on the coins of Boeotia. Some influences must have determined the choice of form, such as I have just suggested in the case of Aegina. Did the first Boeotian Mintmaster shape his coins with the real buckler in his mind's eye? On the reverse of these coins we find the incuse forming a rude X, which is bounded by a circle of dots, whilst in the centre of the incuse is the initial letter of the name of the issuing town, such as ⊕ for Thebes, ⊟ for Haliartus. Does the X-shaped incuse represent conventionally the cross-bars of the frame of the shield seen at the back, the circle dots indicating the outline? The letters on these coins are the earliest inscriptions on the coins of Greece Proper. We can easily see how they came to be placed on the coins, as soon as we remember that there was a Λ on the Lacedaemonian shields, a Σ on the Sicyonian, a M on the Messenian[1]. Why do not we find the initial in the coins placed on the front of the shield, where it must have stood on the real buckler? If as is held by the best authorities the coins of Boeotia formed a federal currency, we see a reason for the practice. As the silver shield replaced the real buckler, the old unit which had been universally employed through Boeotia, no town would have been permitted to put its initial on the shield engraved on the obverse. No doubt the old actual shield of currency was plain, and each purchaser painted the initial of his own country upon it. The Mintmasters accordingly of each town regarding the whole coin as a shield placed the letter of these several states on the reverse. Baumeister (*Denkmäler, s.v.* Wappen) gives pictures of the back of two shields. The frame of the shield consists of a circular rod, with two cross bars. The idea of making the incuse represent the other side of the object given in relief on the obverse seems to be just the stage between a complete representation of the object as in the tunny of Olbia, and that evinced by the early coins of Magna Graecia, on which the reverse gives

---

[1] Photius *Lex. s.v.* Λάμβδα. Eustathius on Homer p. 293. 39 seqq. Xenophon *Hell.* IV. 4. 10 (which shows that the letter was on the front, cf. Pausan. IV. 28. 5).

in the incuse exactly the same form as that in relief on the obverse.

At first sight the result of this great variety of local units apparently places impassable barriers to trade, but a knowledge of the actual facts of barbarous communities and their monetary systems as they exist in our time easily dispels this impression. I quoted above (p. 46) the words of Mohammed Ibn-Omar, wherein he points out that every separate district in the Soudan has its own lower unit or units, whilst everywhere alike the ox and the slave are the higher units; these local units are equated one to the other, so that there is no difficulty in trading. The same holds true of ancient Greece; the tortoise-shell of Aegina may have been reckoned equal to a certain amount of Attic olive oil or to a jar of wine of certain size, which formed the unit of commerce at Thasos and Chios, whilst in its turn a jar of wine was reckoned as equivalent to a package of silphion from Cyrene, a kettle from Crete, or an axe, or certain number of axes, or half-axes from Tenedos, or an ox-hide shield from Boeotia. All were sub-multiples of the ox, and had a fixed value in gold, and later in silver, as weighed against grains of corn. This supposition is in complete accord with the system revealed to us in the Homeric Poems, and is confirmed by the evidence drawn from barbarous races in modern times. It is likewise to be borne in mind that the tendency to place religious and mythological types on Greek coins was one especially developed in the later but not in the earliest period of coinage. No doubt aesthetic considerations played a large part in the adoption of such types, which came especially into prominence when Greek art was at its height. On the early coins one simple type is the rule, whilst at a later stage, besides the old national type, many adjuncts and symbols are added. Contrast the early coins of Athens with the later. The archaic issues have an olive spray and an owl, the later have not merely the owl, but an amphora, and a symbol in the field alluding to the legend of Triptolemus. Again, at Argos the early coins have simply the wolf or half-wolf or wolf's head, with a large A on the reverse, but in the later times the A is accompanied by symbols, such as a crescent and letters. The hare appears on the coins of Rhegium and Messana, having been chosen as a type, according to Aristotle, by the tyrant Anaxilas in commemoration of the introduction of that animal by him into Sicily; but it also appears on a rare coin of Messana, not as a main type, but as caressed by Pan. This does not prove that the hare was a symbol of Pan, but that for artistic purposes the rustic god in the act of caressing the hare is chosen instead of the more commonplace type of the hare all alone. So at Thasos the coins with old Silenus quaffing from a wine-cup do not signify that Silenus was a principal object of worship, but he is simply added for picturesque effect. We can at all events draw one conclusion from the historical origin assigned to both this type and that of the axe of Tenedos, that in the middle of the 4th cent. B.C. the Greeks did not see any religious significance in them, any more than they did in the representation of the mule-car which had won at Olympia, placed on his coins by Anaxilas. If, as has been so emphatically laid down by the leading modern Greek numismatists, the types on Greek coins

FIG. 47. Coin of Messana.

are so essentially religious in origin, it is extremely difficult to explain the extraordinary rapidity with which all such notions as regards their origin must have vanished from the minds of the most learned of the Greeks, at so early a date as the 4th cent. B.C. (hardly more than two centuries after the introduction of the art of coining). The Greeks regarded those types from much the same point of view as we regard St George and the Dragon on sovereigns and crowns, or the Lady Godiva riding *in puris naturalibus* on the Coventry tokens. The effort to turn agonistic into religious types by contending that, as the Olympic festival was of religious origin, so the successful chariot which had won at Olympia was a sacred symbol, can only be regarded as an ingenious effort to attach by even the most slender thread a simple commemorative type to a religious origin.

There is not the slightest reason for treating with incredulity the statement that Anaxilas introduced the hare into Sicily. Pollux[1] tells us that there were no hares in Ithaca, and from the same source we learn that the islanders of Carpathus, wishing to add the animal to the products of their isle, introduced a single pair, the descendants of which became in a short time so numerous that they ruined the crops, a story which finds a singular parallel in the history of the introduction of the rabbit into Australia in our own days. The hare was to the old Greek sportsman (as we know from the Tracts on Hunting of Xenophon and Arrian) what the stag was to the mediaeval baron, and the fox to the modern English squire. If William the Conqueror, as says the chronicler, "loved the tall deer as though he were their father," the tyrant Anaxilas may well have prided himself upon the introduction of the hare into Sicily in much the same manner as modern sportsmen have brought the French partridge into England. When once the type was started, the dislike of any change in coin types is so strong that we need not be surprised at the hare appearing for a long period on the coins of Messana and Rhegium. Besides, the hare was considered by the Greek gourmet as the choicest of viands: all readers of Aristophanes are familiar with "jugged hare" as a proverbial expression for "the best of cheer."

Variation of Silver Standards.

The connection between the types on early silver coins of Greece and the earlier local units of value being probably such as I have indicated, we next approach the question of changes in the weight of the silver coins at various places and at various

---

[1] Pollux, v. 66.

times. Besides the ordinary Euboic and Aeginetic standards we find others such as the Rhodian, and the Ptolemaic, the former so named because the island of Rhodes from the beginning of the 4th century B.C. ceased to strike tetradrachms of the full Attic weight of 270 grs. and coined instead pieces which range in weight from 240 to 230 grs., the latter getting its name from the dynasty of the Lagidae, who quickly dropped the full weight of the tetradrachm (270 grs.) as struck by Alexander, and reverted to the Phoenician silver of 220 grs., which they used not only for silver, but also for gold; it is to this last fact that the name Ptolemaic as given to the standard is really due, for as a standard for gold it was certainly new. But not merely shall we find coins standing so far apart from the usual standards that we are obliged to give them distinctive appellations, but we likewise find various modifications of the Aeginetic in various places, whilst in some parts of northern Greece and Thrace we shall find the so-called Phoenician and Babylonian standards in occupation. It is hardly possible that mere degradation of weight will account for all the phenomena; accordingly the object of this section will be to show that from first to last *the Greek communities were engaged in an endless quest after bimetallism*: we shall find, as we have already indicated, that whilst the gold unit never varies in any part of Hellas until a late epoch, the silver coins exhibit differences not merely between one district and another, but even between one period and another in the self-same city or state. There is incontrovertible evidence to prove that the same trouble was caused by the fluctuation in the relative value of gold and silver as arises in modern times. Xenophon[1] in his treatise *De Vectigalibus* (speaking of the benefit likely to accrue to the state if the silver mines of Laurium were better worked) makes the most interesting remark that "if any one were to allege that gold too is not less useful than silver, that I do not deny, yet this I know that gold, whenever it turns up in quantity,

---

[1] Xenoph. *De Vectigalibus*, iv. 10, εἰ δέ τις φήσειε καὶ χρυσίον μηδὲν ἧττον χρήσιμον εἶναι ἢ ἀργύριον, τοῦτο μὲν οὐκ ἀντιλέγω, ἐκεῖνο μέντοι οἶδα ὅτι καὶ χρυσίον ὅταν πολὺ παραφανῇ, αὐτὸ μὲν ἀτιμότερον γίγνεται, τὸ δὲ ἀργύριον τιμιώτερον ποιεῖ.

becomes on the one hand cheaper itself, and on the other makes silver dearer." This passage alone is sufficient to show how sensitive was the old Greek money market in the beginning of the 4th century B.C., and this statement is amply substantiated on Italian soil by a passage quoted by Strabo[1] from Polybius, from which we learn that after the discovery of a rich gold mine in the land of the Taurisci of Noricum, within the space of two months "gold went down one third in value throughout all Italy." Such being the effect of a discovery of gold, it is evident that either the silver currency must undergo certain modifications in order that a definite round number of silver units may be equal to the gold unit, or on the other hand the gold unit must undergo modification. But as we have shown that the gold unit remained unaltered throughout all Hellas, Asia and Egypt down to the time of the Ptolemies, it follows that whatever changes were necessary must have taken place in the *silver* standards. Of this we have proof in the case of Rhodes itself. Down to 408 B.C. the three ancient cities of Ialysus, Camirus and Lindus issued each a separate series of coins, Camirus on the Aeginetic standard, the other two on the Phoenician. In 408 B.C. all these united in founding the new city of Rhodes, and henceforward there is a single coinage. At first the Attic standard seems to have been employed for silver, as rare tetradrachms of 260 grs. are found, but it must have very soon given place to the so-called Rhodian, the tetradrachm of which ranges from 240 to 230 grs. About the same time (400 B.C.) the Rhodians began to issue gold staters of the so-called Euboic standard, and for a century this double issue of gold and silver continued unbroken. It is plain, from the case of this famous island, that it is only the silver standards which changed. There can be no doubt that the unit by which gold in bullion was reckoned before that metal was coined was the so-called Euboic or ox-unit, but during the archaic period we find both the so-called Phoenician (220 grs.) and Aeginetic (drachms of 92 grs.) being employed for silver in the island, whilst after 408 B.C. gold is issued on the ox-

---

[1] Strabo, IV. 208, συνεργασαμένων δὲ σὺν βαρβάροις τῶν Ἰταλιωτῶν ἐν διμήνῳ, παραχρῆμα τὸ χρυσίον εὐωνότερον γενέσθαι τῷ τρίτῳ μέρει καθ' ὅλην τὴν Ἰταλίαν.

unit, but silver, although at first on this standard, immediately changes to the Rhodian of 240 grs. Evidently then the fixed element is the gold, the fluctuating the silver. The coinage of Rhodes likewise exemplifies the doctrine already indicated, that the employment of religious and mythological symbols seems to mark not the earlier but rather the later stages of Greek coining. Thus Camirus employed the fig-leaf, Ialysus half a winged boar, and Lindus the lion's head with open jaws, but after 408 Helios the Sun-god, from whom all Rhodians alike claimed descent, and to whom the island was sacred[1], becomes the regular type, with the *type parlant* of the Rose (*Rhodon*) on the reverse.

Next let us take the money of Macedonia, where there was an abundant coinage of both gold and silver. The Pelasgian tribe of Bisaltae, and the Thracian Edonians and Odomanti, had during the half century which preceded the Persian wars all struck silver on the so-called Phoenician standard. It is commonly supposed that they obtained this standard from the important town of Abdera, which at the same period employed a like standard, and it is suggested that Abdera had borrowed it from her mother Teos, who had borrowed it from Miletus and the other great towns of the Ionian seaboard, among which it was especially employed for electrum. But unfortunately, whilst the types of Teos and Abdera are the same (a seated Griffin), the staters of Teos weigh only 186 grs., which is the Aeginetic, not the Phoenician (220 grs.) standard. Shortly after the overthrow of the Persian host Alexander I. of Macedon acquired the land of the Bisaltae along with the rich silver mines, which were said to produce for him a talent daily, and he adopted both the types and standard of the Bisaltian silver coinage, only substituting his own name for that of the Bisaltae. During the century which elapsed between Alexander I. and the accession of the famous Philip II. the coinage of Macedon and that of Abdera followed the same course in each case; the Phoenician standard of 230 grs. gave way to the so-called Babylonian or Persian of about 170 grs. Again, it has been suggested that Abdera influenced the neighbouring communities

[1] Pindar, *Olymp.* VII. 58 *sq*.

in this change. But when Philip came to the throne he returned to the Phoenician standard for silver, and when for the first time in Macedon he issued a bountiful coinage of gold staters, they were struck on the ancient gold unit, the so-called Euboic standard of 130 grs. But hardly had Philip slept with his fathers, and Alexander reigned in his stead, when a need was felt for a change in the silver standard. Accordingly the latter in the early years of his reign began, and continued to his death, to strike his silver on the same standard as his gold. Let us now study the lessons to be learned from this history of currency. There can be no reasonable doubt that the ox-unit or *stater* was the unit by which gold was estimated from first to last in that region. Unless it already existed Philip would not have employed it for his gold coinage at a time when he was making changes in his silver, but would have assimilated his gold to his silver standard. But, as before remarked, just because gold was not coined anywhere in Greece until the closing years of the 5th century, and in all transactions it passed as bullion, so much the stronger was the reason for keeping its weight-unit unchanged. But was the standard of 220 grs. really an imported Phoenician, or was it not rather one arrived at in that region by the natives themselves owing to the relations then existing between silver and gold? It is evident from the account given of the Bisaltian silver mines that in the time preceding and immediately posterior to the Persian invasion silver was exceedingly abundant in all that region. It is then by no means unlikely that it required ten silver pieces of 220 grs. each to make the equivalent of one gold unit of 130 grs. With the exhaustion of the silver mines, and perhaps a greater output of gold, silver became dearer, and consequently 10 silver pieces of 170 grs. each were now equal to a gold stater. Abdera on the coast would come perfectly within the sphere of such changed conditions, and her standard would consequently likewise undergo modification. With Philip's accession, fresh conquests and a general development of resources may have temporarily thrown more silver on the market, thus inducing him to revert to the 220 grs. standard, but the exploiting of the famous mines of Crenides increased the supply of gold to such an extent that by the time Alexander mounted his father's throne gold stood to silver in the relation of 10 : 1, and it was found extremely convenient to coin this on the same footing as gold, 10 silver pieces of 135 grs. being exactly equal to the gold stater of like weight. A like explanation applies to the coinage of Thrace. Amongst the Thracian tribes who dwelt near Mount Pangaeum and worked the gold and silver mines of that region the art of coining had been known from the 6th century B.C. and they issued silver coins of about 160 grs. This is regarded by some as debased Babylonian or Persic standard. But it is far more rational to suppose that in that region gold was more plentiful in proportion to silver than it was at that time further west in Macedonia, and accordingly a certain number of silver didrachms of 160 grs. were found to represent the gold stater or ox-unit. It seems most unlikely that a people long acquainted with both gold and silver could not devise for themselves a simple method of making some convenient number of silver pieces be equivalent to one gold, and that, on the contrary, having once obtained a certain standard fixed for silver in Asia Minor, at a time when gold was to silver as 13 : 1, they would blindly cleave to this standard, no matter how great a change took place in the relation of the metals. In face of the statements of Xenophon and Polybius already quoted and the fact that Solon deliberately constructed a new silver standard, it is simply impossible to believe such a doctrine.

On the opposite shore from Thrace lay the flourishing city of Cyzicus. This wealthy community commenced to issue electrum staters and *hectae* in the 5th century B.C., if not earlier, the former being about 252 grs., the latter 41 grs. These electrum staters have been shown by Professor Gardner to have contained gold and silver in about equal proportions[1].

---

[1] *Numismatic Chron.* VII. 185. That the Cyzicene staters were at some time and at some places (Cyzicus itself?) less in value than a Daric is made possible from the new-found Mimiambi of Herondas (VII. 96 *seqq.*); where 4 Darics seem worth more than 5 staters:

ταύτηι δὲ δώσεισ κε[ῖ]νο τὸ ἕτερον ζεῦγοσ
κόσου; πάλιν πρήμηνον ἀξίαν φωνὴν
σεω < υ > τοῦ.
Κ. στατῆρασ πέντε ναὶ μὰ θεοὺσ φο[ί]ρᾳι

This most important fact, taken in connection with the literary evidence derived from Xenophon and Demosthenes, makes it probable that the Cyzicene stater of 252 grs. was counted equal to a Daric of 130 grs. of pure gold[1]. "These coins of Cyzicus," says Mr Head, "together with the Persian Darics formed the staple of the gold currency of the whole ancient world, until such time as they were both superseded by the gold staters of Philip and Alexander the Great[2]."

Not only did they circulate side by side with the Darics, but it is worthy of notice that when the Cyzicenes struck coins of pure gold (*circa* 413 B.C.) they were of Daric type and standard. The earliest silver coins (430—412 B.C.) were small pieces of 32 and 18 grs., whilst the larger coins which come later are on the Phoenician silver standard of 212 grs. (412 B.C.), whilst from 400 B.C. to 330 B.C. the Rhodian standard of 235 grs. prevailed. From the story of her coinage we learn clearly that at Cyzicus the inferior metals bowed to the sway of gold. The electrum stater of 252 grs. is made equal to the pure gold unit, and whilst the silver standard changes from 212 grs. to 235 grs. the gold and pale gold pieces in currency remain inviolate. Once more, it is almost certain that some displacement in the relative values of the metals had caused the raising of the standard from 212 grs. to 235 grs. One thing certainly is beyond doubt, and that is the utter improbability of the introduction of the 235 grs. standard being in any way due to the influence of Rhodes. This remark likewise applies to Chios, where from a very early period (600—490 B.C.) side by side with electrum staters of 217 grs. we find didrachms of silver of 123—120 grs., "a weight peculiar to Chios," says Mr Head, "which was probably the Phoenician somewhat raised." But why was it raised? The real solution is that the relations between gold, electrum and silver at Chios necessitated the striking of

ἡ ψάλτρι' <Εὐ>ετηρισ ἡμέρην πᾶσαν
λαβεῖν ἀνώγουσ'· ἀλλ' ἐγώ μιν [ἐχθα]ίρω
κἢν τέσσαράσ μοι δαρεικοῦσ ὑπόσχηται
ὁτεύνεκέν μευ τὴν γυναῖκα τωθάζει
κακοῖσι δέ[ν]οισ. ει . . . . χρείη.

[1] Xen. *Anab.* v. 6. 23; VII. 3. 10. Dem. *Phorm.* p. 914.
[2] *Op. cit.* p. 449.

silver on a standard a few grains lighter than the gold unit in use (the Persian Daric), and the electrum stater of 217 grs. Space forbids our going through all the cities of the Ionian coast in detail, but the principle which we have laid down and illustrated from the currency systems of several leading states is sufficient to indicate the method by which we would explain the fluctuations in the silver standards employed at different times in various states. The Daric is the universal gold unit of all this region; by its side is the electrum stater usually of 217 grs. and most probably the equivalent in value of the pure gold coin of 130 grs.: along with them we find singular fluctuations in the silver currency; towns that are close neighbours employing different systems contemporaneously.

There is, however, one state which cannot be passed over without more particular reference. At an earlier page I spoke of the gold mines of Thasos, which had attracted the attention of the Phoenicians at a very early time. But, in addition to the mineral wealth of their own island, the Thasians drew a huge annual revenue from their mines on the mainland. Although the first influence in the island was Phoenician, and the Thasians themselves were Ionians from Paros, instead of finding the Phoenician standard employed for its silver coins, we see them striking their archaic coins on the so-called Babylonian system. Under the supremacy of Athens this standard fell so much that it eventually coincided with the Attic (138 grs.) or even was lower. The Thasians, after revolting from Athens in 411 B.C., struck gold coins for the first time; these were on the Euboic or ox-unit standard (consisting of half-staters and thirds). But about the same period they began to coin silver on the so-called Phoenician of 220 grs. It is indeed strange that in the early age, when the Phoenician tradition was still strong, they did not employ the 220 grs. standard, but only resorted to it after employing for a long period the Babylonian and Attic standards. It is evident that in Thasos, as elsewhere, there had existed the same gold unit for untold generations, else at the very time when they revolted from Athens and adopted a new standard for their silver, they would not have struck gold on what is commonly called the Attic or Euboic standard. It is evident that the changes in the silver standards were due to changes in the relation of silver to gold, the fall in standard from 168 grs. to 135 grs. indicating perhaps that silver, which at first was to gold as 1 : 13, had gradually grown dearer.

Commercial Weight System.

We must now turn to the commercial weight system. As elsewhere, one of the chief commodities to come under such a system was copper, and the history of the weighing of this metal, as far as it can be learned, will be of great importance to us. Now we should naturally expect that at Athens, which had in later days but one standard for gold and silver, copper likewise would have been estimated on this unit. But, as a matter of fact, there were two distinct standards in use at Athens, as is proved by two weights preserved in the British Museum, the inscription on one of which is *Mina of the Market* (MNA ΑΓΟΡ), that on the other is *Mina of the State* (MNA ΔΗΜΟ). This mina of the market is the same as that called the *Commercial Mina* on an Attic inscription[1], where its weight is given as that of 138 silver drachms, that is, the weight of an Aeginetic mina of silver. Athens had not coined any money of her own up to Solon's time, but seems to have employed the coins of Aegina. But this standard, although no longer employed for silver, did not fall into desuetude. As already pointed out, all peoples have felt the need of a heavier standard for cheap articles than that which serves for gold. Probably the Aeginetic mina had been used at Athens for copper: accordingly, when Solon made his new silver standard for the weighing of silver, the Aeginetic standard was found convenient for less costly and more bulky wares, and was therefore retained in use as the mercantile or market standard, the name STATE being given to the silver standard.

We have learned already that in the early stages of society copper and iron are not sold or appraised by weight, but rather by measurement. We have also seen that there is every reason to believe that the Greek obol originally was a spike or rod

[1] *Corp. Inscr. Graec.* 125, ἀγέτω ἡ μνᾶ ἡ ἐμπορικὴ Στεφανηφόρου δραχμὰς ἑκατὸν τριάκοντα καὶ ὀκτὼ πρὸς τὰ σταθμία τὰ ἐν τῷ ἀργυροκοπείῳ.

of copper of a definite length and thickness. If we can believe the statement of Ephorus given by Strabo that Phidon of Argos established a weight as well as a measure system for the Peloponnesians (although Herodotus is silent as regards weights), it is not at all improbable that, taking this story in conjunction with the dedication of the old bar money by Phidon in the temple of Hera, we have here a genuine tradition of the superseding of the bars of metal, the value of which simply depended on their dimensions, by a system based essentially on weight. It is plain that, as copper was weighed both at Aegina and Athens by the Aeginetic silver standard, copper most probably was never estimated by weight until after the forming of the separate silver standard in the way already described.

We have previously noticed the fact that the two principal terms applied to silver coins, *drachm* and *obol*, give clear indications that they have been borrowed from an ancient system of copper (just as we shall presently find that the *denarius*, the special term employed for their silver currency by the Romans, owes its origin to the ancient copper *as*). If further proof were required, it is afforded by the name employed for the subdivisions of the obol. The latter at Athens was divided into 8 *chalci* or *coppers* (χαλκοῖ). The smallest silver coin at Athens was the half-obol, but in some places names, *Trichalcum*, *Tetrachalcum*, etc. were given to copper coins. Now, as the Aeginetan obol weighed about $16\frac{1}{2}$ grs. and the Attic $11\frac{1}{4}$, the former is one-third greater than the latter. But we shall see shortly that as the Attic obol has 8 *chalci*, the Aeginetan must have had 12, from which it follows that the ancient copper obol or bar used in Aegina, throughout Peloponnesus, and at Athens, and probably throughout Boeotia, was everywhere the same.

The Sicilian System.

In dealing with the Sicilian and Italian systems we must reverse the order of treatment of the metals, and as it is in the copper that we shall find the closest link between the Greek and those other systems, we shall therefore commence with that metal.

On the Italian Peninsula and in Sicily we find a series of weight and monetary terms totally distinct from any found in Greece Proper. From this alone we may infer that, even before the settlement of any Greek Colonies in Magna Graecia and Sicily, there existed a well defined system, if not of weight, at least for the exchange of copper by fixed standards of measurement. In various Sicilian cities we find small silver coins called *litrae*; these beyond all question are simply the representatives in silver of an ancient copper unit employed by the Sicels, and which they had brought with them into the island. These Sicels were a tribe of the great Italian stock (itself a branch of the Aryan family) closely related to the Umbrians, Latins, and Oscans, had probably formed the van of the Aryan advance into the Peninsula, and had finally crossed the straits and overcome the Sicanians, an Iberic race, who were the earliest inhabitants of the island of whom any historical record exists. The word *litra* is merely a dialectic form of the same original *lidhra*[1], from which the Latin *libra* itself is sprung. But whilst we shall have little difficulty in finding out the weight at which the Latin *libra* was fixed, we have just as great difficulty in discovering that of the Sicilian *litra*, as we have lately found in the case of the ancient Greek copper obol. As copper was only coined at a late period, and the copper coins are merely tokens, or money of account, we are unable to arrive at any conclusion as to the original full weight of the litra from any data afforded by the copper coins of the various Sicilian states, although, from the circumstance that many of these coins bear marks of value, at first sight it might seem far otherwise. Thus at Agrigentum in the period preceding 415 B.C. the copper litra weighed about 750 grs., between 415 B.C. and 406 B.C. 613 grs., and from 340 B.C. to 287 B.C. it was about 536 grs. only. At Himera between 472 B.C. and 415 B.C. it was about 990 grs., but within the same period it fell to 200 grs., whilst at Camarina between 415 B.C. and 405 B.C. it was about 221 grs. Not only therefore is it futile to attempt any statement of the reduction of the litra in Sicily in general, but also to arrive at any sound approximation to its

---

[1] Cf. Wharton, *Etyma Latina*, s.v. *litra*.

full original weight, as far as the weight of the copper coins is concerned. On the other hand, any calculation based on the relative values of copper and silver has been up to the present unsatisfactory, owing to the great uncertainty which still prevails, Mommsen making the relation in the earlier period stand as 288 : 1, whilst Mr Soutzo thinks it never can have been higher than 120 : 1.

The latter view I have already proved to be untenable when we apply the test of the value of cattle, and it was made probable that in the 5th century B.C. silver was to copper as 300 : 1. From this it will be possible to show that the full weight of the copper litra was originally about 4900 grs.

Any effort to determine the original weight of the copper litra by a new method calls for a merciful consideration, even though it too may fail. Whilst the original weight of the litra is still a matter of doubt, we are fortunately completely acquainted with the method of its subdivisions. The litra was divided into 12 parts called Ungiae, Unciae or Onciae, a name which is no other than the Latin *Uncia*. This at once brings us face to face with the Roman copper system, where the *as* was the higher unit, and was divided into 12 unciae (ounces). But there are other striking coincidences of nomenclature. Thus $\frac{1}{6}$ of the *as* was called *sextans*; one-sixth of the litra is called *Hexâs* (ἑξᾶς), and the *Triens* and *Quadrans* are paralleled by the *trias* (τριᾶς) and *tetras* (τετρᾶς) although there is a difference in the application of these terms. Then the five-twelfths of the *as* is *Quincunx*, the same fraction of the litra is *Pentonkion* (πετόγκιον). We have plainly therefore a common Italo-Sicilian copper system, the terms of which were adopted and Graecised by the settlers in Italy and Sicily.

Now we have already adverted to the fact that the earliest Sicilian towns which coined money, Naxos, Zancle and Himera, although Chalcidian colonies, yet employed the Aeginetic standard, whereas we might naturally expect them to follow the Euboic. This would give the maximum of $16\frac{1}{2}$ grs. for the silver obol. Now according to Pollux, Aristotle in his lost treatise on the constitution of Agrigentum says that the litra is worth an Aeginetan obol, and Pollux goes on to say that "one would find in him (Aristotle) in his Constitution of the Himeraeans likewise other names of Sicilian coins, such as *ungia*, which is equivalent to one *chalcus*, and *hexas*, which is equivalent to two *chalci*, and *trias*, which is equivalent to three *chalci*, and *hemilitron* (half litra), which is equivalent to six, and litra which is equivalent to an obol[1]." It is plain from this that Aristotle knew that the Aeginetic obol was divided into *twelve chalci*. Thus the proposition laid down above, that the ancient Greek copper obol was a rod or spike divided into 12 parts, is thoroughly proved. The reason why the Attic obol had only 8 *chalci* is now plain; it was, as we saw, only two-thirds of the Aeginetan and consequently only contained two-thirds of the whole number of pieces of copper into which the ancient copper unit was divided. Now, as we find the Chalcidian settlers of Himera and other places not using their native Euboic standard for coining, but employing the Aeginetic, and as the Aeginetic obol was equal to the Sicilian litra, we are justified in the conclusion, that when the Greek settlers reached Italy and Sicily they found their Italic kinsfolk using a copper unit exactly the same as that employed in Greece; and that finally, when they began to coin, they found it more convenient to strike silver on a standard which was both convenient in reference to exchange with gold, as I have shown above, and had the further advantage of corresponding accurately in value to the ancient copper unit in use among the Sicels. If, as I indicated, silver was to copper as 300:1, the Aeginetic silver obol of $16\frac{2}{3}$ grs. would be worth 5000 grs. of copper (practically the same as the early Roman *libra*). It follows then that if we could only discover the weight of the Sicilian litra we should know that of the old Greek *copper* obol. Is this possible? We have no reason to doubt that the obol was a rod of copper of a certain size, which in the course of time after the introduction of coined money shrank up until the original rod was only represented by what had been its equivalent in silver, or a small copper coin, whose name still survives in the *ob* used in old account books as the symbol for

[1] Pollux, IX. 80.

half-penny[1]. The Greek coinage has preserved for us but faint traces of the various steps in the degradation of the copper obol, but, as we have already seen, we find the Sicilian copper litra in various stages of its decadence from 990 grs. down to 200 grs. Again, whilst no trace has as yet been found of obols at all in the archaic shape of rods, or anything approaching it, we find in Sicily at Agrigentum *litrae* which are in form distinct survivals of an earlier stage when the litra, like the obol, was a rod or bar of copper. These are very strange looking lumps of bronze made in the shape of a tooth with a flat base, having on one side an eagle or eagle's head and on the other a crab, while on the base are marks of value ::, ∴, : (*tetras, trias, hexas*). The *uncia* is almond-shaped with an eagle's head on one side, and a crab's claw on the other[2]. As we found the Chinese knife shrinking up into a shorter and thicker mass until at last it only survives in the round *cash*, so in all probability we here find the Sicilian litra in its mid course from its original full size and shape to that of the ordinary round copper coin of a later age. That the shape of the original copper unit of the Italians was that of a rod or bar we shall now proceed to demonstrate in the case of the Roman *as*.

The Italian System. Bronze.

As the cow formed the highest unit in the monetary system of ancient Italy, so the lowest unit employed was a certain amount of copper called an *as*. We have already found the cow serving the same purpose in Sicily (as late as the time of Dionysius forming the rateable unit at Syracuse). The systems of Further Asia, where the buffalo stands at the head of the scale and the hoe or a piece of raw metal of a certain size stands at the bottom, form a perfect analogy in modern times. As far as its value and divisional system go, we have identified the Sicilian litra with the ancient Hellenic

---

[1] Cf. Shakespeare, *I. Henry IV.* II. 4, 590, in Falstaff's tavern bill: "Item, Anchovies and sack, 6d. Item, bread, Ob. O monstrous! But one halfpenny worth of bread to such an intolerable deal of sack!"

[2] Head, *op. cit.* p. 105.

obol or rod, and we have in turn discovered a very close resemblance between the divisions of the litra and that of the *as*. I now propose to examine into the original nature of this denomination, and the form of the object to which it was applied. This will have been effectually accomplished, if I can succeed in establishing the proposition *that the* as *was primarily a rod or bar of copper, one foot in length, divided into* 12 *parts, called inches (unciae), thus coinciding with the Greek obol in form, as also in its duodecimal division.*

We must, as a preliminary, note carefully several most essential facts connected with the *as*: (1) The term *as* (as used in respect of metals) is never employed for either gold or silver, but is appropriated to *bronze* exclusively; (2) it is not the Roman unit of weight, for that is expressed by the general term *libra*, a word exactly corresponding to the Greek *Talanton*, since it means both the *weight* and the *scales*; (3) the *as* is not confined to weight, but is also employed as the unit of linear measure equal to the foot, and also as the unit of land measure equal to the *jugerum* or acre.

The following table exhibits the subdivisions of the *as*:

As (Pes, Jugerum)

Deunx	$\frac{11}{12}$
Dextans	$\frac{10}{12}$
Dodrans	$\frac{3}{4}$
Bes	$\frac{2}{3}$
Septunx	$\frac{7}{12}$
Semis	$\frac{1}{2}$
Quincunx	$\frac{5}{12}$
Triens	$\frac{1}{3}$
Quadrans	$\frac{1}{4}$
Sextans	$\frac{1}{6}$
Uncia	$\frac{1}{12}$
Semuncia	$\frac{1}{24}$
Sicilicus	$\frac{1}{48}$
Sextula	$\frac{1}{72}$
Scriptulum	$\frac{1}{288}$

Now it has been hitherto assumed by all writers that the

system of division employed in the *as* as a unit of *weight* has been transferred to *measure*. This however is contrary to all experience, for, as we have had occasion constantly before to notice, weight units are derived from measures, e.g. the bushel from the measure of that name, and so on. In the next place as the *as* is not the unit of Roman weight, if even the measure unit was borrowed from the weight, we ought to expect the foot to be called a *libra* rather than an *as*. It is far more likely that a unit originally employed for measure would in time give its name to a weight-unit corresponding in mass to the original measure-unit. There are besides certain pieces of evidence afforded by the nomenclature of the submultiples which point directly to the original as being a measure rather than a weight-unit. The 24th part of the uncia is called the *scriptulum*, *little scratch*, or *line* (*scribo*), which is exactly translated by the Greeks as *gramme* (γραμμή, scratch or line)[1]. Now whilst 24 strokes make an excellent method of dividing the uncia in its capacity of *inch*, they of course have no significance as submultiples of uncia, meaning *ounce*. Moreover, the forms of several of the best known divisions of the *as*, such as triens, quadrans, sextans, which are not easy to explain on the hypothesis that the terminology was primarily applied to weight, on the other hand admit of a ready solution when we take the *as* as originally a unit of measure. For sextans means not a sixth, but that which makes a sixth, triens not a third, but that which divides in three parts, and quadrans not a fourth, but that which makes fourfold, i.e. divides into four, for *quadra* means not a fourth part, but that which has four parts (hence usually a square). If we regard these words as referring to certain lines drawn across a bar of metal, their meaning is obvious. Whilst *sextans uncia*, the ounce which makes a sixth, is nonsense, *sextans linea*, the line which makes a sixth, gives excellent sense, so likewise *triens linea* fits in admirably with the required meaning, whilst *quadrans linea* seems to mean *the line which divides the whole into four parts*.

[1] The forms *scripulum*, *scrupulum*, *scrupulus* are all due to its simply being regarded in later times as a *weight*, and thus falsely identified with *scrupulus*, a small pebble.

THE ITALIAN SYSTEM. 353

The etymology of the word *as* has long been a puzzle. Scholars starting with the assumption that *as* was the Roman abstract term for unity have accordingly searched for an appropriate derivation. Some have identified it with the Greek *heis* one (εἷς through a Tarentine ἅς), whilst the most recent attempt connects it with the first syllable of *elementum*. The same principle has been carried out with regard to *uncia*, which has been treated simply as meaning *unit* and connected with *unus* and *unicus*.

Now it is notorious that the Roman mind was essentially concrete, and found great difficulty in arriving at abstract ideas, and consequently at abstract terms. This alone would make us hesitate to believe that *as* had originally begun as an abstract term meaning unit, and rather incline us to believe that it started in life as a name for some common concrete object. But we have seen above that the numerals in all languages seem originally to have meant certain actual physical objects which served as counters, such as the fingers and toes (*decem* δέκα, *digitus* δάκτυλος), seeds or pebbles. If such has been the origin of the various names for *unit*, we can hardly believe that any term for *unity* can have originated independently of some concrete object. To add to the mists which hang round the origin of the *as*, its division into 12 parts is taken to indicate a Babylonian source. Now the Roman foot was divided, not merely into 16 fingers like the Greek, but also into 12 unciae or inches like our own. The latter is most probably the true Italian system, as it is that found among their cousins and neighbours the Kelts, as well as amongst the Teutonic peoples. With ourselves still the rustic measures inches by his thumb, just as he measures feet by means of his own natural foot. The ancient Irish foot was divided into 12 thumbs or inches (*ordlach*, Lat. *pollex*, the initial *p* being lost in Irish)[1]. The Romans too (as did likewise the Teutonic peoples, *e.g.* Icelandic *tomme*, an inch) used the thumb (*pollex*) as the ordinary measure in practical life[2]. The division then into 12 unciae is simply the result of the fact that

[1] Book of Aicill, p. 335.
[2] Caesar, *B. G.* III. 13.

a certain natural relation exists between the breadth of the thumb and the length of the foot, and as the relation held true just as much for the Kelt as the Chaldaean, there was no need for the ancient Italians to borrow their duodecimal system from the East. Now what are we to say as to the origin of the word *uncia?* Does it mean anything more or less than the breadth of the (thumb) *nail?* The use of *unguis,* a nail, as a measure was common in Latin, as we know from the phrases *transversum unguem* (the thickness of a nail) and *latum unguem* (a nail's breadth) side by side with *transversum digitum* (a finger's thickness) in Plautus. *Uncia* may be simply a derivative from *unguis;* there is no phonetic impossibility, and even if there were any linguistic irregularity, false analogy with *unicus* would amply account for it. The use of a word meaning *nail* to express the divisions of the foot is completely paralleled by the ancient Hindu system, where the *finger-breadth* is termed *angala, i.e.* nail (cognate of *unguis* and ὄνυξ).

Next we come to the word *as* itself, which appears in old Latin as *assis.* It is masculine in gender, which of itself is sufficient to throw doubts on its being a really abstract word. Can it be that we have a close relative of it in *asser* a rod, bar, pole, which is likewise masculine in gender? Whilst one form of the name was specially confined to a small rod or bar of copper, the other was employed in a wide and general way. These two forms *assis* and *asser, -is* are completely analogous to *vomis* and *vomer, -is,* a ploughshare. The meaning *rod* is in complete harmony with what we have said about the Greek obol. All that is now wanting to make our proof complete is some evidence that the primitive Italian *as* was really in the form of a rod or bar. The most archaic specimens of ancient Italian bronze money as yet described are those found at the Ponte di Badia near Vulci in 1828. These consisted (1) of quadrilaterals broken in pieces, weighing from 2 to 3 pounds each, stamped with an ox and trident, (2) cube-shaped pieces of copper without any mark, weighing from an ounce to a pound, and (3) some ellipse-shaped pieces for the most part weighing two ounces[1]. But in the British Museum are preserved a number of pieces of

[1] *Blacas*, Mommsen, I. p. 177.

bronze which are roughly quadrilateral. A cursory examination showed me that, whilst two parallel sides exhibit the marks of a mould, the two remaining sides displayed unmistakable

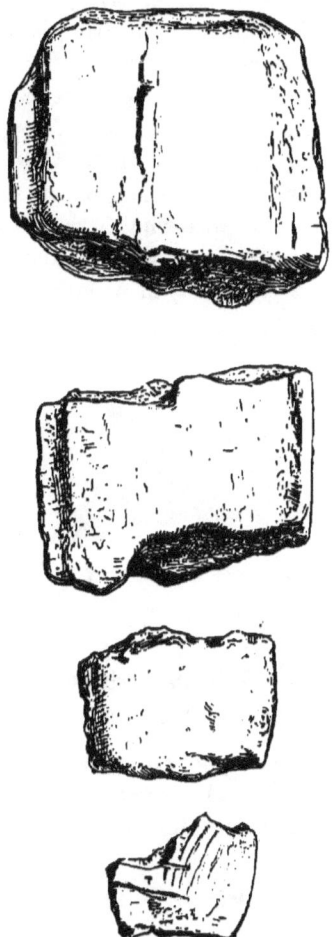

Fig. 48. Aes Rude.

signs of fracture. Several of them are end pieces, showing the volutĳng of the mould on two sides and at one end, whilst the other end shows marks of having been broken (Fig. 48).

Several of them bear stamps, or letters. There can be no doubt that these are pieces of short bars of bronze, which were afterwards cut up, as occasion demanded. The imprints on them prove them to be of comparatively recent date. If therefore the *asses* still retained their bar shape after the art of stamping metal to serve as currency had come into use, *à fortiori* the primitive *as* of Italy must certainly have been nothing more than a plain rod or bar of copper, which passed from hand to hand as the obols in Greece, and the bars of iron and copper pass at the present among savages of Africa and Asia[1]. This was what was called by the ancient writers *the raw copper* (*aes rude*), as distinguished from *the stamped copper* (*aes signatum*) of a later date. The fact that

FIG. 49. Bronze Decussis.

early specimens of *aes signatum*, such as the *decussis*, bearing a cow on both obverse and reverse (Fig. 49), were still made in the shape of a bar, is a further proof that such was the original form.

It will be observed that I can give no positive evidence for

---

[1] It is worth noticing that Plutarch (*Poplicola* 11) translates the *libral asses* of early Rome by the Greek *obolos*; ἦν δὲ τιμὴ προβάτου μὲν ὀβολοὶ δέκα, βοὸς δὲ ἑκατόν· οὔπω νομίσματι χρωμένων πολλῷ τότε τῶν Ῥωμαίων, ἀλλὰ προβατείαις καὶ κτηνοτροφίαις εὐθηνούντων. It is quite possible that Plutarch embodies a genuine tradition that the original *as* and *obol* were the same. Otherwise like Dionysius of Halicarnassus he would have represented the asses by the value in Greek money of his own time. For he can hardly have supposed that at any time an ox was worth only 100 of the obols of his own time.

the length or breadth of the *as*. The pieces in the Museum are all fragments, and, even if there were any of them whole, they would not by any means decide the original *length*, although they would of course represent the *weight*. For as they are late, they would probably have been made at a time when the original rod was shrinking up into a more compact form, just as the Chinese bronze knives get shorter and thicker. But the fact remains that the *as* was identified completely with the Roman *foot* measure, the divisions being the same in each. We therefore may with great probability infer that the *as* was originally a piece of copper a foot in length, and of a known thickness. We have seen that copper and iron are not weighed in the early stages of society, but are appraised by measurement. Why should not the same hold true for Rome? It may be asked, how came it that the *as* was taken as the typical unit for weight and superficial measure, and to express even an inheritance? The answer is not far to seek. To express fractional parts has ever been a great difficulty with primitive people. As the Malays cannot conceive abstract numerals, but must append the concrete *padi* to each of their numbers, so the old Italian found it necessary to employ some concrete object, the subdivisions of which were familiar, to express the fractional parts whether it be of an estate or anything else. The most common unit in use was the rod of copper divided into 12 thumbs. Accordingly, if a Roman wished to say that Balbus was heir to one-twelfth of an estate he expressed this by the homely formula that Balbus had come in for *one inch*, the denominator 12 being mentally supplied, as everyone knew that there were 12 inches in the copper bar. The same principle of taking some familiar object, the ordinary method of dividing which was known to all men, is seen in the method of expressing one-tenth. The Roman *denarius* was divided into 10 *libellae*; accordingly, when Cicero wishes to say that a certain person had come in for a tenth part of an estate he says that he has come in for a *libella* (*heres ex libella*). From this the reader will at once see that we might just as well declare that the word *denarius* is an abstract word meaning *unity* as make the same assertion about the *as*. Again, when the Roman land surveyors elaborated their system of mensuration, they found that the simplest method of expressing the fractional parts of the *jugerum* was to employ the old duodecimal method of the *as*. Nor is this without a parallel elsewhere. As the yard was the common English unit of linear measure, it was applied to the most common unit of land, the quarter of the hide, which was accordingly termed a yard of land, or a virgate (*virga terrae*). The English analogy is even still more complete, for as the *as* or foot-rod became the unit of weight, so in Cambridge the yard of butter is identical with the pound of butter[1].

Our next step will be to trace the process by which the *as* or rod became the general weight-unit, the pound (*libra*). The term *libra* is not the oldest Latin name for *weight*, for *pondus* or its cognate verb *pendeo*, which literally means to *hang*, is the true claimant for that position. *Libra* seems properly to mean the *balance*, as is seen from the legal formula (employed in Mancipatio) *per aes et libram*, by means of copper and the balance. From the fact that its chief use was to weigh *asses* of copper, the mass of an *as* came to be termed the *weight par excellence*, just as the most usual amount weighed in the Greek *talanta* (scales) became the *talanton par excellence*. This process can be illustrated by modern examples. Thus in the south of Ireland potatoes are sold by the unit of 21 lbs., which consequently is termed a *weight*, and instead of speaking of so many stones or hundredweights, everyone speaks of a weight of potatoes. But, as already remarked, it was only at a comparatively late epoch that the bars of copper were weighed. It would be only with the growth of greater exactitude in commercial dealings that the art of weighing, which was employed for all dealings in gold and silver, would be applied to copper. Just as the Malays and Tibetans have been gradually taught by the careful Chinese to employ weights commercially, so the Italian tribes may have been led to do so under the influence of the astute Greek traders from Magna Graecia and Sicily.

---

[1] So the word *mark* means not only a weight but is also used as a linear measure = 48 *alen*, and also as a measure of *area*, as in the term *arable mark* etc. See Appendix.

The system in vogue for gold was that of our old friend the ox-unit. This is proved from the fact that not only is the oldest gold coinage of the Etruscans, the close neighbours of Latium, based upon this standard, but that also in Sicily and Southern Italy there was the small gold talent, the three-fold of the ox-unit. This three-fold of the stater was also used at Neapolis. Although the earliest Greek colonies in Sicily employed at first the Aeginetic standard for silver, we soon find them reverting to the gold or Euboic standard for that metal, whilst the early silver coinage of the Etruscans (before 350 B.C.) is also of the Euboic standard. We may with high probability assume that when the Sicilians and Italians first essayed to weigh their copper rods, they naturally employed the standard already in use for gold and silver. The highest unit of this was the small talent of 3 staters which weighed about 405 grs. The bar was divided into 12 inches, and it was found that an inch of copper rod closely approximated in weight to the small gold talent. The weight of the bar, which was the ancient unit for copper before weight had been employed, now became the standard weight-unit for that metal. It is to be observed that this ounce of 405 grs., though some 27 grs. less than the full Roman *uncia* of later times, is only 15 grs. lighter than the Roman ounce prior to 268 B.C., for it is an ascertained fact that the old Roman *uncia* did not exceed 420 grs.[1] It must be remembered that the weight of the ounce would depend on the standard foot by which the bar was measured. Now, whilst the Roman foot measures 296 millim., there was likewise in use in Campania, and probably in many parts of Southern Italy, a foot of 276 millim. The relation of bars of these lengths and of a given thickness to the Roman libra is not without interest. If we take an ordinary engineer's table of materials we shall find that a copper rod a Roman foot long, and half a Roman inch in diameter, weighs 5040 grs. Now, as the Roman pound weighs 5184 grs. this approximation seems almost too close to be a mere coincidence. If on the other hand we take a rod of a foot of 276 millim. and with a diameter of the corresponding half-inch, we shall get a pound of 4680

---

[1] Many of the Roman unciae in the British Museum are under 410 grs.

grs. and an ounce of 390 grs , which is certainly not far from the weight of the small gold talent. It follows from this that we may expect pounds of different weights in Italy, according as the foot-unit varies in different districts.

In later times, besides the pound of 12 unciae, there were several commercial pounds on Italian soil, the pound of 16 ounces (from which our own avoirdupois is probably descended), that of 18 unciae, and that of 24. The last two are easy of explanation, since one is simply the double, the other one and a half times the Roman pound. But perhaps a different explanation must be sought for the 16 ounce pound. The foot was divided by Greeks and also by Italians into 16 fingers as well as into 12 thumbs. Was therefore the pound of 16 ounces simply derived from the division of the foot bar into 16 fingers, the weight of the finger being however equated to that of the Roman thumb or inch of copper?

The *as*, having been once subjected to weight, its hundredfold, the *centumpondium* or "hundred weight," became the highest Roman weight-unit. Thus the *as* and the *centumpondium* of the Italians correspond to the mina and talent of the Greeks. But it will be observed that the Italians obtained their higher unit by the old decimal system, whereas the Greeks had borrowed the mina and its sixtyfold from Asia. The *centumpondium* must be regarded as a true-born Italian unit, not one borrowed from Greece or Asia, and of this there is further proof. We saw by the ancient Roman law that the cow was estimated at 100 *asses*, the sheep at 10 *asses*. No doubt from time out of mind 100 of the bars of copper, which formed the chief lower unit of barter, made one cow, just as in Annam 280 little hoes make one buffalo (p. 167). When copper came to be weighed, the amount of copper which formed the equivalent of the highest unit of barter, the cow, was taken as the highest weight-unit. From what I have said above it is not improbable that the Roman libra and the Sicilian litra of copper were almost equal in weight. The fact that the Greek writers always employed the Sicilian word litra (λίτρα), to translate the Latin *libra*, likewise indicates that in the Greek mind there was a tradition of their identity. And if the doctrine here put forward of the original nature of the *as* be right, nothing can be more likely than that the Italians who had crossed into Sicily and their kinsfolk who had remained behind employed rods of similar size, and that when they began to weigh the latter, the "weight" (libra or litra), derived from the standard copper rod, should be the same in each region, until certain modifications occasioned by new monetary conditions according to the needs of different communities had caused some divergency in *coin* weights, although as a *commercial* weight the litra remained unchanged. As Aristotle identified the Aeginetic obol and *chalcus* with the Sicilian litra and *onkia*, we may with some plausibility suggest that the ancient Greek copper obol or spike and the Italian *as* or rod were identical in dimensions and in origin.

In Greece the copper obol rapidly fell in weight, for, when once silver currency had been introduced, copper was thrust aside, and it was not till the fourth century B.C. that copper coins came into use. When the copper obol appears as a coin it is but a small piece, being in fact a mere token.

The history of the degradation of copper was seen better in Sicily, where we found the litra still weighing 990 grs., but it

Fig. 50. As (*aes grave*). (Before 2nd Punic War.)

rapidly sank to only 200 grs., evidently in this case also being mere money of account. For as the silver litra was about 13½ grs., unless the 200 grain copper litra was a mere token, silver would have been to copper as 17:1, which is obviously absurd. In the case of the Italian *as* the process is still clearer, for we have every stage of the *as*, from the bars which I have described through the *libral as* (*aes grave*), the *sextantal as*, the uncial and half-uncial, down to the small coin of the empire commonly called "a third brass."

FIG. 51. As (half uncial standard).

FIG. 52. As, 3rd Cent. A.D. ("Third Brass").

Gold and Silver.

Whilst in the infancy of coining the Sicilian silver litra was probably the same as the Aeginetic obol, that is about 16⅔ grs., the Aeginetic didrachm being probably treated as a *decalitron* (ten-litra piece), nevertheless after no long time the common Euboic standard of 135 grs. was employed at Syracuse and elsewhere, and we have the authority of Aristotle for the statement that the *Corinthian stater* was called a *decalitron*. Corinth, as we saw above, used the 135 grain unit for her famous Pegasi, commonly known as "Colts" ($\pi\hat{\omega}\lambda o\iota$), and

FIG. 53. Didrachm of Corinth.

therefore the litra was by this time 13½ grs. Now, in Etruria we find about 400—350 B.C. a silver currency struck on this same 135 grs. standard. These coins bear marks of value, X on coins of 131 grs., Λ on those of 65 grs., ΙΙ' on those of 32 grs., and Ι on those of 14 and 13 grs. It is plain therefore that the stater of 135 grs. was considered to consist of 10 units of 13½ grs. each. In other words, whatever the Etruscans may have called their stater, it was exactly the same in weight and method of subdivision as the *decalitron* of Syracuse. At a later period (350—268 B.C.) we find on coins of like weight the symbols XX instead of X, X instead of Λ, Λ instead of II'. The unit now is exactly half of what it was at an earlier stage, 6¾ grs. instead of 13½ grs.

Not till 268 B.C., just on the eve of the First Punic War, did Rome first coin silver. This coin, called *denarius*, as its name implies, represented 10 *asses*. It was divided into four parts, each of which was called a *sestertius* or 2½, and was marked with the symbol IIS representing that number.

It is very remarkable that the Etruscan coin of the second series, marked 2½, is only very slightly heavier than the Roman sesterce (*sestertius*) which bears a similar mark. Hence it has been very reasonably inferred that when the Romans set about the coinage of silver, they simply adopted with slight modifica-

Fig. 54. Sesterce of first Roman silver coinage.

tion the silver system employed by their neighbours across the Tiber. This is all the more probable, as it is almost certain that, though Rome did not strike silver she like Athens before the time of Solon, and like Syracuse, used freely the coins of other communities for a long time previously. The Etruscan coins would therefore serve as silver currency at Rome. We may then assume that the monetary system must have been much the same on both sides of the river. Accordingly, since in 268 B.C. we find the Romans striking a coin in silver representing 10 copper *asses*, which is almost the same in weight as the Etruscan coin marked X, we may reasonably infer that, if the Romans had commenced coining silver a century earlier, their *denarius* or 10-*as* piece would have been the same weight as the Etruscan.

Now besides the *litra*, which we found to be both a copper-unit and a silver coin in Sicily, there is another term of great interest, especially as it plays an important part in the history of Roman money. The general Latin name for a coin is *numus*, which in the later days of the Republic usually meant a *denarius* when used in the more restricted sense, but in the earlier period it was the term specially applied to the silver sesterce (*sestertius*). This is almost certainly a loan-word, for Pollux is most explicit in warning us that, although the word seems Roman, it is in reality Greek and belongs to the Dorians of Sicily and Italy[1]. It is always a name of a coin of silver in Sicily, being so used by Epicharmus. The coin meant by this poet cannot have been one of great value, for he says: "Buy me a fine heifer calf for ten *nomi*." It was in all probability the Aeginetan obol, for Apollodorus in his comments on Sophron set it down at three half (Attic) obols, that is, almost 17 grs. This is confirmed by the fact that an Homeric scholiast makes the small talent weigh 24 *nomi*, which gives nearly 17 grs. as the weight of that unit. Crossing into Italy, we find that according to Aristotle[2] there was a coin called a *noummos* at Tarentum, on which was the device of Taras riding on a dolphin. This is the familiar type of the Tarentine

Fig. 55. Didrachm of Tarentum.

didrachms which, from their first issue down to the invasion of Pyrrhus (450—280 B.C.), weigh normally 123—120 grs., although one specimen weighs 128 grs. This coin Mommsen recognized as the *noummos* of Aristotle. Professor Gardner afterwards suggested that the diobol, on which occasionally the same type is found, was rather the coin meant. Recently

---

[1] ὁ δὲ νοῦμμος δοκεῖ μὲν εἶναι Ῥωμαίων τοὔνομα τοῦ νομίσματος, ἔστι δὲ Ἑλληνικὸν καὶ τῶν ἐν Ἰταλίᾳ καὶ ἐν Σικελίᾳ Δωριέων.

[2] Pollux IX. 84.

Mr A. J. Evans has almost proved this hypothesis impossible by showing that all the diobols yet known are probably later than the time of Aristotle[1]. As, however, this rests on negative evidence, and is liable to be overthrown at any moment by the discovery of an archaic diobol, it is advisable to cast about for some more positive criterion. Heraclea of Lucania, the daughter-city and close neighbour of Tarentum, as we know from the famous Heraclean Tables (which scholars are agreed in regarding as written about the end of the 4th cent. B.C.), employed as a unit of account a silver *nomos*. It is so probable that the *nomos* employed at Heraclea (*circ.* 325 B.C.) would be the same in value as that employed at Tarentum in the time of Aristotle (*ob.* 322 B.C.), that if we can prove the *nomos* of Heraclea to be a *didrachm* and not a *diobol*, we may henceforth hold with certainty that the *nomos* of Tarentum was the larger coin.

On the Heraclean Tables it is enacted that those who held certain public land should pay certain fines in case they had failed to plant their holdings properly; four olive trees were to be planted on each *schoenus* of land, and for each olive tree not so planted a penalty of 10 *nomi* of silver was to be exacted, and for each *schoenus* of land not planted with vines the penalty was two *minae* of silver[2]. The *schoenus* is identical with the Roman *actus* (half a *jugerum*), being the square of 120 feet. Four olive trees were the allowance for each *schoenus*. Now if we can determine the number of vines which were planted on a *schoenus*, we shall be able to get a test of the value of a *nomos*. Two minae of silver contained in round numbers 110 Tarentine didrachms of 123 grs. each, or 675 diobols of about 20 grs. each. Olives were many times more valuable than the vine, so that any result which will make the vine about the same value as the obol will be absurd.

[1] Evans, *Horsemen of Tarentum*, pp. 9—11.
[2] *Tabulae Heracleenses* (Boeckh *Corp. Inscrip. Graec.* 5774—5; Cauer, *Delectus* 40, 41) I, 122. αἱ δέ κα μὴ πεφυτεύκωντι κατὰ γεγραμμένα, κατεδικέσθεν πὰρ μὲν τὰν ἐλαίαν δέκα νόμως ἀργυρίω πὰρ τὸ φυτὸν ἕκαστον, πὰρ δὲ τὰς ἀμπέλως δύο μνᾶς ἀργυρίω πὰρ τὰν σχοῖνον ἑκάσταν.

Now Mr A. J. Evans, when in Southern Italy, at my request kindly ascertained that vines, when trained on poles on vineyard slopes, are usually about 3 yards apart, whilst when trained on pollard poplars (as is much more usual in Campagna), they stand about 6 yards apart. In the case of the former about 150 vines would go to a *schoenus* (1600 sq. yards), whilst in the latter case barely 50. We cannot doubt that the distance between the vines must have been much the same in ancient as in modern times.

If now we take the *nomos* to be a *diobol*, each vine is worth $4\frac{2}{3}$ *nomi*, or 14 *nomi*, according as there are 50 or 150 vines to the *schoenus*. Now, as the valuable and slow growing olive is only worth 10 *nomi*, and it is impossible to believe that the relative values of olive and vine could have ever been such as those arrived at on the assumption that the *nomos* is a diobol, we must turn to the alternative course and take the *nomos* as a didrachm. The penalty for a *schoenus* of vines is two minae or 110 didrachms. If 150 vines go to a *schoenus*, each will be worth about $\frac{2}{3}$ didrachm, 15 vines being equal to one olive, or taking 50 vines to the *schoenus*, each vine will be worth about two didrachms, 5 vines being worth one olive. This result is so rational that we need hesitate no longer to regard the well-known Tarentine didrachm as the *nomos* (*noummos*) of Aristotle.

There is such a difference between the *nomos* of Sicily, identical with the Aeginetan obol, and that of Tarentum that we are forced to conclude that the term *nomos* is not specially applied to any particular coin unit. In Sicily we found the native unit, the litra, identified in certain cases, at least in earlier times, with the Aeginetan obol as well as with the *nomos*. Why two names *nomos* and *litra* for the same unit? Is one Sicilian and the other Greek? This at least gives a reasonable explanation. The Dorians then in Sicily gave the name to their earliest coins, *nomos*, with them indicating the unit of currency established by law, just as did *nomisma* among other Greeks. As in Sicily the Aeginetic obol was the *legal coin* (*nomos*) *par excellence*, so at Tarentum, where didrachms were the first coins to be struck, the term (*nomos*) was applied to that unit. We may therefore expect to find the term *nomos* applied to various kinds of coins among the Italiotes and Italians, according to the particular coin chosen by each state as its own unit of account.

Accordingly we find the term *nomos* applied to certain bronze coins struck on the sextantal (two ounce) and uncial standards, at Arpi and other towns, which are inscribed N II (the double *nummus*), N I (*nummus*), ... (*quincunx*), ... (*triens*), ... (*quadrans*), .. (*sextans*), . S (*sescuncia*), . (*uncia*), and Σ (*semuncia*). The divisions being those of the *as*, it is clear that the *nomos*, or current coin in those places, was the reduced *as*. Finally, when the Romans first use the term *nummus*, it means the silver *sestertius* ($2\frac{1}{2}$ asses), the one-fourth of the *denarius* or ten-*as* piece, which weighed a scruple (*i.e.* $18\frac{1}{2}$ grs.) at the time of the first Roman coinage of silver. Here we have all our positive evidence for the *nomos*. As diobols of 18 to 17 grs. are found in the coinages of various towns in Magna Graecia, such as Arpi, Caelia, Canusium, Rubi, and Teate, it has been plausibly held that such a diobol was the *nomos par excellence* of these states, and that it was from contact with them that the Romans learned both the use and the name of such a monetary unit. But Rome may have been influenced by her Etruscan neighbours, for, as we have seen, the smallest denomination in the second silver series of Etruscan coins (of which the coins weigh 129 grs., 32 grs. and 17 grs. respectively) is just the weight of the Roman sestertius, and bears the symbol ΛII ($2\frac{1}{2}$), just as the latter bears IIS ($2\frac{1}{2}$). Taking into consideration these facts, it looks as if the Romans and Etruscans grafted on to a native system the diobol, or current silver coin of Southern Italy, the Romans (and for all we can tell the Etruscans likewise) adopting at the same time the name *nummus*. Finally, we observe that this *nummus* is identical with the Sicilian *nomos*, which in turn was found to be none other than the Aeginetic obol. The Roman *sestertius* being a *scriptulum* ($17\frac{7}{12}$ grs.) in weight, we thus find a direct connection between the latter and the Aeginetic obol ($16\frac{2}{3}$ grs.). This need not surprise us, for it is most natural that in the welding of a weight system (partly foreign, and on the native side only employed for gold and silver) and of a system of measurement employed for bronze, certain features derived from the special silver units in use would be introduced into the new system, which afterwards became universal for weighing all commodities. The term *Sicilicus*[1] employed for the quarter-ounce is good evidence for this hypothesis. Its name seems to mean simply *Sicilian*. In weight it was about 108 grs. Now, didrachms struck on such a foot are found in the Greek cities of south-western Italy, at Velia, Neapolis and at Tarentum, after the time of Pyrrhus. Did the Romans, who must have carried on by weight all dealings in silver up to 268 B.C., treat such coins as quarter-ounces, and ultimately take the name of the coin (wrongly connecting it with Sicily) to designate the quarter-ounce? In like fashion it was probably discovered that the Aeginetic obol of the Greek colonists was about equal in weight to the line (*scriptulum*) which is one-twenty-fourth of the inch (*uncia*) of copper. Thus as there are 24 *nomi* in the Sicilian talent, so there are 24 *scriptula* in the Roman *uncia*. These considerations help to explain the relations which existed between the *nomos* (Aeginetic obol), *sestertius*, and *scruple*.

Mr Soutzo[2] gives a very different account of the *nomos*. Starting with the Egyptian hypothesis he makes all the Italian weight systems of foreign origin. He thus makes the Roman libra the $\frac{1}{100}$ of a Roman *talent*, which he seems to identify with a light Asiatic talent[3]. Starting with the talent he supposes that on Italian soil it was divided into 100 *librae* instead of 60 heavy or 120 light minae, as in the East. Each of these *librae* or *pounds* was divided into 12 *ounces*, and each *ounce* into 24 fractions. He holds likewise that the Italians adopted from the East the use of bronze "comme matière première de leurs échanges," at the same time as they obtained the first germs of civilization and their first weight

---

[1] Boeckh, *Metrol. Unters.* 160, takes the *Sicilicus* as originally the Silician *quadrans* in the Roman silver reckoning. Cf. Mommsen, *Blacas*, I, 243. Hultsch, *Metrol.* p. 145.

[2] *Étude des monnaies de l'Italie antique*. Première partie, pp. 8 and 16.

[3] *Ibid.* p. 29.

standards. The *centumpondium* or 100 weight therefore he takes as his prime unit. But besides the talent and the mina and the *centumpondium* and *libra* or *as*, according to Mr Soutzo, "all the Italian peoples availed themselves of an intermediate weight unit: this was the *nomos* or *decussis*[1]. This unit was the *libral nomos*, the twelfth of the heavy talent, being worth ten *minae* or *librae*, and the *libral decussis*, the *tenth* of the *centumpondium*, weighing 10 *librae*." The monetary *nomos* and *decussis*, he thinks, played an important part in the history of Italian coinage. He admits however that no specimen of either *nomos* or *decussis* of libral standard is known, the heaviest being a *decussis* of the Roman triental (one-third) standard, whilst the pieces from Venusia and Teanum Apulum marked N I and N II (*nomos* and double *nomos*), representing 10 and 20 minas respectively, belong to a still much more reduced standard. The simple multiples of the *as* (libra) and litra, such as the *tripondius* and *dupondius*, were just as rarely cast in the libral epoch. The *mina* or the *as* with their fractions, on the contrary, were the kinds most employed: originally the series was ordinarily composed of the *as* (marked I or sometimes ....), the *semis* (S), the *triens* (....), the *quadrans* (...), the *sextans* (..), the *uncia* (.) and *semuncia* (Σ). In some series the *as* is rare and the *semis* is wanting, but in addition to the other denominations here given the *quincunx* (:··:) and the *dextans* (S...., 1 *semis* + 4 *unciae*) are found. The presence or absence of these pieces characterizes certain Italian and Sicilian monetary systems[2]. All the evidence virtually which can be produced by Soutzo for this hypothetical *nomos* is that at Syracuse the Corinthian stater of 135 grs. was called a *decalitron*, that the Tarentine didrachm of 128 grs. (max.) was similarly divided into 10 *litras*, that the Romans employed the tenfold of the *as* (*decussis*) and when they coined silver called their silver unit a *denarius* as representing 10 copper *asses*, and the fact that certain copper coins such as those of Arpi, called *nomi*, were evidently regarded as containing 10 units, the half being the *quincunx*. But, as we have already seen, the real explanation of these coins seems to be that they represent

[1] *Ibid.* p. 30. [2] Soutzo, *ibid.* p. 31.

reduced *asses*. We must remember that the heaviest Roman *as* yet known is only 11 ounces, whilst the great proportion of the earliest specimens are only 10 *unciae* or (*dextantals*). When the idea of a real copper currency for local purposes gained ground, and it was found that it was not necessary to have the *as* of account of full weight, and at the same time to enable the state to make a profit of this copper currency which was solely for home use (just as our Mint makes a large profit of our silver coins), the first stage in reduction was to take off an ounce, or much more frequently two full ounces. I have already pointed out the vitality and universality of the *uncia* as an unit, and have given the reasons for this. Hence arose *asses* or *bars* of 10 ounces. The number 10 had of course great advantages, and presently, when further reductions in the copper currency took place, certain communities clave fast to the decimal system and, instead of taking off some more whole ounces, simply reduced the ounce itself, and retained the denomination, continuing to place the marks of value as before. In those Hellenized states of Apulia just referred to this reduced copper *as* or *litra* was the *legal* unit, and therefore denominated a *nomos*, especially as it probably corresponded in value (at least as money of account) to the silver unit or *nomos* in circulation in each district. But whilst Mr Soutzo seems wrong in his view of the *nomos*, there can be no doubt that there was a consensus among the Sicilians and Italians in favour of making an intermediate unit between 1 and 100, the tenfold of the *litra* and *as*, into a higher unit. The Syracusan *decalitron* and the Roman *decussis* and *denarius* are incontrovertible facts. For the latter at least a most interesting connection with a unit of barter can be proved. We saw that by the Lex Tarpeia (451 B.C.) a cow was counted at one hundred *asses* (*centussis, centumpondium*) whilst a sheep was estimated at 10 *asses* (*decussis*). The reader will observe that, even if the theory were true that the Roman *centumpondium* is the starting-point of the Roman weight system, and that it was borrowed from the East, the cow all the same plays a most important part in the founding of the system. It would be another instance to prove the impossibility of framing a weight standard independent of the unit of barter, just as we have already seen that the Irish, when borrowing a ready-made weight system from Rome, found it absolutely necessary to equate the cow to the ounce of silver, and as Charlemagne had to adjust the *solidus* by the value of the same animal. If again the *centumpondium* and *as* grew up independently as *weight* units on Italian soil, and copper was weighed there before gold, the cow is evidently the basis of the system; whilst again, on my hypothesis that *copper* went by bulk in bars of given dimensions, and was not weighed until long after the scales had been employed for gold, the cow is directly connected with that unit of weight (the gold ox-unit of 135 grs.) which ultimately forms the basis of the uncia (as *weight*) and libra. On every hypothesis alike the cow must be retained as the chief factor in the origin of the Roman weight system. It will be observed that Mr Soutzo offers no explanation why the Romans, instead of retaining the sexagesimal division of the talent which they are supposed to have imported, subdivided it according to the decimal scale. It cannot be alleged that they had any deep-rooted antipathy to the duodecimal system, seeing that the *as* was divided into 12 *unciae*, and the ounce into 24 scruples. The fact that the Romans resisted in this respect the Greek influences, which were so potent a factor in their civilization, is strong evidence that the employment of the tenfold and hundredfold of the *as* was of immemorial native origin, and most intimately connected with the animal units, which must certainly be held to be autochthonous. As we found in Further Asia and Africa hoes or bars of metal as the lowest unit of currency, so many hoes being worth a kettle, so many kettles a buffalo, so in ancient Italy 10 bars (*asses*) of copper made a sheep, and 10 sheep made a cow. It is exceedingly probable that the same system prevailed among the Sicels and Sicilian Greeks, 10 litras going to the sheep, 10 sheep to the cow. For we saw on an earlier page that at Syracuse down to the time of Dionysius the cow remained the unit of assessment, just as at the present moment the buffalo is the unit of assessment among the villages of Annam; and, just as with the latter the buffalo is the unit of value, so we may well infer that with the Sicilians the cow played the same rôle. It may therefore be assumed with considerable probability that the employment of the *decalitron* and *decussis* as monetary units was originally due to their connection with the value of the sheep.

As Soutzo has observed, the degradation of the local copper series moved on most unequal lines, and no doubt in some places the *decussis* did not represent perhaps one half the value of its archetype, the sheep, whilst at the same moment the copper unit in another community stood at almost its original weight and value. Where silver was coined the degradation of copper went on all the quicker; there was a tendency more and more to get rid of the old cumbrous copper coins, and to employ those of a lighter and more portable size. Moreover the inter-relations between copper and silver made the coinages in these metals act and react upon each other. Thus the state after reducing the copper would reduce likewise the silver, so as to make the two series correspond. This was probably facilitated in some cases at least by the change in the relative value of these metals. Italy was not a silver-producing region, whilst it was rich in copper. Naturally with the increase of commerce and the development of silver mines in neighbouring countries such as Spain, silver became more abundant and the price of copper rose accordingly. We have had occasion already to remark that the abundance or scarcity of gold or silver is indicated by its being employed or not for coinage. In the case of gold we know that it is only when the supply of that metal is in excess of its demand for purposes of ornament that it is or can be employed in the form of coined money. The history of the coinage of Persia, Lydia, Macedonia, Rhodes and elsewhere in ancient times, as well as the history of mediæval gold coining, make this evident, whilst modern Hindustan teaches us the same lesson. Of course in times of great financial straits under the pressure of war a gold coinage was sometimes issued, as perhaps at Athens[1] in 407 B.C. and as

[1] If we take the καινὸν κόμμα of Aristophanes (*Ranae* 720) to refer, as the scholiast *ad loc.* asserts on the authority of Hellanicus and Philochorus, to a gold issue in B.C. 407, which was much alloyed. As Mr Head says it is quite

at Rome during the second Punic war in 206 B.C. Backwardness in the coinage of silver among certain peoples is probably to be accounted for in the same way. The employment of iron money at Sparta (and Byzantium) was probably due to the dearth of precious metals rather than to any ordinance of Lycurgus against the employment of the latter. If accordingly we find that Rome did not coin silver until 268 B.C. we are justified in concluding that it was from want of silver she had been so long in following the example of the Etruscans and the Greeks.

It is certainly most significant that within four years after the capture of Tarentum (272 B.C.) and the subjugation of all Southern Italy we find her issuing a well-matured silver currency. Doubtless by her conquests she obtained a vast supply of the precious metal, for we know from the records of Livy and Pliny that great masses of foreign coins and bullion flowed into the treasury after every fresh conquest. We may therefore reasonably assume that previous to 272 B.C. silver had been much dearer in relation to copper.

But to return. We have seen that with the imprinting of some device on the primitive bars of copper, the tendency to reduce their weight would quickly evince itself. Accordingly it was possible that in certain places when the coinage of silver began, and there was still a desire to make the silver unit equal to the copper, the latter having been already reduced, the silver would be proportioned thereto. Thus when silver was first coined in some towns in Sicily, the silver Aeginetic obol of $16\frac{1}{2}$ grs. was regarded as the equivalent of the copper litra, but when Syracuse started a coinage of Corinthian staters, a piece of silver of $13\frac{1}{2}$ grs. was accounted as the litra.

But in other parts of Italy the process was somewhat different. For we find the silver unit when once fixed remaining the same in weight, but simply having its denomination altered to meet the requirements of certain changes in the

---

possible that Aristophanes alludes to the new bronze coinage issued the year before the Frogs was acted (*Hist. Num.* 814). No such base gold coins of Athens are known, and as her gold coins are of excellent quality, it is better to refer them with Head to 394 B.C., the period of her restored prosperity, when Conon and Pharnabazus brought aid from the great king.

bronze series. Thus the Etruscan silver staters of the period prior to 350 B.C., which weigh 130 grs., are marked X, whilst the coins of the same weight at a later epoch are marked XX, showing that the copper unit had undergone a change. This Soutzo thinks was simply a reduction from the triental to the sextantal foot, and in no wise due to any change in the relative value of silver and copper. That however both influences may have aided in the change will be made clear from the history of the reduction of the Roman *denarius* and *as* in the second Punic war. Finally when the Romans coined their first *denarii* in 268 B.C., the *libella* or tenth of the *denarius*, which represented in silver the copper *libra*, was only 7 grs., an indubitable proof that the *as* was but then a mere fraction of its former self. Yet all the same it is clear that this silver *denarius*, which represented a reduced *decussis* of bronze, had its ultimate source in nothing else than the 10 libral *asses* which represented the value of a sheep. Are we not then justified in suggesting that the Etruscan stater of 135 grs. marked X had a like origin, that the 10 litra piece or *noummos* of Tarentum of almost the same weight, and the Syracusan 10 litra piece of 135 grs., had also a similar origin, whilst at an earlier period 10 Aeginetic obols (the *nomi* of the poems of Epicharmus and Sophron) were the equivalent of the same animal? Ten *nomi* were the price of a calf in the time of Epicharmus, and as we have seen already the value of a sheep and a young calf is always about the same, even down to the present day.

Roman System.

Although it is not our concern to go into the history of Roman money, it is nevertheless necessary to give the reader a short sketch of its principal features in order to make the history of the Roman weight standards intelligible.

First came oxen and sheep, which according to their age and sex bore definite relations to each other, and by which all other values were measured. From an early period (at least 1000 B.C.) copper was in use, not yet however weighed, but estimated by the bulk, as I have already described. Side by side with it ingots of gold and silver passed from hand to hand. Such ingots are mentioned by Varro under the name of *bricks* (*lateres*)[1]. Though this mention refers to a later period, we can yet infer from it with certainty that the practice of trafficking in small ingots of gold and silver prevailed in Italy as elsewhere. With gold came the art of weighing, which was also applied to silver. We have given reasons for believing that the weight-unit employed was the same as that which I have termed the ox-unit. We found the Etruscans, the close neighbours of the Romans, and who had access to the gold fields of Upper Italy, employing this unit as their standard from the commencement of their coinage in the 5th century for both gold and silver. Any of the towns of Southern Italy which struck gold, such as Metapontum, coined on the same standard, which was likewise employed for silver, sometimes a little reduced, by many communities, such as Tarentum. The standard ingot of gold would bear a known relation to that of silver, to the bar of bronze, the cow, and the sheep. We have given absolute proof of the relation between cattle and bronze in the 5th cent. B.C., and we may well infer similar constant relations between cattle and bronze, and the other metals. With greater exactness in commercial dealings the bronze rod was next weighed by the standard already in use for gold, and it was found that each of the 12 parts or unciae into which it was divided weighed just three times the ox-unit, that is, the weight of the small talent which we have found likewise in Macedon, Sicily, and Lower Italy, and which may have itself represented originally the conventional value of a slave, which was three cows among the Celts, the close kinsfolk of the Italians, and probably about the same among the early Greeks. As soon as the rods or *asses* were exchanged by weighing, they would quickly lose their original form, which was only required so long as it was necessary that they should be of certain fixed dimensions. Under the new system it mattered not whether an *as* was

---

[1] Varro ap. Non. p. 356 nam lateres argentei atque aurei primum conflati atque in aerarium conditi. *Lateres* is used in this sense by Tacitus, *Annals*, XVI. 1.

·8 inches long, and three inches thick, provided only it was of full weight when placed in the scale. These are the pieces which are known as *aes rude*; as yet they are mere lumps of metal, without any stamp or device. Gaius well describes this stage: "For this reason bronze and the balance are employed (in *mancipatio*) because formerly they only employed bronze coins, and there were bars (*asses*), double bars (*dupondii*), half-bars (*semisses*) and quarters (*quadrantes*), nor was there any gold or silver coin in use, as we can learn from a law of the Twelve Tables, and the force and power of these coins depended not on their number but on weight. For as there were bars (*asses*) of a pound weight, there were also two pound bars (*dupondii*), whence even still the term *dupondius* is used, as if two in weight[1]. And the name is still retained in use." The half-bars likewise and quarters were no doubt proportionately adjusted to weight. It will be observed that the omission of all mention of the *decussis* as a standard seems to throw additional doubt on Mr Soutzo's hypothesis. The plain fact is that a mass of bronze ten pounds in weight would have been extremely cumbrous and unhandy for purposes of manufacture into the implements of everyday life.

When and by whom a stamp was first placed on the bars, it is of course impossible to say. Tradition however seems unanimous in assigning it to the Regal period. Pliny's account of the Roman coinage is as follows[2]: "King Servius first stamped bronze. Timaeus hands down the tradition that aforetime they employed it in a rough state at Rome. It was stamped with the impressions of animals (*nota pecudum*), whence it was termed *pecunia*. The highest rating in the reign of that king (Servius) was 120,000 asses, and accordingly this was the first class. Silver was struck A.U.C. 485 (B.C. 268) in the Consulship of Q. Ogulnius and C. Fabius, five years before the first Punic war, and it was enacted that the *denarius*

---

[1] Gaius I. 122. This passage is unhappily corrupt. The Verona MS. runs asses librales erant et dupondii——unde etiam dupondius. As *dupondius* is really a masculine adjective used as a noun, a masculine noun must be understood, this can only be *as*. Dupondius then is simply a two-pound bar.

[2] xxxiii. 3. 13.

should pass for ten pounds of bronze, the *quinarius* for five, and the *sestertius* for two and a half. Now the libral weight was reduced in the First Punic war, as the state could not stand the expenditure, and it was appointed that *asses* of the weight of a *sextans* (2 *unciae*) should be struck. Thus there was a gain of five-sixths, and the debt was cleared off. The type of that bronze coin was on the one side a double Janus, on the other a ship's beak, whilst on the *triens* and *quadrans* there was a ship. The *quadrans* was previously termed a *teruncius* from *tres unciae* (three ounces). Afterwards under the pressure of the Hannibalic wars in the dictatorship of Q. Fabius Maximus, *asses* the weight of an ounce were coined, and it was enacted that the *denarius* should be exchanged for sixteen *asses*, the *quinarius* for eight, the *sestertius* for four; thus the state gained one half. Nevertheless in the soldiers' pay the *denarius* was always given for ten *asses*. The types of the silver were *bigae* and *quadrigae* (two-horse and four-horse chariots), hence they were termed *bigati* and *quadrigati*[1]. By

FIG. 56. Romano-Campanian Coin.

and by in accordance with the Papirian law half-ounce *asses* were struck. Livius Drusus when tribune of the Plebs alloyed the silver with an eighth part of bronze. The *Victoriatus*

FIG. 57. Victoriatus.

[1] Before striking silver at Rome the Romans had struck silver coins with type of quadriga and ROMA in Campania. Hence it is that Pliny regarded these the *quadrigati* and *bigati* as the oldest issue instead of the coins with the Dioscuri (Fig. 54). The *biga* came next, after it the genuine Roman *quadriga*.

was struck in accordance with a law of Clodius, for previously this coin brought from Illyria was treated as merchandize. It was stamped with a Victory and hence its name. The gold piece was struck sixty-two years after the silver on such a standard that a scruple was worth twenty sesterces, and this on the scale of the then value of the sesterce made 900 go to the pound. Afterwards it was enacted that 1040 should be coined from gold pounds, and gradually the emperors reduced the weight, most recently Nero reduced it to 45."

This statement of Pliny is supported in various details by several disjointed passages of Varro and Festus. Thus the former says that "the most ancient bronze which was cast was marked with an animal (*pecore notatum*)[1], and elsewhere he says that the ancient money has as its device either an ox, or a sheep, or a swine[2]," a statement repeated by Plutarch and other later writers. Festus (*s. v. grave aes*) says "*aes grave* was so called from its weight because ten *asses*, each a pound in weight, made a *denarius*, which was so named from the very number (i.e. *deni*). But in the Punic war, the Roman people being burdened with debt, made out of every *as* which weighed a pound (*ex singulis assibus librariis*) six *asses*, which were to have the same value as the former." We have also a statement in the fragment of Festus (4, p. 347, Müller) that afterwards the *asses* in the *sestertius* were increased (*i.e.* to 4 from 2½), and that with the ancients the *denarii* were of ten *asses*, and were worth a *decussis*, and that the amount of bronze (in the *denarius*) was reckoned at XVI *asses* by the Lex Flaminia when the Roman people were put to straits by Hannibal[3]. Again, Festus says: "*Asses* of the weight of a *sextans* (two ounces) began to be in use from that time, when on account of the Second Punic war which was waged with Hannibal, the Senate decreed that out of the *asses* which were then libral (a pound in weight)

---

[1] Varro, *R. R.* II. 1. 9.

[2] Varro ap. Non. p. 189 *aut bovem aut ovem aut vervecem habet signum*. Probably *uerrem*, not *ueruecem*, is the true reading, since Plutarch says that the coins were marked with an ox, a sheep or a *swine* (βοῦν ἐπεχάραττον ἢ πρόβατον ἢ ὗν). *Popl.* 11.

[3] Festus fragm. p. 347 Müller *s.v. Sextantari asses*.

should be made those of a *sextans* in weight, by means of which when payments began to be made, both the Roman people would be freed from debt, and private persons, to whom a debt had to be paid by the state, would not suffer much loss[1]."

FIG. 58. Sextans (Aes Grave).
(The two globules mark the value.)

Varro likewise is worth hearing: "In the case of silver the term *nummi* is used: that is borrowed from the Sicilians. *Denarii* (were so named) because they were worth ten (coins) of bronze each, *quinarii* because they were worth five each, *sestertius*, because a half was added to two (for the ancient *sestertius* was a *dupondius* and a *semis*). The tenth part of a *denarius nummus* is a *libella*, because it was worth a *libra* of bronze in weight, and being made of silver was small. The *sembella* is half the *libella*, just as the *semis* is of the *as*. *Teruncius* is from *tres unciae*; as this is the fourth part of the *libella* so the *quadrans* is the fourth part of the *as*."

As so much difficulty and controversy surround the various questions connected with the beginnings of Roman currency, I have thought it best to give at full length the scanty data afforded by the ancient authorities. Let us now state the principal facts revealed by those extracts. (1) The Romans in the Regal epoch employed *aes rude*, but according to the testimony of Timaeus (an Italian Greek historian who wrote about B.C. 300), they had already before the days of the Republic stamped bronze with figures of cattle. (2) Silver was first coined five years before the beginning of the First Punic war: (3) Some time during that war the *as* was reduced from a pound to two ounces; (4) In the Second Punic war under like circumstances the *as* was reduced from two ounces to one ounce; (5) The *denarius* when first struck represented ten

---

[1] v. 173 Müller.

libral *asses*, or a *decussis;* (6) In the Second Punic war when the *as* was reduced, the *denarius* was ordered to pass for 16 instead of 10 *asses;* (7) In spite of this reduction, the *denarius* continued to be regarded as containing only 10 *asses* when employed in paying the soldiers.

Considerable numbers of *asses* and the parts of *asses* have come down to us, many of them bearing marks of value as before described. There is undoubted evidence of a constant reduction of the *as*. The question arises, did the reduction take place *per saltum* or by a gradual process? Mommsen thinks that the *as* continued to be of libral weight until shortly before 264 B.C. and that it was then without any intermediate steps reduced to the triens (4 ounces). Mr Soutzo on the other hand maintains with vigour that from 338 B.C., the date at which he fixes the first coinage of *asses* at Rome, to 264 B.C., the degradation was a gradual process, and he arraigns Mommsen on a charge of disregarding the ancient authorities, who state, as we have seen, that the change was from libral to sextantal *asses*. Mr Soutzo is thus compelled to state that all the *asses* within that period (338—264 B.C.) although they have a range from almost full libral weight to only 3 ounces were treated as libral *asses*. Now this of course is a very reasonable hypothesis on the principle which I have adopted that bronze money was in fact merely token currency, used only for local circulation and not for extraneous trade. But Mr Soutzo is precluded from adopting such a position unless he gives up the basis of his whole work. He has laid down that the bronze money was not a mere conventional currency, but always was actual value for the amount which it represented. On this assumption he obtains his relation of 1 : 120 between copper and silver. Assuming that the sextantal reduction was contemporaneous with the issue of the first *denarius* (which is in direct defiance of the historians), he found that the *denarius* of 70 grs. = 2 ounces (840 grs.) of bronze; therefore silver was to bronze as 120 : 1. Again, when the financial crisis took place during the Second Punic war and the *denarius* was reduced (as we learn from the actual coin weights) to 62 grs., and it was made to pass for 16 *asses* instead of 10 *asses*, he finds that since 62 grs. of silver = 16 *asses* of 432 grs. (*unciae*) silver was to bronze as

112 : 1. But in the latter case he omits to explain why it was that the *denarius* in paying the troops only counted for *ten asses*. It is evident that if the relation between copper and silver was really as 1 : 112, there could have been no need for making this difference. But as the soldiers were serving outside Rome, and Roman local token currency would not be taken in payment, it was necessary to pay them according to the market value of bronze. At Rome the *denarius* was made to pass for 16 *asses*, or three-fifths more than its actual value. It appears therefore that the data given us by Pliny are not sufficient to allow us to come to any definite conclusion as regards the relative value of silver and bronze at that time. Moreover there is no evidence to show that the *denarius* was reduced from 70 grs. to 62 grs. by the Lex Flaminia. It is on the whole more likely that this reduction took place when the first gold coinage was issued (62 years after the first silver) in 206 B.C., since there was every inducement to make such a change in the silver as would admit of a convenient relation between the gold *scruple* and 20 *sestertii*. This again raises just doubts as regards the accuracy of Mr Soutzo's calculation. With reference to the reduction of the *as* to the sextantal standard we have seen that the truth of his deductions rests entirely on the assumption that the degradation took place *before* the First Punic war at the same time as the issue of the first silver coinage. This of course is directly contradicted by the historians. But even granting that it was correct, it is difficult to see why we should assume that the Roman *as*, which according to Soutzo's own principles had been nothing more than a token, should suddenly have been treated as though it really was of the actual value which it represented. There was no reason why, even though the unit of account was the sextantal *as*, the *as* should have been anything else than a token in its relation to the silver currency: certainly it is strange that, if the Romans after treating the *as* as a token down to 268 B.C. then suddenly gave it its full monetary value, they did not continue to carry out their new principle. For as a matter of fact there are very great differences in the weight of the sextantal *asses*, and after the reduction to the uncial standard, the same process of degradation went on without ceasing, as

Soutzo himself has shown[1]. All these facts point to the conclusion that the bronze coinage at Rome was only a local token currency, such as is our own silver and bronze series at the present day.

Let us now see if we can give a consistent explanation of the statements of the ancient writers which I have quoted above. *Aes rude* or bronze in an unstamped or unmanufactured state was originally in use at Rome, according to Timaeus. This period corresponds to that time when, as I have endeavoured to show, *asses* or *bars* of given dimensions intended to be made into articles for use or ornament passed from hand to hand, as do the brass rods mentioned above at the present moment in the Congo region of Africa. Then came the stamping of the *asses* towards the close of the regal period (according to Timaeus), when figures of animals were placed thereon. We have seen above (p. 354) that such figures are actually found on certain rough quadrilateral pieces of bronze found in some parts of central Italy. With the use of weight instead of measure for appraising their value, the shape of the *asses* would become modified, getting shorter and thicker. Finally, they assume the round shape of ordinary coins, and bear certain well-defined symbols on both sides, such as the Janus head and Rostrum on the *as*, that of Mercury on the *sextans*. But as few of these round *asses* are found to weigh more than 10 *unciae*, it would seem that the process of degradation had already set in before their issue. Gold and silver at the same epoch passed by weight either after the ancient fashion in ingots, or as the coined money of the Greek cities of the South or of the Etruscans. The unit of account continues to be the *as* of *full weight*. Thus all penalties due to the state would be paid not in reduced *asses* of only 5 or 4 ounces, but in full libral *asses* as weighed in the balance. On the other hand although reduced *asses* were used by the state in paying debts to private individuals, they were only received as tokens, and no doubt the state was bound if called upon to pay a full pound of bronze for

---

[1] Deux. Partie p. 41. "Le poids normal de l'as oncial est de 27 gr. 25, mais il alla en s'affaiblissant progressivement du commencement à la fin de la periode."

every stamped reduced *as* presented to it, but in ordinary times this made no practical difference, for the bronze currency was purely local all over Italy and Sicily, as we have seen above. It was far too cumbrous to be used as a medium of international trade.

When the Romans after defeating Pyrrhus and taking Tarentum had reduced all Southern Italy and hence obtained great quantities of silver, they proceeded five years before the beginning of the First Punic war to issue silver *denarii* or ten *as* pieces. Are these pieces real representatives of the *as* of account, or do they rather simply represent the value of the then normal *as* of currency, which was probably not more than a *triens* or four ounces or perhaps not more than a *quadrans* or three ounces? The latter is the more likely hypothesis. They had been long accustomed to a bronze token currency, and it was most likely that the new silver currency would be adapted to it. It is then likely that the *denarius* equalled ten *asses* of at least 3 ounces each, in which case silver was to bronze as 180 : 1. In transactions inside the state the balance would be commonly, and in dealing with strangers invariably, employed in all monetary transactions, ancient states being very jealous of alien mintages. This is exemplified by Pliny's statement that the Victoriates brought from Illyria were treated simply as merchandize. Then came the First Punic war, which lasted for two-and-twenty weary years, during which the resources of the Republic were almost drained dry. The state became virtually a bankrupt and simply paid in modern phraseology 3*s*. 4*d*. in the pound. It was effected thus: up to the present the *as* of full weight was the unit of account, although the coined *asses* had by this time come to be simply tokens of about 2 ounces each. The state accordingly enacted that the *as* of currency should become the unit of account, and paid the state debt by these coins, and at the same time made it legal for private individuals, who were bound under the old order of things to pay their debts in libral *asses*, to discharge their obligations by sextantal *asses*. Thus Pliny is perfectly right in saying that the state made a profit of five-sixths. The influx of silver after the conquest of Southern Italy and the requirements of large quantities of bronze for the building of fleet after fleet, and for military equipment, may have very well tended to appreciate the value of bronze at this period. As the reduction in the size of the *as* continued, though the unit of account was two ounces, under the pressure of the Second Punic war they repeated the same process. The *as* was now not more than an ounce, so they decreed that the *as* of currency should again be the *as* of account, and the state thus gained a half, this time paying ten shillings in the pound.

The *ounce* and *libra* had been long well defined at Rome before the silver coinage first appeared, and whilst we saw that the *sextula* or one-sixth of the *uncia* was the lowest weight employed for bronze, the fourth part of this weight, the *scriptulum*, had been regularly employed in weighing silver and gold; as we have seen it owed its origin to the fact that the Aeginetan silver obol was found to be about the weight of the 24th part of an *uncia* or inch of bronze. The first *denarii* were the weight of a *sextula* or 4 *scriptula* (70 grs.) of the older weight. The *scriptulum* and *sestertius* were thus identical, and hence in later days the unit of account was the *sestertius* and not the *as*. Accordingly when the gold coinage of 206 B.C. was issued, it was based on the *scruple*, and consisted of pieces of 1, 2, and 3 scruples.

We have now traced the origin of Roman currency sufficiently for the purposes of this work. After various fluctuations in the weight of the gold pieces under Sulla, Pompey, Julius Caesar and others, Constantine the Great finally fixed the

Fig. 59. Gold Solidus of Julian II. (the Apostate).

weight of the *aureus* or *solidus* at 4 scruples in 312 A.D., and so it remained until the final downfall of the Empire of the East in 1453. From this famous coin the various mintages of

THE ROMAN SYSTEM. 385

mediaeval and consequently of modern Europe may be said to trace their pedigrees. The *solidus* was divided into *thirds* or *tremisses*, for the scrupular system had been abandoned, the *solidus* being regarded simply as a *sextula* or one-sixth of the *uncia*, and not as a multiple of the *scruple*. The *tremissis*

FIG. 60. Gold Tremissis of Leo I.

therefore weighed 24 grs. Troy, or 32 wheat grains. When the barbarian conquerors of the Roman Empire began to coin silver they took as their model the gold *tremissis*. In the earliest stage of the Anglo-Saxon mintage we find so-called gold pennies of 24 grs. occasionally appearing. These are nothing else than *tremisses*. But silver henceforward was to form for centuries the staple currency of Western Europe, and the silver penny of 24 grs. (whence comes our own penny-weight) became virtually the unit of account. As its weight shows, the penny was based on the gold *tremissis*.

The first regular coinage of gold in Western Europe began with the famous gold pieces of Florence in the beginning of the 14th century. These weighed 48 grs. or 2 *tremisses*. From their place of mintage the name *florin* (fiorino) became a generic term for gold coins. Accordingly when Edward III. issued his first gold coins of 108 grs. each, although differing so completely in weight from their prototype, they too were called *florins*. In reality however Edward's coin was $1\frac{1}{2}$ solidus (72 + 36). The first attempt did not prove satisfactory, and with the issue of the famous noble, first of $136\frac{1}{2}$ grs., and afterwards of 129 grs., the series of English gold coins may be said to begin, of which the latest stage is the sovereign of $120\frac{1}{4}$ grs. Troy.

I have already explained at an earlier stage the origin of the Troy grain; before we end let me add a word on the origin of the Troy ounce. The Troy pound like the Roman has 12 ounces, but whereas the Roman ounce had 432 grs. Troy or

576 grs. wheat, the Troy ounce has 480 grs. Troy or 640 grs. wheat. How came this augmentation of the ounce?

It is in Apothecaries' weight that we find the key. This standard runs thus

$$\begin{aligned} 20 \text{ grs.} &= 1 \text{ scruple,} \\ 3 \text{ scruples} &= 1 \text{ drachm,} \\ 8 \text{ drachms} &= 1 \text{ ounce,} \\ 12 \text{ ounces} &= 1 \text{ pound.} \end{aligned}$$

Now note that there are 24 scruples in the ounce, and 288 scruples in the pound, exactly as in the Roman system. But there is an element foreign to the old Roman system as seen in the drachm of 60 grs. Now Galen and the medical writers of the Empire used the post-Neronian *denarius* of 60 grs. as a medicine weight. What more convenient weight unit could be employed than the most common coin in circulation? The *drachma* and *denarius* had long since been used synonymously in common parlance. But as there were 18 grs. (Troy, 24 wheat grs.) in the old scruple, and there were 60 grs. in the drachm or *denarius*, they were not commensurable, and accordingly to obviate this difficulty the physicians for practical purposes raised the scruple to 20 grs., in order that it might be one-third of the drachm. The number of scruples in the ounce remaining 24 as before, the ounce became augmented by 48 grs. (24 × 2) and accordingly rose to 480 grs. We saw above that the Troy grain is the barley-corn. Why is the latter so closely connected with 'Troy weight'? When the scruple was raised from 18 grs. Troy, 24 grs. of wheat, to 20 grs. Troy, it no longer contained an even number of wheat grains, for the new *scruple* contained $26\frac{2}{3}$ grs. wheat. As this was inconvenient, and on the other hand the new scruple weighed exactly 20 barley-corns, the latter henceforth became the lowest unit of this system.

Conclusion.

It now simply remains to sum up the results of our enquiry. Starting with the Homeric Poems we found that although certain pieces of gold called *talents* were in circulation among the early Greeks, yet all values were still expressed in terms of cows. We then found that the gold *talent* was nothing else than the equivalent of the cow, the older unit of barter, and we found that the *talent* was the same unit as that known in historical times under the names of Euboic stater or Attic stater, and commonly described by metrologists as the light Babylonian shekel. Our next stage was to enquire into the systems of currency used by primitive peoples in both ancient and modern times, and everywhere alike we found systems closely analogous to that depicted in the Homeric Poems, and we found that in the regions of Asia, Europe and Africa, where the system of weight standards which has given birth to all the systems of modern Europe had its origin, the cow was universally the chief unit of barter. Furthermore gold was distributed with great impartiality over the same area, and known and employed for purposes of decoration from an early period by the various races which inhabited it. We then found that practically all over that area there was but one unit for gold, and that unit was the same weight as the Homeric Talanton. Next we proved that gold was the first object for which mankind employed the art of weighing, and we then found that over the area in question there was strong evidence to show that everywhere from India to the shores of the Atlantic the cow originally had the same value as the universally distributed gold unit.

From this we drew the conclusion that the gold unit, which was certainly later in date than the employment of the cow as a unit of value, was based on the latter; and finally we showed that man everywhere made his earliest essays in weighing by means of the seeds of plants, which nature had placed ready to his hand as counters and as weights. Then we surveyed the theories which derive all weight standards from the scientific investigations of the Chaldeans or Egyptians, and having found that they were directly in contradiction to the facts of both ancient history and modern researches into the systems of primitive peoples, we concluded that the theories of Boeckh and his school must be abandoned.

Next we proceeded to explain the development of the various systems of antiquity from our ox-unit, taking in turn the Egyptian, Assyrio-Babylonian, Hebrew, Lydian, Greek and Italian. New explanations of the origin of the Talent and Mina and also of the earlier types on Greek coins and of the varieties of standard employed for silver by the Greeks were offered, and finally in dealing with the systems of Sicily and Italy arguments were advanced to show that the Roman *as* was originally nothing more than a rod or bar of copper of definite measurements, and was in weight and method of division the same as the Sicilian Litra and the Greek Obol.

In how far the propositions here put forward have been proved, it must remain for others to decide.

Laus Deo, Pax Vibis,
Requies Mortuis.

APPENDIX A.

The Homeric Trial Scene.

Κεῖτο δ' ἄρ' ἐν μέσσοισι δύω χρυσοῖο τάλαντα,
Τῷ δόμεν, ὃς μετὰ τοῖσι δίκην ἰθύντατα εἴποι.

Il. xviii. 507—8.

I WOULD not return to so well-worn a theme, were it not that editors like Dr Leaf (*ad loc.*) still state that there is nothing in the *language* of the last line to hinder us from taking it either of the litigant or of the judge.

Scholars have fixed their attention so closely on the words δίκην εἴποι that they have completely overlooked the qualifying ἰθύντατα. In modern courts of law we do not expect to hear the *straightest* statement of a case from advocates, but rather from the judge. The ancient Greek would never dream of expecting a litigant to give a *straight* statement of his case. The following passages will show that ἰθύς, ἰθύνειν, εὐθύνειν, ὀρθός are always applied to a judge (the converse σκολιώς being used of unjust judges). The metaphor is from the carpenter's rule (cf. ἐπὶ στάθμην ἰθύνειν *Od.* v. 245).

Pind. *Pyth.* IV. 152 καὶ θρόνος, ᾧ ποτε ἐγκαθίζων Κρηθεΐδας ἱππόταις εὔθυνε λαοῖς δίκας.

Solon 3. 36 εὐθύνων σκολιὰς δίκας.

Il. XVI. 387 οἳ βίῃ εἰν ἀγορῇ σκολιὰς κρίνωσι θέμιστας.

Hesiod *Opp.* 221 σκολιῇς δὲ δίκῃς κρίνωσι θέμιστας.

Hes. *Opp.* 222 (Δίκη) κακὸν ἀνθρώποισι φέρουσα
οἵ τέ μιν ἐξελάσωσι καὶ οὐκ ἰθεῖαν ἔνειμαν.

Arist. *Rhet.* I. 1 οὐ γὰρ δεῖ τὸν δικαστὴν διαστρέφειν εἰς ὀργὴν προάγοντας ἢ φθόνον ἢ ἔλεον· ὅμοιον γὰρ κἂν εἴ τις, ᾧ μέλλει χρῆσθαι κανόνι, τοῦτον ποιήσειε στρεβλόν.

Pind. *Pyth.* XI. 15 ὀρθοδίκαν γᾶς ὀμφαλόν.

Aesch. *Persae* 764 εὐθυντήριον σκῆπτρον.

No one can then doubt that the words δίκην ἰθύντατα εἴποι can only refer to the judge.

The following account of a trial on the Gold Coast so well illustrates the principle of payment having to be made to the judges that I think it worth quoting. (*Eighteen years on the Gold Coast of Africa*, by Brodie Crookshank, Vol. I. p. 279, London, 1853.)

"When the day arrived for the hearing of Quansah's charge, a large space was cleanly swept in the market-place for the accommodation of the assembly; for this a charge of ten shillings was made and paid. When the Pynins (elders) had taken their seats, surrounded by their followers, who squatted upon the ground, a consultation took place as to the amount which they ought to charge for the occupation of their valuable time, and after duly considering the plaintiff's means, with the view of extracting from him as much as they could, they valued their intended services at £6. 15s., which he was in like manner called upon to pay. Another charge of £2. 5s. was made in the name of tribute to the chief, and as an acknowledgment of gratitude for his presence upon the occasion. £1. 10s. was then ordered to be paid to purchase rum for the judges, £1 for the gratification of the followers, ten shillings to the men who took the trouble to weigh out the different sums, and five shillings for the court criers. Thus Quansah had to pay £12. 15s. to bring his case before this august court, the members of which during the trial carried on a pleasant course of rum and palm wine."

APPENDIX B.

What was the Unit of Assessment in the Constitution of Servius Tullius?

Th. Mommsen in his Roman History (I. 95—96 English Trans.) has laid down that land was the basis of assessment, on the analogy of the Teutonic *hide*. He makes the members of the First Class those who held a whole hide; and the remaining four classes were made up of those who held proportionally smaller freeholds. When Mommsen has once spoken, it is presumptuous to raise doubts. If however it can be shown that the Italians rather based their assessments on cattle, and that furthermore the statements of the later historians point to an original rating which harmonizes well with such an original condition, it may have been worth while to start enquiry once again in a case where the data are so scanty and obscure.

Pliny *H. N.* xxxiii. 3. 13. Maximus census cxx. assium fuit illo rege, ideo haec prima classis. This is confirmed by Festus (*s.v. infra censum*, p. 113 Müller) infra classem significantur qui minore summa quam centum et viginti millia aeris censi sunt.

Livy I. 42 says the rating of the *prima classis* was Centum millia aeris, of the *secunda classis* was infra centum assium ad quinque et septuaginta millia. *Tertia classis* quinquaginta millia, *Quarta classis*, quinque et viginti millia. *Quinta classis*, undecim millia.

Dionysius of Halicarnassus (IV. 16—17) puts the rating of the 1st class at 100 minae (of silver) or 10,000 drachms; of the 2nd at 75 minae, of the 3rd at 50 minae, of the 4th at 25 minae, and that of the 5th at 12 minae.

All are agreed that it is absolutely incredible that the original rating of the first class was 120,000 *libral* asses of bronze. The cow was worth 100 *libral* asses at Rome in 451 B.C. Therefore the rating of 120,000 asses would have been equivalent to 1200 cows. It is impossible to believe that there could have been a numerous body of men in early Rome possessed of such vast capital. Boeckh's explanation is that with the reduction of the *as* from its original weight of a *libra* to two ounces, and one ounce, there was a corresponding raising of the amount of the rating of the several classes.

Mommsen on the other hand thinks that the rating was originally on *land*, and that the change in the method of rating from land to bronze took place at a time when land had greatly risen in value, and that accordingly 120,000 *asses* of the First Class are libral *asses*. Such a change as Mommsen supposes must have taken place before 260—241 B.C., for the *as* was reduced to two ounces during the first Punic War. Yet we cannot easily suggest any period before that date when there was likely to have been so great a rise in the value of land, as is necessary to account for the large rating of 120,000 *asses*, which according to Mommsen's reckoning would be worth about 400 lbs. of silver (or according to Soutzo 1000 lbs. of silver).

Boeckh's hypothesis seems to fit better the conditions of the problem. Much of the importance of the rating of the various classes passed away when Marius (104 B.C.) changed the whole military system and chose the troops from the *Capite censi*, as well as from the five property classes.

The *as* had been reduced to a single *uncia* in the 2nd Punic War (cf. p. 377). Thus 12 *asses* of the *uncial* standard were required to make up the weight of the old *libral as*. Accordingly 120,000 *asses* of the 2nd century B.C. would be equal to 10,000 *libral asses* of the earlier days. But as by the Lex Tarpeia 100 *asses* is the value of a cow, 10,000 *libral asses* = 100 cows. This would be by no means an unlikely number of cows, to form the minimum of the wealthiest class of a pastoral community. There is another curious piece of evidence which seems to confirm my hypothesis. One of the provisions of the Licinian Rogations (367 B.C.) was that no one should hold more than 500 *jugera* of the Public Land, or should be allowed to feed more than *one hundred* large cattle or 500 small cattle on public pastures. μηδένα ἔχειν τῆσδε τῆς γῆς πλέθρα πεντακοσίων πλείονα, μηδὲ προβατεύειν ἑκατὸν πλείω τὰ μείζονα καὶ πεντακοσίων τὰ ἐλάσσονα. Appian, *Bell. Civ.* I. 8.

If 100 large cattle were the number which qualified a Roman for the first class, there was every reason why Licinius and Sextus should have taken 100 as the *maximum* number of cows which a citizen could keep on the public pastures.

Next I shall show that the method of rating by cattle and not by land was that actually practised in Sicily. That island stood in such close relations to the Italian Peninsula both geographically and ethnologically that we may reasonably infer that the method of rating in use there was also in use in Italy.

Now we learn from Aristotle's *Oeconomica* (II. 21) that when the tyrant Dionysius oppressed the Syracusans with excessive exactions, they ceased to keep cattle:

Τῶν δὲ πολιτῶν διὰ τὰς εἰσφορὰς οὐ τρεφόντων βοσκήματα, εἶπεν ὅτι ἱκανὰ ἦν αὐτῷ πρὸς τοσοῦτον· τοὺς οὖν νῦν κτησαμένους ἀτελεῖς ἔσεσθαι, πολλῶν δὲ ταχὺ κτησαμένων πολλὰ βοσκήματα, ὡς ἀτελῆ ἑξόντων, ἐπεὶ ᾤετο καιρὸν εἶναι, τιμήσασθαι κελεύσας ἐπέβαλε τέλος, κ.τ.λ.

If the citizens of Syracuse, a great Greek trading city, were still rated in cattle in the time of Dionysius (405—367 B.C.), *à fortiori* we may expect the same primitive method of assessment to prevail among the pastoral peoples of Central Italy in the 6th and 5th centuries B.C.

Among the Kelts, the close kinsfolk of the Italians, the same system probably prevailed. Thus in the ancient Irish laws, where the various classes of freemen are described, there are a number of them called *Bo-aires*[1], cow-freemen.

As modern research has shown that everywhere among the Aryans land was originally held in common, and that separate property in land sprung up only at a comparatively late period, we may with some confidence infer that in Italy likewise in early days a man's wealth was reckoned in his cattle, and not in lands, such as I have shown to have been the practice among the Greeks of the 'Homeric times' ('The Homeric Land System,' *Journal of Hellenic Studies*, 1885).

[1] *Ancient Laws of Ireland*, Vol. I. p. 61. O'Curry, *Manners and Customs of the Ancient Irish*, Vol. I. pp. 100 seq.

APPENDIX C.

Keltic and Scandinavian Weight Systems.

It is always dangerous to deal with things Keltic. So much difficulty is there in getting at any facts amidst masses of wild assertions and loose conclusions, that a prudent man may well shrink back. However, as it is worth while to give some *facts* respecting the actual weights of gold rings and other ornaments, I have thought it best to print the following pages.

Attempts have long ago been made to find the standard of the so-called ring money. Sir William Betham, followed by John Lindsay[1], after weighing many examples, arrived at the conclusion that they are based on the ounce Troy. Now as the ounce Troy is entirely unknown to the Brehon Laws, and was only brought into Ireland by the English settlers, it is needless to argue further against that doctrine. Dr Petrie's[2] discussions about Irish coins are similarly vitiated by his treating as Troy grains the grains of wheat mentioned by the authorities.

1. *Irish.* Let us work back from the known to the unknown.

The system in the Brehon Laws is as follows:

    1 Cumhal (ancilla) = 3 Cows.
    1 Cow   = 1 Unga (uncia of silver).
    1 Unga   = 24 Screapalls.
    1 Screapall   = 3 Pinginns.
    1 Pinginn   = 8 grs. of wheat.

Unga = 576 grs. of wheat.

[1] *Survey of the Coinage of Ireland*, p. 3.
[2] *Ecclesiastical Architecture of Ireland*, p. 213 seqq.

The ounce seems to be the highest unit of weight, and just as in the Brehon Laws an *unga* of silver is equated to a cow, so in early times an *unga* of gold seems to have been the regular value of a slave, the most valuable of living chattels. At least we may so infer from a curious story of St Finnian of Clonard:

Life of St Finnian (of Clonard, Co. Meath).

(Book of Lismore, fol. 24 b, c.)

Tainic iar sin Finnen cu Cilldara co Brighit, cu m-bui ic tiachtuin leiginn ocus proicepta fri re. Ceilebrais iar sin do Brigit ocus dobreth Brighit fainne oir dho. Nir 'bho santach som imon saegul: ni roghabh in fainne. " Ce no optha," ar Brigit, "roricfea a leas." Tainic Finnen iar sin on Fotharta Airbrech. Dorala uisce do. Roinnail a lamha asin usci[1]: tuc lais for a bhais asan uisci in fainne targaidh Brighit dó.

Táinic iar sin Caisin, mac Naemain, co faelti moir fri Finden. Ocus coneadhbair fein dó ocus roacain fris ró Fotharta ic cuinghidh oir fair ar a shaeire. "Cia mét," ar Finnen, "conaidheas?" "Noghebhudh uingi n-oir," ar Caisin. Rothomthuis sé iar sin in fainne [ocus frith uingi oir[2]] ann. Dorat Caisin hi ar a shaeriri.

[1] Folio 24 c.
[2] The bracketed words are interlined in a recent hand; but the final word shows that they were a portion of the text.

Translation.

"After that came Finnian to Kildare to Brigit and he was engaged in teaching and preaching for a time. He takes leave afterwards of Brigit and Brigit gave a ring of gold to him. He was not covetous regarding the world: he accepted not the ring. "Though thou refusest," said Brigit, " thou wilt require it." Finnian came after that to Fotharta Airbrech[3]. [On his way] he met water. He washed his hands with the water [and] brought on his palm from out the water the ring that Brigit offered to him.

After that came Caisin, son of Naeman, with great joy to [visit] Finnian. And he offered himself to him and complained to him that the king of Fotharta was demanding gold from him for his liberation. "How much," said Finnian, "asketh he?" "He would accept an ounce of gold," said Caisin. He [Finnian] weighed after that the ring (and there was found an ounce of gold[4]) in it. Caisin gave it for his liberation."

[3] Near Croghan Hill, in the north of King's Co.
[4] See note on Irish text.

I am indebted for this valuable reference, which also enables us to form an idea of the relative value of gold and silver in early Ireland, to the Rev. B. Mac Carthy, D.D., of Youghal.

But there is another weight called crosoch (crosóg or crosach), found in the most ancient poems. For instance in Cuchulaind the brooch of Queen Medbh, "My spear brooch of gold which weighs thirty ungas, and thirty half ungas, and thirty crossachs and thirty quarter [crossachs]." (O'Curry, *Manners and Customs*, Vol. III. p. 102.) The weight of a crosoch we learn from a gloss quoted by O'Donovan (Supplement to O'Reilly's Dictionary) from *MS. R. I. A.*, No. 35, 5. 49.

da pinginn agas cetrime pinginne isin lacht caerach i, crosóg[1].

"Two pinginns and a fourth of a pinginn are a milk of a sheep, i.e. a crosog." Since 1 pinginn = 8 grs. wheat therefore a crosóg = 18 grs. wheat or 13·5 grs. Troy.

There are accordingly 32 crosochs in the unga of the Brehon Laws.

Inspection at once shows that the crosoch must have belonged to a different system, on which either the system of ungas and screapalls was grafted or *vice versa*. The expulsion of the crosoch from the later Irish shows that the first alternative is the true one.

Again, it is certain that the unga and screapall were borrowed from the Roman system, probably before the time of Constantine, as after his time the solidus became universal throughout the Empire, and has left its impress everywhere.

The crosoch therefore must be non-Roman, *i.e.* belong to the native population.

Above we saw that it was used along with ungas and half ungas in describing Medbh's Fibula. Here is historical evidence of its use in the weighing of gold ornaments.

There were certainly 32 crosochs in the ounce of the Brehon Laws, but if we can show in another system of north-western Europe a weight exactly the same as the crosoch, with an ounce which is its thirty-fold, we may hesitate to lay down that the full Roman ounce with its 432 grs. Troy (576 grs. wheat) was the earliest form of Irish *unga*.

There is no mention of screapalls in the weight of Medbh's brooch. It is quite possible that under ecclesiastical influences the full Roman ounce and its division into screapalls may have been introduced at a comparatively late period. The contact between Kelts and Scandinavians in early times has of late excited much interest.

---

[1] O'Donovan has omitted *caerach* of the MS.

2. Let us now turn to the old Norse system. It is as follows:
   1 pening = 13·5 grs. Troy
   10 penings = 1 örtug = 136·7 grs.
   3 örtugs = 1 öre = 410 grs.
   8 öres = 1 mark = 3280 grs.

Let us deal first with the mark. As its name signifies, it in all probability was originally not a *weight*, but a *measure*. The use of *mark* as a land measure is well known in the Teutonic languages. It is also used as a measure of length. Thus a mark of cloth consists of 448 *alen* or *ells*. After what we have learned about the history of the Roman *as* (p. 354) we need not be surprised if a term originally used as a measure of some article which was not as yet sold by weight, came in similar fashion to be incorporated at a later period into the weight system as a higher unit. If the mark was originally a given measure of bronze or iron, we can readily see how it came later on to be used as a weight, and ultimately to be the chief unit of account among our Anglo-Saxon forefathers, until it was at last driven out by the *pound*.

That silver was cast into bars which weighed a mark is rendered highly probable by the fact that three of the silver bars found at Cuerdale weigh respectively 3960, 3954, and 3950 grs. Troy; that is, just the weight of 160 pennies of the reign of Alfred. 160 pennies are two-thirds of a pound of 240 pennies, or in other words a *mark*.

The practice of running silver into ingots of such a weight may well have arisen from an earlier practice of employing bronze or iron bars of such a weight. It is at all events certain that the mark is native Teutonic and is not borrowed from Rome. That the Kelts at least used bars of iron as money is made not unlikely by a famous passage of Caesar which I shall quote later on. A various reading states that the Britons used iron rods as money (*ferreis taleis*). Even without this we may reasonably infer from what we have learned of the practice of primitive peoples in dealing with iron or copper, that the Teutons and Kelts must have used these by measure. It is well known that the Swedes used ingots of copper as currency down to comparatively recent times. It is then most likely that the *öre* or ounce of 410 grs. was the highest original weight unit, just as the *unga* is in the ancient Irish system. The weight of this *öre* is of great interest. If we found the Roman pound of 12 ounces in Scandinavia, we should at once say that the *öre* of 410 grs. was the reduced Roman ounce (432 grs.). But as the native mark evidently got its position before the influence of Rome was felt in the North, we may well consider the *öre* to be pre-Roman. The reader will remember that I identified the ancient Roman *uncia* with the small talent of Sicily and Macedonia. The latter weighed 3 ox-units or about 405 grs. I also suggested that it originally represented the value of a *slave*, and was thus the original highest unit used for gold or silver. I showed on an earlier page (141) that the Norse *örtug*, the one-third of the *öre*, was the price of a cow. If three cows were the price of a slave in Scandinavia as they were in Ireland, and probably in Homeric Greece, an *öre* of gold was the price of a slave. The passage from the life of St Finnian given at once shows that an ounce of gold was the regular price of a slave in early Ireland, and probably a good Scandinavian scholar could soon find similar evidence for the value of the old Norse slave.

The meaning and derivation of the term *örtug* have been much discussed. It occurs in the forms *örtog*, *örtug*, *ertog*, *œrtug*. Cleasby's Lexicon makes nothing out of the first part of the word, but takes the second part (-tog -tug = tugr = 20), because *örtug* had the value of 20 *penningar*, though *tugr* means 10. But as a matter of fact there were, as we saw above, 240 *penningar* in the mark, and therefore there were 10 *penningar* in the *örtug*. Holmboe[1] goes more deeply into the origin of *örtug*. He says, "As *á*, pl. *ær*, signifies a ewe, and *tug-r* as a derivative of *ten* both by itself and in compounds signifies *ten*, *ertug* seems originally to have signified 10 *ewes*, just as the weight *ertug* betokens the weight of 10 *peningar*, and *peningr* itself also means a *sheep*. It may be regarded as questionable to assume the plural *ær* to form the first part of the compound, yet *ær* must at an early period have been used in the formation of compounds, since both the folkspeech of Norway has the form *ær-saud-ewe*, sheep, technically a *ewe-with-lamb*, and the folkspeech of Denmark has *ær lam* in the sense of *ewe-lamb*[2]." Another suggestion is that *örtug* comes from *arta* = a pea-*formed knob*, so that örtug = örtu-vog, the weight of a pea. The objection to this would be that the pea would weigh 13·5 grs. Troy, which seems far too much.

In spite of the philological difficulty in making *örtug* = 10 ewes,

---

[1] *Norges Mynter*, IV—V.
[2] I am indebted to Mr E. Magnússon for the translation of Holmboe.

it is very remarkable that this value corresponds so accurately with the value of a cow, which I independently found for it. I have already pointed out that 10 sheep were the usual value of a cow. So it was at Rome in 451 B.C. and so it is with the Modern Ossetes. The ox fit for the yoke was probably worth 20 lambs or 5 sheep in Lusitania[1], and as we saw that in the Welsh Laws the ox when fit for the yoke was worth half a full-grown cow, the Lusitanian cow was worth 10 sheep. So also at Athens, when Plutarch[2] says an ox was worth 5 sheep, he probably means an ox fit for the yoke, the cow being worth 10 sheep. In the Brehon Laws 8 sheep go to the cow, but as I have already pointed out the insulated position of Ireland would tend to cause a variation in prices from those on the mainland of Europe. Thus we see from the story of St Finnian that gold must have been worth only three times its weight in silver in Ireland in the early centuries of our era. For the price of a slave was an ounce of gold, whilst in the Brehon Laws it is 3 ounces of silver. It might be said that we cannot prove that this was the value of a slave in gold and silver at any one time, and that silver may have been much cheaper at an earlier date. When we recollect that silver has never existed in any quantity in Ireland, and that where it does exist it can only be obtained by systematic mining, a thing impossible in the eternal turmoil of Ireland, and also bear in mind that when Japan was opened to Europeans in this century gold was exchanged for three times its weight in silver, we need not think such a relation at all unlikely in ancient Ireland. The paucity of silver ornaments in the Royal Irish Academy Museum confirms this opinion. But the evidence from the Penitentials shows that silver was scarce at a comparatively still early date in Ireland[3]. Thus XII altilia vel XIII sicli praetium unius cuiusque ancillae.

I have already shown the universality of making gold ornaments after a fixed weight. The passages given above show that a similar practice existed among the ancient Irish.

Let us turn to the numerous gold rings, commonly called Ring Money, of which there are some 50 in the Museum of the Royal Irish Academy of various weights and sizes. I give these weights. Let us examine them, and see if we can find any indications gained inductively of a weight standard.

[1] Polybius xxxiv. 8.
[2] *Solon* 23, see p. 324 *supra*.
[3] Wasserschleben, *Die Bussordnungen d. Abendlandisch. Kirchen* (De disputatione Hibernensis Sinodi et Gregori Nasaseni sermo), p. 137.

400   APPENDIX C.

As by inspection we see that the smallest rings weigh 13 and 14 grs. Troy, and the next three 29, 31, 32 respectively, which look like the double of the smaller, I shall group the rings according as they approximate to the multiples of 15.

Multiples of 15	Actual Ring Weights (Royal Irish Acad.)	Multiples of 15	Actual	Rings	Weights
15	13, 14	180	179	345	——
30	29, 31, 32, 36	195	199, 203	360	——
45	40, 46	210	206, 209	375	372
60	54, 56, 58, 59, 61, 65, 65	225	220	370	
75	69, 73	240	247		
90	84, 84, 88, 96	255	259		
105	98, 104, 111	270	——		
120	121, 124	285	*288, 283		
135	——	300	——		
150	144, 144, 147, 147, 150, 151	315	322		
165	171, 172	330	332		

A glance at the foregoing table shows that the most numerous group of rings occurs at the fourfold (60), no less than seven specimens ranging themselves at that point, next we find six specimens at the tenfold (150), whilst next in order comes the sixfold with four examples. There are three cases of the double (30). On the other hand it is worth noticing the absence of the ninefold, whilst there are three instances of the sevenfold, and the absence of the eighteenfold (2 × 9) likewise, whilst we have the elevenfold, twelvefold, thirteenfold, fourteenfold. However from the absence of the twentyfold (2 × 10) we cannot lay great stress on this. The heaviest specimen (372) closely approximates to the twenty-five fold (375).

I add the weights of the ancient Irish gold rings preserved in the British Museum.

Irish small plain ring money. Some are without localities but may be assumed to be Irish. Marked thus *.

*103, 563, *389, *121, *29½, 218, 224, 323, 295 injured, 218, 122, 90, 28, 56, 215 copper plated with gold (injured), 299, 148, 98, 366, 89 piece

APPENDIX C. 401

cut from a larger bracelet ?, 48½ hollow and open ? plating of bronze ring ? (banded), 422, 410 (ounces), 288 (injured).

Irish fluted ring money. * *No precise locality, but presumably Irish.*
*106, *123 (worn), 30, 59, 90, 66, 59½.
With disks, 249, 806 (2 oz.), 595, 283, 169, 665, 139, 119.
Dots, no lines, 32.

The weights of these rings show many points of agreement with those in the Irish Museum. Thus we get 28, 29½, 30, and 32 grs. corresponding to 29, 31, and 32 grs. of the second group in the Irish Table. Again, 56 and 59½ where we get 54, 56, 58, 59, 61 in the Irish, and 66 corresponding to 65, 65; 98 to 96 and 98; and 89 corresponding to 88 and 90; 119, 121, 122 and 123 to 121 and 124; 139 to 144, and 144 and 148 to 147 and 147; then 169 to 171 and 172. Then comes a break, and we get 215, 218, 218, 224 corresponding to 220, and 249 to 247, and 283 to 283 and 283; and 323 to 322, and 360 to 366. But the British Museum gives us in the higher weights three very important specimens: for 410 grs. is the ounce corresponding exactly to the old Norse *öre* of 410 grs., and the ring of 422 grs. looks like the later ounce rising towards the full weight of 432. The ring of 806 grs. is plainly 2 ounces of the standard of 410 (806 ÷ 2 = 403).

The occurrence of several specimens so constantly all of the same weight, as for instance those about 220 grs., points beyond doubt to the conclusion that when the rings were being made a given quantity of gold was weighed out for the purpose. The story of St Finnian proves that for any transaction in which rings were employed as money, the scales were employed.

There is a set of leaden weights in the Royal Irish Academy Collection, found at Island Bridge, Dublin, in 1869, when Ancient Irish and Scandinavian remains were found together. As they are more or less corroded, it is not advisable to lay much stress on their present weights.

		grs.
1.	Semicircular weight	1852
2.	Animal's head	1550
3.	Circular	1221
4.		958
5.		634
6.	Oblong	539
7.		459
8.	Quadrangular	414 (oz.)
9.		395 (oz.)
10.		220

R.

There are certainly some interesting points of agreement between the weights and the gold ornaments, *e.g.* the weights of 220, 390, 414, 630, have corresponding weights in gold. The largest weight may be $4\frac{1}{2}$ oz. of 410 grs.

Let us now return to the Irish monetary system, and see if we can determine more accurately its relation to that of Rome.

    8 grains of wheat = 1 pinginn.
   24   ,,   ,, = 3 pinginns = 1 screapall.
  576   ,,   ,, = 72   ,, = 24 screapalls = 1 unga.

As regards *unga* and *screapall* we have spoken already. Of their origin there is no doubt. The pinginn on the other hand is not so easy. The name is certainly Teutonic, said to be ultimately a loan word formed from *pecunia*. It seems to have been employed as a general term for the smallest form of currency. Hence we find the Saxon form (*pendinga*) applied to the 240th part of the lb., and of about 32 grs. wheat, and the Norse *peningr* used for the 240th part of the *mark*, whilst in Ireland the cognate form is applied to the 72nd part of the ounce, and is of the weight of 8 grains *wheat.*

The Irish employed the system of Uncia and Scripula. Shall we say then that this system was in vogue in Britain likewise before the time of Constantine and yielded slowly before the later one?

Since then it was common to the Kelts on both sides of the Irish Sea, and we find that in Ireland it was grafted upon an earlier system, of which the *crosoch* is a survival, we may reasonably infer that the Kelts of Britain had likewise a native system analogous to the *crosoch*. But further, of this we have strong evidence of two kinds. Caesar *B. G.* v. 12, when describing the British Kelts and their manners, says; pecorum magnus numerus. Utuntur aut aere aut nummo aureo aut annulis ferreis ad certum pondus examinatis pro nummo[1]. The passage has been mutilated by Editors, but this is the reading of the best MSS. Caesar thus tells us that they had

---

[1] Beside the difficulty about *numo aureo* there is a further variant between *anulis ferreis* and *taleis ferreis* (bars of iron). Can Caesar have in reality written both? May the original reading have been: utuntur aut aere aut numo aureo, aut aureis anulis, aut taleis ferreis etc.? Caesar speaks of the Britons having iron of their own, and it is highly probable that they employed ingots or bars of it as money, as the wild tribes of Annam and Africa do at present. They probably used their gold or bronze rings and armlets as money also.

APPENDIX C.   403

a system of weights of their own. Secondly the evidence of the actual British Coins (cf. Evans, *Coins of Ancient Britons*) which are of a standard not Roman.

Now we have seen above that the Irish gold rings were weighed on a standard of almost 13·5 grs. Troy. Let us now see if the larger gold ornaments preserved in our Museums confirm or disprove the evidence of the rings. I shall first give the weights of those in the Royal Irish Academy:

[1] *Crescent shaped ornaments:* 1539, 434 (ounce of Brehon Laws?), 733, 1008, 255, 2013, 489, 552, 660, 1081, 98, 432 (ounce of Brehon Laws), 339, 400 (early ounce=Norse *öre*?), 187, 390 (old ounce?), 797 (2 ounces, $2 \times 398\frac{1}{2}$).

The following are not in Wilde's Catalogue: 472, 505, 542, 540, 630, 647, 667, 687, 720, 722, 737, 1092, 4331.

Torques: 476, 1013, 1527, 3126, 3168, 4722, 5941, 6007, 10268.

Not in Wilde: 154, 342, 1946, 2715, 4172, 5207, 5275, 6012, 6881.

Armlets: 144, 158, 182, 329, 401 (small pre-Roman ounce), 421 (ounce), 487, 510, 684, 757, 894, 989, 1037, 1369, 1630 (4 ounces of 407 grs.?), 1716 (4 ounces of 426 grs.?), 2089 (5 oz. of 418 grs.?), 5635 (14 oz. of 402 grs.?), 6265 (15 oz. of 417 grs.).

Not in Wilde: 130, 145 ($\frac{1}{3}$ of oz. of 432 grs.?), 178, 184, 187, 199, 208, 215 (half oz. of 432 grs.?), 241, 289, 301, 303 ($\frac{3}{4}$ oz. of 405 grs.?), 345, 396 (oz.?), 487, 509 ($1\frac{1}{4}$ oz.?), 547 ($1\frac{1}{4}$ of oz.), 606 ($1\frac{1}{2}$ oz. of 405 grs.?), 630 ($1\frac{1}{2}$ oz. of 420 grs.?), 740, 753 ($1\frac{3}{4}$ oz.), 1093 ($2\frac{1}{4}$ oz.?), 1190, 1210 (3 oz. of 405 grs.), 1267 (3 oz. of 422 grs.?), 1322, 1641 (4 oz. of 410 grs.), 1730 (4 oz. of 432 grs.?), 1836, 1836 ($4\frac{1}{2}$ oz. of 410 grs.?), 1940 (5 oz. of 388 grs.? or $4\frac{9}{4}$ oz. of 410 grs.?), 1980 (5 oz. of 396 grs. or $4\frac{9}{4}$ oz. of 410 grs.?), 2201, 6144 (15 oz. of 410 grs.?), 13557 (33 oz. of 410 grs.?).

Fibulae: 56 (4 crosachs), 179, 180 ($\frac{2}{3}$ oz. of 400 grs.?), 415 (oz.), 600 ($1\frac{1}{4}$ oz. of 400 grs.?), 1231 (3 oz. of 410 grs.), 1345 ($3\frac{1}{4}$ oz. of 432 grs.), 1596 (4 oz. of 399 grs.?), 2301 ($5\frac{1}{4}$ oz. of 400 grs.), 2536 (6 oz. of 422 grs.), 17200 (43 oz. of 400 grs.?), 8092 (20 oz. of 404 grs.), 19440 (48 oz. of 405 grs.).

Not in Wilde: 61, 106 ($\frac{1}{4}$ oz.), 170, 170 ($\frac{2}{3}$ oz. of 425 gr.), 191, 196 ($\frac{1}{2}$ oz.?), 207, 209 ($\frac{1}{2}$ oz.), 248, 275 ($\frac{2}{3}$ oz. of 411 grs.), 315 ($\frac{3}{4}$ oz.?), 379 (oz.), 542 ($1\frac{1}{3}$ oz.?), 557 ($1\frac{1}{4}$ oz.?), 586 ($1\frac{1}{2}$ oz.?), 649 ($1\frac{1}{2}$ oz. of 432 grs.?), 1187 (3 oz. of 396 grs.?).

Gorgets: 1160 (3 oz. of 387 grs.?), 2020 (5 oz. of 404 grs.?), 3091 (8 oz. of 386 grs.?), 3444 (8 oz. of 430 grs.?).

[1] These are taken from Sir W. Wilde's Catalogue, but for the weights of articles acquired since 1862 I am indebted to the kindness of the Curator, Major Macenery.

The result of an examination of the foregoing weights is to show that in all probability the vast majority of them were made on a standard much lighter than the Roman ounce of 432 grs., which was in full use in mediæval Ireland. We saw that the Roman ounce had been only 420 grs. down to the Second Punic war, and I suggested that originally it was of the same weight as the Sicilian talent 390—405 grs. Can we observe a similar increase in the Irish ounce? The ounce of 400—410 seems to point to a time when Kelt and Scandinavian had a common higher unit of similar weight corresponding to the value of a slave[1], just as the Sicilian and Macedonian talent of three ox units represented the same slave unit.

I shall now give the weights of the various ornaments of gold found in England, Wales and Scotland which are preserved in the British Museum. For these I am indebted to the great kindness of Mr F. L. Griffith of the Anthropological department.

Torques with rings.

Boxton, Suffolk, torque band twisted. 1·038 (2½ oz. of 415 grs.) with double ring. Weight 24·8 grs.

(A ring of 8 parallel sections, bronze plated with gold, injured, weighs 111 grs.; the locality is not known, but it seems connected with this class. Probably Irish, one in Wilde's catalogue of 7 sections.)

Another double ring, Devonshire, weighs 563 grs. (1⅓ oz. of 420 grs.).

Lincolnshire torques; 1454 grs. (3½ oz. of 415 grs.), coiled band 119½. Quadruple ring, 93½ (¼ oz.?), another similar 93.

Cambridgeshire torques (not in B. M.) 1944 (5 oz. of 387? or 4¾ oz. of 410), rest in B. M. viz.:—bracelet 613 (1½ oz. of 412 grs.), two treble rings linked together, combined weight 358, double ring, weight 132 (⅓ oz.), another 131½, two others similar but smaller are each 68 (⅙ oz.).

Wales. Two plain bracelets, near Beaumaris, Anglesea, 1028 (2½ oz. of 410 grs.); 420 (1 oz.), crescent-shaped gorget, Caernarvon, 2861 (7 oz. of 410 grs.).

Scotland. Noard, near Elgin, torques formed of a plain twisted band, 207 (½ oz.): 215 (½ oz.): 192 (½ oz.): 119 grains.

The evidence points to an ounce of 420 grs. It is worth noting that this is just 5 times the weight of the latest British coins, 84—82 grs.

---

[1] My friend Mr F. Seebohm has shown me that as a *weight* the Swedish *Jungfrau* is equal to the Irish *Cumhal*.

Whence then did the Britons obtain this pre-Roman standard? Was it of native development or borrowed from some other people? By Britons we must be careful to express not all the natives of Britain. They fall most certainly into at least two groups. I. The Kelts in the East and South East. II. The barbarous inhabitants of the interior, who subsisted by hunting and fishing, and who were probably of that Iberic race, which spread over all Western Europe before the advance of the Aryans. It is only with the first group that we are immediately concerned. They almost exclusively possessed the art of coining, as is shown by the area over which British coins are found. Furthermore Caesar tells us of the close relationship of the first group to the Gauls, as is shown by their tribal names, language and customs. In addition their coinage is similar. Now there can be no doubt as regards the source from whence the Gauls derived their coinage. As they got the art of writing from the Phocaeans of Massilia (founded circ. 600 B.C.), so likewise did they gain the art of money-stamping from the same famous town, as has been completely demonstrated long since. People are inclined at once to assume that the Gauls and Britons got their weight standards also from Marseilles. There is certainly some evidence to support this belief. Thus the gold torque lately found in Jersey weighs 11500 grs., which is exactly the mina of the Phocaic system at a time when $57\frac{1}{2}$ grs. went to the drachm. Again we have seen that there were a considerable number of gold ornaments in Ireland and Britain which weigh 224—216 grs. This is the Phocaic (or Phoenician) stater. But the question is not so simple as it might appear at first sight in relation to the weight system, as will appear most readily by a short survey of the history of the monetary system of Massilia.

I. The earliest coinage consists of silver, small divisions of the Phocaic drachm (58—54 grains Troy). These have various symbols on the obverse, but have uniformly the incuse square on the reverse. These may be placed after 500 B.C. "Notwithstanding their archaic appearance, it does not seem that these little coins are much earlier than the middle of the 5th century."

II. Next comes a series, chiefly obols for the most part with head of Apollo on obverse, and a wheel on reverse, the latter probably a development of the earlier incuse square. They are mostly obols of 13—8 grains.

III. About the middle of the 4th century the drachm first

appears with the head of Artemis on obverse and a lion on the reverse, weighing 58—55 grains.

Now over all Gaul, and far into Northern Italy, and the valleys of the Alps, as far as the Tyrol, the coinage of Massilia made its way and was abundantly imitated. In fact these imitations formed the entire medium of those regions until the Roman conquest. The imitations of the little coins with Apollo and the wheel as reverse are found right into the north of France, and in England.

Did the Kelts borrow their $13\frac{1}{2}$ grain unit from the 13 grain obol of Massilia, or is it of far earlier growth? The Etruscans used a unit of $13\frac{1}{2}$ grs. in the 4th century B.C., and we find the Massaliotes having almost the same. Is the true answer this? All over Western Europe the ox unit of 135 grs. of gold was subdivided into 10 parts each of $13\frac{1}{2}$ grs. These 10 parts corresponded to 10 sheep, the regular value of a cow. There was also a higher unit from Greece to Gaul and Britain corresponding to the slave. There were fluctuations in their worth in various times and places, but on the whole there was a tendency to raise the weight of the higher unit (ounce). But it is natural that the Kelts may have taken over into their system certain units from the Phocaic system which they used as multiples of their own smaller units, just as the Teutonic peoples took the Roman pound into their own system, and the natives of West Africa made the Spanish dollar the multiple of their own native weights, based on seeds. Some idea of the relative ages of Keltic gold ornaments may perhaps be got from applying the criterion of weight standard to them.

INDEX.

Abdera, 340
Abraham, 112, 113, 197
Abrus, 172
Absalom's hair, 120, 275
Abyssinian gold in beads, 82
Actus, 365
Aegina, 211, 328
Aeginetan measures, 306
— obol, 366
— standard, 9, 21, 311
— — its origin, 217
— — used for copper, 345
— system, 307
Aelian, 144
Aes, 86
Aes grave, 378
Aes rude, 355, 376
Agariste, 212
Agathocles, 138
Agerept, 150
Agonistic types, 337
Agrigentum, 347, 350
Aicill, Book of, 353
Airgid, 63
Alalia, 180
Alamanni, 140
Alaska, 47
Alexander, 29, 198, 342
Alexandrine talent, 244
Alfred's penny, 180
Al-li-ko-chik, 15
Alphabet, the, 227
Alps, gold of, 88
Altin (=gold), 70
Alyattes, 71
Amber, 227
— beads, 46
— golden, 110; red, 110
Anaxilas, 336
Angala, 354
Annals of Four Masters, 31
Annam, 23
— barter system of, 164
Ant coins, 22
Ants, gold-digging, 66
Apis, worship of, 50
Apollo, 107

Apulia, 370
Aquileia, 87
Arab weights, 179, 182
Arabia, gold of, 75
Archimedes, 36, 100
Argippaei, 68
Argos, 215, 335
Arimaspians, 66, 68
Aristaeus, 314
Aristeas, 108
Aristotle, 96, 106, 131, 138, 213, 318, 323, 336
— Polity of Athenians, 305
Armlets, 42
Arpi, 367
Arrows, 24, 43
Arrugia, 101
Artabri, 97
Arverni, 90
As, 350
— derivation of, 353
— divisions, 351
— land measure, 351
— linear measure, 351
— of empire, 362
— reduction of, 380
— sextantal, 362
— symbol of, 369
— used only of bronze, 351
As libralis, 135
Assam coinage, 177
Asser, 354
Asses, sacrifice of, 107
Assis, 354
Assurbanipal, 201
Assyrian weights, 183, 199, 249
Astronomy, 199
Asturia, 101
Astyra, 71
Aternian law, 134
Athene, statue of, 211, 220
Athenian coinage, 124, 306, 372
Athens, Polity of, 214, 305
Attic choenix, 214
— didrachm, 5
Aulus Gellius, 135
Aura (old Norse), 63

Aurès, 183, 254
Aurum, 87
Ausum (aurum), 61
Axe, 318
Axes, Tenedos, 50
— West African, 40
Aymonier, 23, 161
Aztec money, 192
— numerals, 192
Aztecs, 17, 59

Babylonian metric system, 251
— standard, 78, 163, 206, 261, 387
— system, 197
Bactria, coins of, 126
Baetis, 97
Bag of rice, 162
Bahnars, 23
Ball, V., 68
Balux, 101
Bamboo-joint, 163, 171
Bar, 39, 158
— (Assyrian), 185, 285
— of silver, 25
Barley, 178
Barleycorn, 177, 179
— = Troy grain, 181
Barrel, 115, 175
Bars, 371
Barter, age of, 11, 114, 196
Bassak, 161
Baug-brotha, 37
Baugr, 37
Beaver, 314
— skin, 12, 153, 323
Beag, 37
Bear skins, 16
Bee, 320
Bekah, 277
Belgic tribes, 94
Bells, 43
Bereniceum, 297
Bermion, 71
Bes, 351
Betzer, 36
Bhascara, 177
Bigae, 377
Bigati, 377
Bimetallism, 338
Bisaltae, 340
Blanket currency, 17
Bo, 33
Boar, 332
Boeckh, 1, 238, 365
Boeotia, 77
Boeotian shield, 331
Bonny River, 40
Boroimhe, 32
Bortolotti, 241
Bosman, 185

βοῦς ἐπὶ γλώσσῃ, 8
Boyd Dawkins, 110
Bracelets, 35
Brahmegupta, 177
Brandis, 129, 195, 294
Brandy, 323
Brass rods, 41
Brassey, Lady, 330
Britain, gold coins, 93
' Britons'' money-system, 179
Bronze in Italy, 368
— in Northern Europe, 86
Brugsch, 122, 195, 196
Buffalo, 24, 164
— value of, 154
— worth a stick of gold, 168
Buffaloes, 25
Bull, 322
— on coins, 321
Bull's-head weight, 282
Burgundians, 141
Bushel, 115
— how fixed, 191

Cacao seeds, 17, 193
Cadmus, 71, 227
Caesar, 179
Calculus, 192
Caldron, 25
Caldrons, Irish, 32
Caldwell, W. H., 152
Calf, 374
Calves' heads, 322
Camarina, 347
Cambodia, 25, 160
Cambridge, 182
Camirus, 339
Campania, 216
Candarin, 158
Cappadocae, 78
Carchemish, 202
Carmania, gold in, 74
Carob, 181
Carthage, 288
Carthaginian coinage, 131, 289
— gold unit, 130
— trade in gold with West Coast of Africa, 83
Cartload, 175
Cash, 157
Cat's eyes, 21, 27
Cattle at Rome, 31
— chief wealth of Britons, Gauls, Italians, etc., 51
— in Avesta, 27
Catty, origin of, 162, 174
Cauer, 365
Cayley, Prof., 231
Centupondium, 136, 360
Centussis, 370
Ceramus, 82

Chabas, M., 239
Chabinus, 76
Chalci, 346
Chalcis, 227, 361
Χαλκός, 86
Chariot of Hera, 116
Chariots in Veda, 26
Charlemagne, 34
Charuz, 60
Chautard, 225
Chauter, 45
Chinese coinage, 10
— shell-money, 21
— weight-system, 156
Chios, 322, 343
Chisholme, 199
Χρυσός, 60
Chrysûs, 277
Cicero, 134
Cilicia, silver of, 286
Cloth, 35
— silken, 22
Cnidus, 321, 322
Cocoanut, 162, 171
Coinage, invention of, 203
— of gold, 125
— of silver at Rome, 136
Coins, early Lydian, 293
— normal weight of, 218
Coin-standards, 210
Colaeus, 62, 96
Colchis, 70
Colebrooke, 176
Colpach, 33
Commercial weights, 344
Comparetti, 314
Compensation for wounds, 30
Concha, 328
Conchylion, 329
Constantine, 384
Constantine's solidus, 181
Conti, Nicolo, 27
Convention, 47
Coomb, 115
Copper coins in Greece, 361
— — in Britain, 94
— in Greece, 312
— in Meroe, 78
— in relation to gold, 77
— native, 58
— of Haidas, 17
— rings, 22
— standards, 348
— wire, Calabar, 40
Corcyraean wine jars, 106
Corinthian standard, 362
— system, 311
Corn sold by measure, 115
Cotton as money, 45
Counters, 192, 228
Coventry tokens, 336

Cow, 2 seqq., 370
— among Ossetes, 30
— at Delos, 5
— at Syracuse, 31
— equal centumpondium, 360
— Hebrew, value of, 148
— in Avesta, 26
— in Rig Veda, 25
— in Scandinavia, 35
— in Welsh Laws, 32
— names for, Sanskrit, Zend, etc., 51
— on coins of Eretria, 5
— suckling calf, 321
— unit of assessment at Rome and Syracuse, 393
— value of, in Gaul and Germany, 140
— — in Greece, Italy, 133
— — at Rome, 135
— — in Scandinavia, 141
— — in Sicily, 137
— — Persian, 151
— — Phoenician, 143
— — (Table), 153
— — the same over wide area, 52
Cowell, Prof., 176
Cowries, 13, 177
— as counters, 229
Cows among Madis, 43
— in Darfour, 44
Crab's claw, 350
Crab's eyes, 186
Crawfurd, John, 170
Crenides, 74, 341
Croesus, 204, 297
Crosach, 36; crosóg, 396
Croton, 328
Cubit, royal, 265
Cucurbita, 258
Cumhal, 33
Cunningham, 55, 117, 127
Curtius, E., 201, 212
Cuttle-fish, 327
Cyathus, 258
Cyrene, 326
Cyzicene staters, 342
Cyzicenes, 301
Cyzicus, 316, 342

Damba, 186
Damleg, 45
Danes, 321
Danube, 106
— flows into Adriatic, 107
— source of, 107
Dapper, 43
Darfour, 44
Daric, 126, 277, 297
— as talent, 6
— derivation of, 300
— = Homeric talent, 7

Datum, gold mines, 74
Debae, 75
Decalitron, 362
Decimal system, 203, 228, 371
—— in Homer, 308
Decussis, 356, 369, 370
Deecke, 130
Degradation, 226
— of coin weights, 223
— of weight, 338
Delian priests, 108
Delphium, 106
Delos, 215
Demareteion, 297
Demeter, 327
Denarius, 357, 363
Deunx, 351
Dewarra, 20
Dextans, symbol of, 369
Dhalac, 330
Digitus, 353
Dinar, 63
Diodorus, 81
Dionysius, 31, 225
— of Halicarnassus, 134
— of Syracuse, 224
Dioscuri, 377
Dirham, 148, 182
Dodona, 215
Dodrans, 351
Dogs, 94
Dollar, Maria Theresa, in Soudan, 56
— Mexican, 24; Spanish, 44
Double Unit, 267
Doukha, 45
Drachm at Athens, 324
— Corinthian, 311
— origin of, 214, 310
Draco, 5
Dragon's eye, 22
Dublin, 321
Duck weight, 83
—— suggested origin, 247
Duck weights, 199, 245
Dungi, 248
Duodecimal system, 371
Dupondius, 376
Dyer, Dr Thiselton, 186
Dyrrachium, 322

Earring, 35
Ebusus, coinage of, 290
Echinus, 328
Egypt, coinage of, 219
— gold in, 78
Egyptian gold-mines, described by Diodorus, 79
— measures, 122
— Monad, 129
— records, 236
— weights, 122

Egyptian weight system, 237
Electrum, 98, 204, 290
— at Carthage, 289
— Lydian, 70, 294
— why coined, 207
Elephant, price of, 24
Elephant's tusk, 25
Ellis, 187
Emporiae, 290
English coinage, 224
— Imperial weights and measures, 266
— penny, 225
— weights, 186
Ephorus, 211
Epicharmus, 137, 364
Eretria, 322
Erman, 146, 242
Erytheia, 110
Eryx, 144
Esterlings, 225
Etruria, 374
Etruscan gold coins, 130
— gold unit, 359
— silver, 363
— standard, 130
Etruscans, 64
Etymology, danger of, 65
Euboic-Attic system, 311
Euboic standard, 9, 210
—— origin of, 222
Eustathius, 125
Evans, A. J., 271, 365, 366
— Dr J., 94
Exagium, 183
Ezekiel, 121, 282

Falgo, 45
Fanam, 173
Fee, 4, 34
Felkin, 43, 263
Fen Ditton, 182
Fertyt tribe, 46
Festus, 134
Fetiches, 187
Fibulae, 41
Fifteen-stater standard, 286
Fiji, 21
Fines, 135
Fiorino, 385
Fish-hooks, 28
Florin, 385
Foot, Roman, 359
Foucart, 219
Fractions, 357
Frankincense, 6
Frazer, J. G., 30, 320
French metric system, 1
Fuel sold by bulk, 115

Gades, coinage of, 290

Gaius, 8, 376
Galetly, A., 30
Gallaecia, 101
Gardner, Dr, 126, 342
— P., 222, 313, 364
Gaul, 325
Gaulish gold unit, 131
Gauls, 332
— in Italy, 61
— value of cow with, 140
Gaus, 51
Gelon, 142
Gerah, 277
Germans, 131
Geryon, 110
Gill, 23, 296
Gold, 57 seqq.
— alone weighed in Homer, 117
— among Salassi, 89
— at Vercellae, 88
— bat, 163
— Coast, 105
— coinage, 372
— coinage, Athens, 124; Macedon, 125; Thasos, 125; Cyzicus, 125
— coinage, Roman, 362
— coins, Athens, 372
— distribution of, 65
— equal distribution of, 114
— first coinage at Rome, 378
— first of all articles weighed, 114
— from India, 257
— in Bactria, 67
— in California, 58
— in China, 22
— in Gaul, 90
— in Meroe, 78
— in Noricum, 87
— in quills, 17, 186, 192
— in Rig Veda, 25
— in rings from Sennaar, 82
— in Swiss lake-dwellings, 85
— in Thibet, 66
— in Wales, 94
— measured, 168
— measured by quills, 186
— mining, methods of, 101
— not weighed, 187
— nuggets of, 75
— of Tolosa, 92
— ornaments of Gauls, 92
— Irish, 402
— placer, 98
— poured into jars, 259
— relation of, to silver in Etruria, 140
— relation of, to silver and copper in Italy, 139
— relative value, and silver, 75
— scarce in Greece, 221

Gold, standard, 211
— Talent of, 3
— unit, the same everywhere, 133
— unit of Attopoeu, 163
— units, table of, 132
— Ural-Altai, 67
— wedge of, 270
— weighed in Veda, 122
— weighing, 167, 172
— white, 97
Golden Bough, 320
— Fleece, legend of, 70
Goliath, 120
Gortyn, 314
Gourds, 43, 258
Greek (old) standard, 306
— standard (table), 310
— system, 304
— weights, 181
Griffins, 68, 70
Guadalquivir, 97
Gunjâ, 176, 178
Gygadas, 206
Gyges, 71, 201, 204, 293

Hachachah, 45
Haddon, 105
Hair weighed, 275
Hakon the Good, 34
Haliartus, 334
Hamilcar, 289
Handfuls of rice, 170
Hanno, voyage of, 83
Hare, 336
— hunting of, 337
Hares at Carpathus, 337
Hare-skin, 13
Harich, 45
Harpoon, 105
Harris papyrus, 239
Hasdrubal, 289
Haxthausen, 4
Head, 130, 138, 196, 314, 316
Hebrew system, 269
— system, tables, 283
Hectae, 342
Hectare, 1
Helbig, 36, 84
Helix, 36
Helvetii, 90
Heraclea, 365
Herakles, 107, 227
— road of, 111
Hercynian forests, 106
Herodotus, 107, 258, 260
Herondas, 342
Hexas, 348
Hide (of land), 301
Hides, 51
— as money, 332
Hierapolis, 202

INDEX.

Himera, 142, 347
Hindu weights, 177
Hiranya-pindas, 26, 258
Hissarlik, 73
Hittites, 202
Hoe money, China, Annam, 22
Hoes, 45, 165, 312, 371
Hoffmann, 36
Homeric Greeks, analogy of, to modern barbarians, 50
— Poems, 2
— Trial Scene, 8, 389
Honey, 34, 122
Horapollo, 129
Horse, value of, 147
Hottentots, 42
Hucher, 131
Hultsch, 95, 129, 202
Hyksos, 50
Hyperborean maidens, 109
Hyperboreans, 107
Hyperoché, 109

Ialysus, 339
Iceland, 18
Icelandic proclamation, 18
Illyria, 378
Incas, weight, 193
Incuse on coins of Magna Graecia, 334
— square, 333
India, mediaeval, 27
Indian weight standards, 176
Ireland, gold in, 95
Irish currency, early, 31
— weights, 180, 401
Iron in Homer, 117
— ingots, 25, 163
— money, 373
— needles of, 27
— plates, 43
— rings, 40
Issedones, 68
Istir, 148
Istropolis, 107
Italian system, 350
Ivory tusks, 42

Jade, 48, 105
Janiform head, 318
Japanese Bean money, 295
Jars in Annam, 24
Jersey torque, 405
Job, 35
Jol, 288
Jones, Quayle, 186
Jordan, 112
Josephus, 277
Jugerum, 358
Juno Moneta, 215

Kaibel, 306

Karnak, 239
Kat, 238
Keller, Dr, 85
Kelts, 31
— their early knowledge of gold, 104
Kenrick, 143
Kenyon, 306
Keseph, 270
Kesitah, 270
Kettle, 31
Kettles, 24
Kid, 33
Kikkar, 264, 279, 309
King's weight, 275
Klaproth, 69
Knife money, 156
Knives, 312
Koehler, 219, 317
Kolben, 43

Lacedaemonian shield, 334
Lachish, 258
Lady Godiva, 336
Lais, 330
Lake dwellings, 84
Lamb, 271
Laodicé, 109
Laos, weight system of, 161
Larins, 28
Lassen, 66
Lateres, 375
Latham, R. G., 57
Laurium, 99
— mines of, 59
Layard, Sir A. H., 85
Leake, Col., 313
Lebetes, 314
Lehmann, 195
Leinster, king of, 32
Lelantum, 222
Lemnos, 323
Lenormant, 129, 242
Leocedes, 212
Lex Flaminia, 378
— Tarpeia, 31
Libella, 357, 374
Libra, 347, 358
Lindus, 339
Linguistic Palaeontology, 60
Lingurium, Greek derivation of, 110
Lion and Bull, 296
— on coins, 321
— weights, 199, 245
Litra, 347
— its subdivisions, 348
— silver, 361
— translation of libra, 360
Litre, 1
L. M. R., 330
Load, 173, 263
— as unit, 172

Load, Greek, 309
Lupinus, 278
Lusitania, 97
Lycia, 332
Lydia, 201
Lydian coinage, 299
— coins, 321
— electrum, 296
— system, 293
Lynx, 110
Lyre, 329
Lysias, 301, 324

Macedonian standard, 346
— talent, 125, 304
Machpelah, 246
Madagascar, 187
Madden, 240
Madi tribe, 43, 263
Maine, Sir H. S., 8
Maize, grain of, 166
Makrizi, 182
Malay weights, 171
Malays, 309
Manā of gold, 26, 122
Mancipatio, 121, 358, 376
Mancus (of silver), 34
Maneh, its origin, 256
Mansous, 46
Manu, 177
Maris, 203
Mark, 358, 397
Marquardt, 181
Marsden, W., 172
Marseilles, inscription at, 142
Massilia, 62
— court of, 111
Mathematical hangmen, 231
Measure of corn or oil, 324
Medbh, 36
Medimnus, 324
Melitaea, 323
Melkarth, 227
Men, 327
Mene, mene, tekel, upharsin, 247
Mennan, 33
Mentores, 106
Mermnadae, 205
Meroe, gold, copper, iron in, 78
Mesha, 272
Mesopotamia, cattle in, 50
Messana, 336
Metals, first objects to be weighed, 114
— relations of, in Greece, 219
— their discovery, 57
Metapontum, 319, 327
Metre, 1
Metric systems, 198
Midas, 71
Miletus, 205, 210, 226, 296
Milk of cow, 33
— of goat, 33

Milk of sheep, 33
Millies, 157
Mill-sail incuse, 334
Mina, Greek, 309
— Hebrew, 274
— in Ezekiel, 284
— origin of, 258
— use of, 309
Mines of Spain, 97
Mithcal, 183
Moda, 46
Modius, 121
Moeun, 162
Mohurs, 35
Moïs, 24
Mommsen, 88, 134, 205, 348, 364, 380
Money, development of, 48
Monro, D. B., 226
Moriae, 324
Moschus, 137
Moura, 160, 175
Movers, 143
Muk, in Annam, 24
Murex, 330
Mycenae, 72
— rings at, 77
Mytilene, 322

Naaman, 280
Nails, 159, 312
Naucratis, 241
Naxos, 348
Nehemiah, 280
Nejd, 29
Nero, 378
New Britain, 20
New Carthage, 289
— — mines of, 99
Niebuhr, 135
Nile, source of, 107
— water, 242
Nineveh, 85
Nissen, 195, 239
Nomads, 75
Nomisma, 366
Nomos, 369
— bronze, 367
— of Heraclea, 364
— Sicilian, 364
Nouممos of Tarentum, 364
Nub (gold), 60
— its derivation, 78
Nubia, 78
Numerals on coins, 130, 363
Nummus, 131, 137
Numus, 364

Oats, 34
Ob, 349
Obol, 346
— Attic, Aeginetic, 346
— copper coin, 361

Obol, its subdivisions, 349
— origin of, 310
Oenone, 211
Olbia, 67, 316
Olive trees, 365
Olives, 324
Olympic victor, 324
Oncia, 348
Onesicritus, 74
Onions, 45
Oppert, 183
Oppian Law, 139
Or (gold in Irish), 61
Orang Glai, 25
Orchomenus, 72
Ordlach, 353
Öre, 397
Ornan's threshing-floor, 148
Örtug, 397
Ossetes, 4, 30
Ostiaks, 4
Ostracism, 329
Ostrakon, 329
Owls, 225
Ox, fore part of, on coins of Samos, 313
— in *Capitularium Saxonicum*, 34
— name of coin, 4
— on coins of Eretria, 313
— value of, in Egypt, 146
Oxus, 204

Pactolus, 70, 206
Padi, 192
Paeonia, gold mines of, 74
Pahlavi texts, 148
Paille, 101
Palacrae, 101
Palae, 98, 101
Palestine, 269
Pallegoix, 161
Pangaeum, 71, 220
Panormus, 130, 289
Parkyns, Mansfield, 82
Parthenon, 310
Pauli, 89
Pausanias, 212
Pea, scarlet, 172
Peach, 78
Pecunia, 4, 376
Pegasus, 362
Pendeo, 358
Pening, 397
Penny, its cognates, derivation, 64;
  weight, 385
Pentacosiomedimni, 324
Pentoncion, 348
Pericles, 215
Perseus, 107
Persian Gulf, 27
— silver standard, 261
— standard, 300, 303

Persian tribute, 129
— wars, 220
— weights, 179
Persians coin money in Egypt, 219
Pertz, 141
Peru, 193
Petrie, W. M. F., 216, 240, 241, 258
Phanes, 320
Pharaoh, 113
Pheidias, 211, 310
Pheidon, 211, 311
Pheidonian weights, 213
Philip II., 74, 341
Philippi, 74
Philippus stater, 140
φλ/ορι, 61
Phocaea, 205, 322
Phocaean standard, 210
Phocaeans, 62, 96, 110, 130, 132
Phoenicia, 86, 200
Phoenician inscription of Marseilles,142
— standard, 206, 261
— — origin of, 286
— system, 285
— weights, 201
— — from Jol, 288
Phoenicians, 117
Phtheirophagoi, 70
Picul, 263, 309
— origin of, 174, 190
Pig, 25
Pindar, 170, 211
Pinginn, 33
Pipilika, 67
Plutarch, 135, 378
Po, 110
Pollex, 353
Polo, Marco, 14, 146
Polybius, 62, 139
Polygamy, 54
Pondus, 358
Poole, R. S., 271
Posidonius, 91, 97
Pottery, in shape of gourds, 258
Pound, English, 266
— of 16 ounces, 18 ounces, 24 ounces, 360
— of silk, 259
Powell, 20
Priam, 71
Propontis, 210
Ptolemaic coinage, 299
— standard, 338
— stater, 279
Pump, Egyptian, 99
Pylus, 214
Pyrenees, 99
Pytheas, 257
— his voyage, 83

Qesitah, 270
Quadrans, 348, 352

INDEX. 415

Quadrigae, 377
Queen Charlotte Islands, 17
Queensland blacks, 152
Queipo, 179, 200
Quills of gold, 17
Quincunx, symbol of, 369

Rakat, 172
Rameses II., 128
Ratti, 127, 176, 186
Red Sea, 76
Regenbogenschüsseln, 140
Reindeer, 4
Relation of gold to silver, to copper, 135
Rhegium, 336
Rhinoceros, horn of, 25
Rhoda, 290, 322
Rhodes, 132, 322, 339
Rhodian standard, 338, 339
Rhys Davids, 29
Rice, 178
— bag of, 162, 172
— grains, 187
Rig Veda, 25, 59, 122, 257
Ring money, 35, 394
Rings, Egyptian, 242
— gold, 34, 128
— — of Egypt, 129
— in Homer, 36
— Mycenaean, 36
— of tin, 44
Road, sacred, 111, 216
Robes, in Homer, 49
Roman coins of Campania, 216
— foot, 359
— (later) weights, 181
— pound, 234
— system, 374
Romans, use of weights by, 121
Rose, 320
Rotl, 46
Royal standards, 250
Rubat, 45
Ruding, 180
Rupee, 4
— purchasing power of, 152
Rye, 34

Saggio, 23, 146
Salamis, 142, 272
Salassi, 89
Sallet (von), 317
Sallust, 110
Salt, 45
Samhaise, 33
Samos, 222
Samoyedes, 3
Sapec, 24, 157
Sarah, 113
Sardes, 206
Sassanide kings, 151

Saxon coins, 321
Sayce, 202
Scales of silver, 193
— used, 226
Scandinavian currency, 34
Scapte Hyle, 78
Schliemann, 129, 231
Schoenus, 365
Schrader, 60, 69, 70, 92
Scillinga, 39
Sciron, 331
Screapall, 33
Scriptulum, 351
Scripulum, 135
Scrupulus, 352
Scythians, 67
— use gold, but not copper, 69
Seal, 322
Sedâcy, 44
Seebohm, F., 404
Sembella, 379
Semis, 369
Sequani, 332
Servius, 376
Sestertius, 363, 379
Sexagesimal system, 198
Sextantal as, 362
Sextans, 348
Sextula, 351, 384
Shakespeare, 349
Shayast, 150
Sheep, 38, 324, 370, 374
— as coin type, 272
— as unit, 272
— weights, 271
Shekel, 35
— as unit of Hebrew system, 273
— earlier than mina, 246
— heavy, 259
— light, heavy, 201
— of Sanctuary, 273
Shekels, 269
Shell money, 14
Shells of silver, 22
Shield, 331, 334
— in Homer, 331
Shilling, 37
Siamese bullet-money, 28
— coins, 161
Sicanians, 347
Sicels, 347
Sicilian gold unit, 131
— silver coinage, 359
— system, 346
— talent, 131, 187, 304, 359
Sicilicus, 368
Sicily, 31
Siculo-Punic coins, 289
Sicyonian shield, 334
Sidonians, 117
Sierra Leone, 39
Siglus, 261

Silenus, 323
Siliqua, 182
Silphiomachos, 326
Silphium, 314, 325
— on coins of Cyrene, 50
Silver, 57
— at Rome, 139, 373
— coinage, Roman, 362
— coins, origin of Greek, 315
— discovery of, 98, 100
— found in Cilicia, 146
— furnaces for, 98
— in Cilicia, 286
— in Gaul, 93
— in Greece, 310
— in Palestine, 147
— not weighed in Homer, 117
— relation to bronze, 380
— scarce in Egypt, 146
— standard, 260
— standards, table, 209
— — variation of, 337
— value of, 146
Silverlings, 269
Silvestre, 157
Sipylus, 71
Six, M., 321
Sjögren, 70
Slave-boy, 326
Slave, foreign, more valuable, 55
— Hebrew, value of, 148
— in Homer, 30
Slaves, 11, 323
— constancy of price, 54
— in Congo, 42
— in Darfour, 46
— in Wales, 32
— male, female, 54
Soanes, 70
Solidus, 33, 181, 384
Solomon, 147
Solon's coinage, 306, 324
— standard, 306
Sophocles, 204
Sophron, 364
Sophytes, 127
Soteria, 327
Soudan, 312
Soul, weighing of, 150
Soumyt, 46
Soutzo, M., 134, 203, 347, 368, 380
— view of relation between the metals, 136
Spain, mines of, 96, 97
Spata, 84
Spear-brooch, 36
Spices weighed, 276
Spirals, 36
— Keltic, 38
— Scandinavian, 37
Squirrel skin as unit, 4
Stater, use of, 308

Sterlings, 225
Stiver, 186
Stockfish, 18, 316
Strabo, 71, 97
String of cash, 24
Sumatra, 172
Sun's diameter, 203
Suvarna, 127, 178
Svoronos, 314
Swine, 378
— with Gauls, 333
Symbol as mark of worth, 324
Syracusan standard, 362
Syracuse, coinage of, 225
Szins, 25

Taberdier, 158
Tacoe, 186
Tael, 158
Taku, 186
Talanton, 228, 304
Talent, 244
— Homeric, 2 seqq.
— Macedonian, 125, 304
— origin of, 262
— Sicilian, 304
Tantalus, 71
Tapaks, 167
Taras, 364
Tarbelli, 92
Tarentum, 364
Tarncih, 44
Tarshish, 97
Tartessus, 96, 97
Taurisci, 87, 339
Tax, hut, 25
Tea as money, 23
Teanum, 369
Tectosages, 90
Temples as banks, 215
Tenedos, 318
Teos, 210, 340
Testudo, 329
Tetl, 192
Tetras, 348
Teutonic peoples, 34
Thasos, 220, 323, 344
— mines of, 73
Thebes, 334
Theocritus, 137
Theseus, 331
Thomas, 176
Thothmes III., 128
Thracian coinage, 342
Thracians, 340
Thucydides, 72, 211
Thumb, 353
Thurii, 322
Tibetan currency, 23
Tical, 29
Timaeus, 51, 379
Time, measurement of, 198

INDEX.

Timoleon, 225, 289
Tin, 97, 173
— Cornish, 83
— discovery of, in Sumatra, 100
— coins, 225
— rings of, 44
Tiryns, 84, 231
Tjams, 24
Tmolus, 70
Tobacco, 45
Tola, 177
Tolosa, 90
Tomme, 353
Torres Straits, 105
Tortoise, 313, 333
— Island, 331
— (sea), 215
— shell, 328
— — currency, 21
— — masks, 105
Tortoises of terra cotta, 329
— of wood, 330
— — and earthenware, 330
Toukkiyeh, 44
Trade routes, 105
Tremissis, 385
Trias, 348
Trichalcum, 346
Triens, 348
Tripods, 314
Troy grain, origin of, 181; of ounce, 386
Tschudi, 70
Tunny coins of Olbia, 317
— fish, 315
— — Cyzicus, 50
— — Olbia, 50
Turdetani, 97
Turkey rhubarb, 83
Turti, 97
Types parlants, 322
Tyre, 200
— fall of, 141
Tylor, 229

Umbrians, 64
Uncia, derivation of, 353
— Roman, 359
Unga, 33
Unguis, 354
Ur, 197
Ural-Altaic range, 204
— region, 68
Uten, 203, 238

Varro, 375, 378
Venusia, 369
Victoriatus, 377
Victumulae, mines of, 88
Vieh, 4
Vines, distance apart, 366

Vomis, 354
Vulci, 354

Wadai, 44
Wade, Sir T., 158
Wai wai, 105
Wales, 31
Wall paintings, 128
Walrus hide, 47
Wampum, 14
Weapons, 35
Weighing of the soul, 150
Weight, its origin, 12
— of potatoes, 358
— unit, how fixed, 168
Weights, false, 241
— in connection with currency, 271
— in form of animals, 153, 401
— — — oxen, 128
— in shape of cows, 243
Weissenborn, 212
Welsh currency, 32
West, E. W., 148
Whale's teeth, 21
Wheat, 122
— corn, 179
— corn in Assyria, 183
— corns, 180
— ear, 327
— grain, 182
Wheaten straw, 109
Wicklow, gold in, 334
Wife, payment for, 44
— price of, 44, 105
Wilamowitz, 306
Wine, 323
— cup, 323
— jar, 323
— trade, of Carthage, of Gauls, 326
Wolf, 335
Wood as currency, 42
Woodpeckers' scalps, 15
Wool merchants, 117
— weighed in Homer, 118
— weighing of, 116

Xenophanes, 205, 293
Xenophon, 337
— *De Vectigalibus*, 338

Yard, English imperial, 266
— of butter, 358
— of land, 358

Zancle, 348
Zechariah, 148
Zend Avesta, 149
— — physicians' fees, 26
Zulus, 2, 42

SOME PUBLICATIONS OF

The Cambridge University Press.

The Types of Greek Coins. By PERCY GARDNER, Litt. D., F.S.A. With 16 Autotype plates, containing photographs of Coins of all parts of the Greek World. Impl. 4to. Cloth extra, £1. 11s. 6d.; Roxburgh (Morocco back), £2. 2s.

The Engraved Gems of Classical Times, with a Catalogue of the Gems in the Fitzwilliam Museum, by J. HENRY MIDDLETON, M.A., Slade Professor of Fine Art. Royal 8vo. Buckram, 12s. 6d.

Illuminated Manuscripts in Classical and Mediaeval TIMES, their Art and their Technique. By J. H. MIDDLETON, M.A. With Illustrations. Royal 8vo. [*Nearly ready.*

An Introduction to Greek Epigraphy.
Part I. The Archaic Inscriptions and the Greek Alphabet.
By E. S. ROBERTS, M.A., Fellow and Tutor of Gonville and Caius College. Demy 8vo. With Illustrations. 18s.

A Catalogue of Ancient Marbles in Great Britain, by Prof. ADOLF MICHAELIS. Translated by C. A. M. FENNELL, Litt. D., late Fellow of Jesus College. Royal 8vo. Roxburgh (Morocco back), £2. 2s.

Essays on the Art of Pheidias. By C. WALDSTEIN, Litt. D., Ph.D., Reader in Classical Archæology in the University of Cambridge. Royal 8vo. 16 Plates. Buckram, 30s.

The Literary Remains of Albrecht Dürer, by W. M. CONWAY. With Transcripts from the British Museum MSS., and Notes by LINA ECKENSTEIN. Royal 8vo. 21s. (*The Edition is limited to* 500 *copies.*)

The Growth of English Industry and Commerce during THE EARLY AND MIDDLE AGES. By W. CUNNINGHAM, D.D., Fellow of Trinity College, Cambridge. Demy 8vo. 16s.

The Growth of English Industry and Commerce in MODERN TIMES. By the same Author. Demy 8vo.
[*Nearly ready.*

London: C. J. CLAY AND SONS,
CAMBRIDGE UNIVERSITY PRESS WAREHOUSE,
AVE MARIA LANE.

www.ingramcontent.com/pod-product-compliance
Lightning Source LLC
Chambersburg PA
CBHW030541300426
44111CB00009B/817